Seeing New Worlds

SCIENCE AND LITERATURE
A series edited by George Levine

Seeing New Worlds

Henry David Thoreau
and Nineteenth-Century Natural Science

Laura Dassow Walls

The University of Wisconsin Press

The University of Wisconsin Press
114 North Murray Street
Madison, Wisconsin 53715

3 Henrietta Street
London WC2E 8LU, England

5 4 3 2 1

Printed in the United States of America

Library of Congress Cataloging-in-Publication Data
Walls, Laura / Dassow, 1955–
Seeing new worlds: Henry David Thoreau and nineteenth-century natural science / Laura Dassow Walls
318p. cm.
Includes bibliographical references and index.
ISBN 0-299-14740-1. — ISBN 0-299-14744-4 (pbk.)
1. Thoreau, Henry David, 1817–1862—Knowledge—Natural History.
2. Literature and science—United States—History—19th century.
3. Natural history—United States—History—19th century.
4. Thoreau, Henry David, 1817–1862—Knowledge—Science. 5. United States—Intellectual life—19th century. 6. Science in literature.
7. Nature in literature. I. Title.
PS3057.n3D37 1995
818'.309—dc20 95-7401

To my parents,
John and Ethel Dassow

It may appear singular, but yet it is not the less correct,
to attempt to connect poetry, which rejoices every where
in variety of form, color, and character,
with the simplest and most abstract ideas.
Poetry, science, philosophy, and history
are not necessarily and essentially divided;
they are united wherever man is still in unison
with the particular stage of his development,
or whenever, from a truly poetic mood of mind,
he can in imagination being himself back to it.

—Wilhelm von Humboldt, quoted by
Alexander von Humboldt, *Cosmos*

CONTENTS

ACKNOWLEDGMENTS

Many people have helped me over the course of this project's long inception and development. Long ago, at the University of Washington, Martha Banta and Robert E. Abrams first started me thinking about Thoreau. When I returned to graduate school at Indiana University, Lee Sterrenburg opened new worlds to me by showing me how I might bring together the exuberant, unpredictable, and self-aware world of nature with the many cultural constructions of it in history. His classes in literature and science suggested to me how literature might be the medium for studying these intersections of nature and mind, and affirmed for me that primary research and good teaching can inspire each other and hand to the willing student the reins of great possibility.

Many others at Indiana University have helped me along the way. Particular thanks go to Kenneth Johnston, James Justus, Christoph Lohmann, and Cary Wolfe for reading the manuscript in its earlier stages, and offering generous and helpful advice all along the way. Frederick Churchill aided and abetted my Humboldt research at a crucial stage. Steve Watt, Pat Brantlinger, Kathryn Flannery, and Beverly Stoeltje all gave steady support and encouragement. Cynthia Jordan was a patient and deeply sympathetic advisor who quickly became a dear friend; both she and Wallace E. Williams helped start this journey, and they both ought to have been a part of it to the end. Finally, without my confederates in SLAG (or the Science and Literature Affinity Group, with thanks and apologies to Donna Haraway), this would be a thinner book; maybe not a book at all. Thanks to you all, especially to Richard Nash and Nancy Rutkowski, and to Anka Ryall.

Many people beyond Indiana University have offered aid and encouragement. I owe special thanks to George Levine, who seemed to know what I was about even before I did; and to Bob Sattelmeyer, whose generous advice and encouragement have helped me find my way as a Thoreau scholar, and whose suggestions have made this book far stronger than it would otherwise be. Bill Rossi has directly and indirectly provided me with many insights and the kind of encouragement and wisdom so important to one first starting out, especially in a field as multilayered as this one. Dan Peck, Lisa New,

Frederick Garber, Wes Mott, David Robinson, Richard Grusin, Ed Schofield, Brad Dean, Bob Richardson, Beth Witherell and her assistants at the Thoreau Edition have all given, at various points and in many ways, crucial advice and assistance. My colleagues at Lafayette College have been patient, sympathetic, and supportive of a project that must often look distinctly nonliterary; particular thanks must go to James Woolley, and to Jeffrey S. Bader and the Committee on Advanced Study and Research, who provided the financial means to bring this project to its completion.

Finally, those closest to me know already how much they have been a part of this project in its very inception, and even more so as its life has become, over the years, my own. My deepest thanks go to Robert Walls, who constantly has believed in me, has pushed me farther than I would have dared to go alone, and has always been there to listen and to remind me, as a husband, a social historian, and a folklorist, that high literary and scientific abstractions ultimately take on meaning only from the texture of everyday life. And first and last of all, I owe more than thanks to my parents, John and Ethel Dassow. Their lives—my mother's as a writer and editor, my father's as a chemist and field biologist—and their aspirations are inscribed into the premises of my life's work, and to them this book is dedicated.

ABBREVIATIONS

Works by Henry David Thoreau

CC *Cape Cod*
CO *Correspondence*
CP *Collected Poems*
DI *Dispersion of Seeds* (in *Faith in a Seed*)
EE *Early Essays*
FB *Thoreau's Fact Book*
LN *Thoreau's Literary Notebook*
MW *Maine Woods*
NHE *Natural History Essays*
RP *Reform Papers*
WA *Walden*
WK *A Week on the Concord and Merrimack Rivers*

Works by Ralph Waldo Emerson

CW *Collected Works*
JMN *Journals and Miscellaneous Notebooks*

Seeing New Worlds

Introduction

Thoreau devoted the last ten years of his short life to studies that have puzzled generations of his commentators. What was the "transcendental" author of *Walden* doing out in all weathers, counting tree rings, listing plant species, measuring stream depths? These are not, on the face of it, very transcendental activities. It is difficult to imagine Emerson, for instance, scouting woodlots in the autumn rain, entering tree ring counts into a field notebook. And it is easy to imagine these activities as fatal distractions from the great task of writing the successor to *Walden*, and thus to marginalize them, in our disappointment, as the product of a declining and tragically misled talent.

Yet there are at least two overriding reasons for attending carefully to these studies. The first is Thoreau's sheer joy in physical engagement with the woods, fields, and waters of Concord, evident still on every page of the late *Journal*. "Each new year is a surprise to us," he writes in 1858, many years into his project; a certain yew tree strikes him as "a capital discovery" (X:304, 306; 3/18/58).[1] He is full of questions, astonishments: "How long?" "Who striped the squirrel's side?" (X:507–8; 6/24–6/25/58). "How interesting now . . . the large, straggling tufts of the dicksonia fern above the leaf-strewn greensward, the cold fall-green sward!" (XII:370; 10/4/59). Already "surprised at the abundance" of the year's acorn crop, he is surprised again to see it destroyed by early frost; "It is a remarkable fact" that the labor of the oaks should thus be lost. As for that seemingly most deadly study of all, counting those endless tree rings: "Thus you can unroll the rotten papyrus on which the history of the Concord forest is written" (XIV:149, 152; 10/19/60). Second,

Thoreau himself felt he was on, not a retreat, but a real and affirmative quest, which was intrinsic to the totality of his career, the attempt to read and tell a history of man and nature together, as and in one single, interconnected act.

The effort to read nature "whole" was shared by many of Thoreau's contemporaries: Goethe, Coleridge, Emerson, Carlyle, Ruskin; Schelling, Paley, Whewell, Humboldt, Darwin. In an age steeped in Enlightenment ideals as well as Platonic idealism, and astonished by the recent successes of natural philosophy in explaining the fundamental forces of nature, such an effort must have seemed just possible. But *how* to go about it? On this there was no general agreement, and it is hardly accidental that the "scientific method" was established and modern science institutionalized during Thoreau's lifetime. Central to this book is the assertion that there is, in addition to the one narrative usually told about romanticism, a second competing narrative. The usual narrative finds the determining metaphors for romantic literature and science in the opposition of organicism and mechanism, a split which resulted by mid-century in the deliverance of emotive and "organic" literature from dry and "mechanical" science. But this narrative divides Thoreau from himself. Along with most of his contemporaries, Thoreau too conceived of nature as one great whole. However, he experienced two very different ways to approach and understand that whole. The first, which I call "rational holism," conceived the mechanico-organic whole as a divine or transcendent unity fully comprehended only through thought. In various forms, this holism characterized the Anglo-American tradition of natural theology, as well as Goethe and the German *Naturphilosophen*, Coleridge and British transcendental morphology, and finally Emerson and Agassiz. The second, "empirical holism," was an emergent alternative which stressed that the whole could be understood only by studying the interconnections of its constituent and individual parts. It was developed and exemplified by Goethe's friend Alexander von Humboldt and his friends and followers, including, most significantly for this study, Charles Darwin. I suggest that Thoreau's growing interest, through the late 1840s and 1850s, in particularized nature distinguishes him as a Humboldtian empirical naturalist. As a Humboldtian, Thoreau saw his task to be the joining of poetry, philosophy, and science into a harmonized whole that emerged from the interconnected details of particular natural facts. By acting as a Humboldtian naturalist, Thoreau participated in and helped to advance an alternative tradition of romantic science and literature that looked toward ecological approaches to nature and that was suppressed, then forgotten, by later organicist interpretations.

Recovering this alternative tradition enables a new understanding of the problematical studies which fill the later years of Thoreau's *Journal*, which are also the years of his greatest literary productivity. Certainly there was a major shift in Thoreau's career, but I wish to redefine that shift: Thoreau transformed not from an Emersonian transcendental poet to a fragmented empirical scientist, but from a transcendental holist to something new which combined transcendentalism with empiricism and enabled innovative, experimental and postsymbolic modes of thinking and writing.[2]

This book offers an argument for a double romantic paradigm, attempting first to locate Thoreau within the dilemmas created by rational holism, then to follow as he encounters and embraces empirical holism, exploring in turn the dilemmas and the possibilities offered by a vision that sought to heal the growing split between poetry and science. For it was a *vision* that Humboldt offered—not a new world, but a new way of seeing that showed the old world to be fresh and unimagined. Under its spell, Thoreau would write:

> How novel and original must be each new mans [*sic*] view of the universe—for though the world is so old—& so many books have been written—each object appears wholly undescribed to our experience—each field of thought wholly unexplored— The whole world is an America—a *New World.* (4:421; 4/2/52)

Thoreau's labor in these years was cut short by his early death, but by 1860 he was shaping interlinked clusters of essays, drawing details of rural nature—acorns, autumn leaves, wild apples, huckleberries—into explorations of perception, epistemology, economics, and morality. In effect, what came after *Walden* was a deep concern with what comes after: with principles of succession, continuity, daily sustenance, and the ongoing, chaotic processes of life. "The sun climbs to the zenith daily high over all literature & science," continues Thoreau in the passage above; "—astronomy even concerns us worldlings only—but the sun of poetry & of each new child born into the planet has never been astronomized, nor brought nearer by a telescope. So it will be to the end of time. The end of the world is not yet."

Before proceeding, it might be helpful to add a few words on the interdisciplinary nature of this project. This rereading of Thoreau's career requires a rethinking of the disciplinary barriers that have kept Thoreau in an uneasy intermediary position, straddling a widening chasm between the hostile camps of literature and science.[3] The "two cultures" may seem a truism today, but in the 1840s and 1850s, the relationship between "literature" and "science," while complex, was hardly hostile. Science was neither so monolithic nor so intimidating as now. Educated readers turned with ease to the primary works of scientists, and responded directly to the arguments advanced therein; scientific and technological advances were seen as signs of the times, part of the buzz and flux of the newspapers, parlors, and periodicals, right alongside—often the subject of—poems and stories and gossipy fillers.[4] Even the nature and degree of their differences were not the same as today, making it necessary to recover what Thoreau meant by "philosophy," "poetry," and "science," rather than impose our meanings anachronistically. This problem is complicated by the fact that Thoreau's own discourse shifts, both as his concerns change and as these terms shift meaning within a broader social context.

As a student at Harvard in the 1830s, Thoreau studied mechanics, astronomy, optics, and electricity under the rubric of "natural philosophy," and the title of his zoology text, an "eighteenth-century classic of physicotheology" by William Smellie, was *The Philosophy of Natural History*.[5] Though the word "scientist" had been coined by William Whewell in 1833, by analogy with "artist," it was not yet in general use and would not be for some decades, largely because of resistance to the narrow professionalization it implied.[6] Meanwhile, "natural philosophy," "natural history," and "science" were virtually interchangeable, and a person practicing in any of them was usually called a "man of science" or a "philosopher."[7] Similarly, "poetry" could designate not just verse or belles lettres but the creative human spirit generally, Wordsworth's "breath and finer spirit of all knowledge," a spirit which could manifest itself in science quite as well as verse.[8] But all of these terms and concepts were actively changing. By 1862, the year of Thoreau's death, "natural philosophy" was rapidly becoming obsolete on both sides of the Atlantic. Louis Agassiz was aggressively promoting "science" at Harvard's Lawrence Scientific School, and the forces that were feeding the success of the American Association for the Advancement of Science were meanwhile demoting "natural history" to merely amateur status. Soon scientists were writing exclusively for each other, and a new class of popularizers arose to interpret their work to the lay public. Our familiar, present-day boundaries were beginning to be drawn and defended.

Implicit in the broad, old-fashioned terms still current in the 1830s was the assumption that all knowledge formed a unified whole, within the common context of natural theology.[9] A canonical work for its time was John Herschel's *Preliminary Discourse on the Study of Natural Philosophy* (1830), which defined and illustrated what would come to be called the "scientific method." For Herschel, man's world is "a system disposed with order and design" by "a Power and an Intelligence superior to his own" (4), and science is a means to its contemplation and use. Before tabulating the progress of the branches of "physics," he impresses upon his readers that "natural philosophy is essentially united in all its departments, through all which one spirit reigns and one method of enquiry applies" (219). But we are unable to study it as a whole, "without subdivision into parts" (219); thus by what he elsewhere calls a "division of labour" (131), the "parts" arise: in the physical sciences, acoustics, optics, astronomy, geology, crystallography, mineralogy, chemistry; in biology, physiology (the study of "organization and life"), zoology, botany. Collective human endeavor can grasp the whole, but individual limitations necessitate division into disciplines. Hierarchies are forming: geology is second in sublimity only to astronomy; experimental sciences like mechanics are making rapid progress; observational sciences must make "slow, uncertain, and irregular" progress unless and until they too become susceptible to experiment (77–78).

Specialization and disciplinary boundaries ultimately prevailed because they created the conditions for an astonishing productivity. But while natural philosophers and poets together hailed the growth of knowledge, they also viewed its consequent subdivision and fragmentation with dismay. Voice after voice calls for knowledge to be made whole again, for the great system to be conceived entire, for all to remember and reiterate, as does Herschel, that the essence of the universe is not division but *unity*. Emerson's 1837 address "The American Scholar" voices the hope and the fear: there is but "One Man,—present to all particular men only partially . . . you must take the whole society to find the whole man." In the "*divided* or social state," functions must be parceled out to individuals; each must, to "possess himself," sometimes return from his labor to embrace the others. "But unfortunately, this original unit, this fountain of power, has been so distributed to multitudes, has been so minutely subdivided and peddled out, that it is spilled into drops, and cannot be gathered." The amputated "members" of society are so many "walking monsters,—a good finger, a neck, a stomach, an elbow, but never a man" (CW I:53). Instead of men, they are so many *things*.

It was shortly after this address that Thoreau came under Emerson's mentorship, to be confirmed and assisted in his own effort to be such a "whole" individual. For serious intellectuals of his time, such an effort might embrace human knowledge, attained and attainable, in whatever form: arts and manufacture; poetry and the creative spirit; science and philosophy. Thoreau took it upon himself to achieve in all three. As his writing matures, he is not so much conflicted as experimenting with new ways to keep fields and activities united that we now perceive to have been diverging permanently. To follow his lead requires the historian to work across disciplinary boundaries as they exist today. Indeed, what discipline can safely be left out? This study draws from literary theory, the history and philosophy of science, and the sociology of scientific knowledge. Thoreau was also a classical scholar and an engineer, a critic of capitalist politics and economies, attentive to aesthetic theory and the visual arts, a contributor to folklore and to environmental science. Of course no study of his time dare ignore social or political history, the construction of race or gender, nor, as should already be clear, theology.

If competence is required in all these fields, then responsible interdisciplinarity is doomed, an oxymoron. But I proceed nevertheless, under the conviction that what now exists as a fractured kaleidoscope of academic specialties existed in Thoreau's day as a more or less experientially unified field. This study, then, looks toward a reconstruction of nineteenth-century interdisciplinarity which asks how, out of a continuous if not unified "field of strategic possibilities," differences emerged to fracture it.[10]

Theoretical formulations are emerging which can underpin and enable such a study, particularly one that works across the "two cultures" of literature and science. N. Katherine Hayles discusses disciplinary convergences and differences in terms of a shared cultural matrix, in which parallels operate through the general cultural field, "a diffuse network of everyday experience," and dissimilarities through distinct disciplinary traditions which guide inquiry and shape thought. "The dual emphasis on cultural fields and disciplinary sites implies a universe of discourse that is at once fragmented and unified." Isomorphisms arise in different discourses in response to a shared cultural environment, while disciplinary differences are maintained through the perpetuation of economic infrastructures; Michel Serres' concept of equivocation, the effort both to celebrate "noise" locally and to suppress it "in global theories," offers a way for the theorist located at the "crossroads of disciplines" to mediate between them. The result is an attempt to create "an equivocal site at which both disciplines can have a voice." Her vision is not of science influencing literature (or vice versa) "but of literature and science as two mingled voices."[11]

That said, I must of course own up to my own disciplinary site: my inquiry is guided and my thought shaped by traditions within the discipline of English and American literary history. But I focus on a canonical literary figure precisely because those disciplinary traditions do not account for his activity. My disciplinary site both determines and enables my own "line of sight," by establishing my concerns in the context of literary criticism, theory, and history; but for Thoreau, deeply concerned with both the enabling and the limiting character of site and sight, a purely literary account proves inadequate. Thoreau, significantly, did develop and employ a working scientific methodology. Thus I hope to show that Thoreau, too, is such a "theorist" at the "crossroads of disciplines," mediating between the disciplinary economies of "literature" and "science" through particular linguistic and conceptual structures, in an effort to see them as fundamentally coincident even as they were, historically, dissociating each from the other. The "two mingled voices" of literature and science are distinctly heard in his own work, though his own favorite metaphor for this mingling is not aural but visual: he experiments with what he calls "intentionality of the eye." If the field is fundamentally unified, then perhaps perception and cognition join to create from it particular and varying views. Disciplinary differences are evoked by approaching the "uni-verse" with consciously varied intentions. Each discipline becomes in effect a mental lens, a mode of perception as well as of discourse.

Hayles's use of terms like "discourse" and "voice" intimates that an underlying connection between disciplines exists in language. This is made explicit in George Levine's discussion: literature and science are, indeed, "modes of discourse," languages which both differ and converge: "science and literature reflect each other because they draw mutually on one culture, from the same sources, and they work out in different languages the same project." This view draws on antifoundationalist arguments (such as those of Feyerabend and Latour), which counter the conventions of naive realism by reversing most of its assumptions: science does not objectively discover hidden truth, but creates theories whose coherence and social valence determine both the data and the degree of acceptance on which their authority supposedly rests. This analysis of science, Levine notes, moves into the "literary fold," but importantly does not subvert science but rather "historicizes and humanizes it." Once science is no longer seen as clearly separable from other humanistic enterprises, literature and science involve each other: "Literature becomes part of the history of science. Science is reflected in literature. And the tools of literary criticism become instruments in the understanding of scientific discourse." For the common medium is, necessarily, language.[12]

Levine continues to distinguish them even on this ground: they are "modes" of discourse, not the same mode; "different languages," not identical. For Thoreau, science was, precisely, a "language," and it is for this reason that he became so interested in it: What were its expressive possibilities? As a language, science was enabling, a site which gave him "sight." This characteristic is most explicit in his love of scientific names, which made visible modes of reality that otherwise went unseen. But at times Thoreau experimented with a more radical formulation: literature and science were not different languages, but in their purest and highest form inseparable, both "simply some human experience" (VI:236–37; 5/5/54). As personal experience, the best science, and the best poetry, would come together and would "read" the same. This position is echoed today by the radical interdisciplinarity of the "strong programme" of the sociology of scientific knowledge: in Steve Woolgar's words, once science and society can no longer be studied as "separate analytic objects . . . these conceptions collapse into one another." That conventional triad, "science, technology, and society," is traditionally diagrammed as interconnected but not overlapping, but "the radical view maps these entities one on top of the other," comprising "a single domain." Science does not have "'social aspects'" but "is itself constitutively social." A scientific theory, then, must be spoken of not as objectively true, but as a more or less "stable construction."[13] With such a view all the barriers are down. Once the demon of "social construction" is let loose, a single social field—or a multitude of variously contested fields—is the only logical outcome. All disciplines collapse back into the "one culture," or the unifying "cultural matrix," out of which Hayles and Levine see the emergences of different discourses.

It is tempting to equate Thoreau's own view with this radical formulation, because it corresponds in some degree both to the nineteenth-century view of the world as a cosmic whole, and to Thoreau's more radical attempts to embrace poetry, philosophy, and science simultaneously. But as Donna Haraway worries, if all nature is a social construction, a projection onto a "screen," it is, essentially, dead. Haraway's solution is to acknowledge, and give voice to, the multiplicity of living agents, not in a global or totalized space but locally, in what she calls "situated knowledges." Haraway asserts the redemptive value of "sitedness"—of knowledge that has the enabling properties of being sited, in discipline, place, history, gender, personality: "The only way to find a larger vision is to be somewhere in particular." Thoreau's conscious experimentation with intentional and partial vision centered in personal, empirical experience anticipates Donna Haraway's quest for a "feminist objectivity" which defies the transcendent God-vision of the unmarked gaze by insisting on the "embodied nature of all vision," through eyes which are "active perceptual systems."[14] Both Thoreau and Haraway look to the same end: sight that is *answerable* to what it sees. Her stress on "situated knowledge" has particular relevance to Thoreau's project, sited, with peculiar attention, in the disciplines of philosophy, science, and literature; in Concord, Massachusetts; in the empirical/imperialist nineteenth century; in his own complexly gendered, aggressively idiosyncratic self. To "situate" knowledge also means acknowledging the limitations of being locally sited: surrender of the power of universality. One must allow others their voices and not attempt to globalize the local, to break falsely through limitations by universalizing them.[15]

Thoreau was faced with the problem of inhabiting both the local and the global, or how to move between the particulars of his time, place, and temperament, and what he often called "higher law"—or between history and myth—without sacrificing either extreme. Along a slightly different axis, he valued both what he called "extravagance," or proliferation, unpredictability, and excess whether in rhetoric or in nature; and "purity," the simple, lawful truth that kept extravagance from fracturing into mere mannerism. The latter he frequently encoded as "science" or "philosophy," the former as "poetry" or "language."[16] Though he sought them both, simultaneously, in "the field" —of his own culture, of the Concord landscape, and of his own sense of self—his search was not conducted according to any of the single models available to him. That is, he finally could not accept the idealist move, to reach the universal by annihilating the restraints of the local and particular; nor did he accept the limited and methodically realized aims of the scientist's methods. Or rather, he *did* accept them both—by a process of reconciliation modeled for him in much of the discourse of the time, which sought to bring together polar opposites into new, progressive, higher unities.

This common feature of nineteenth-century thought was systematized and theorized by William Whewell, who named it "consilience": the advancement of knowledge that occurs when apparently unconnected classes of facts are seen to have "*jumped together*" to form a new, simpler, and more unified theory. This new knowledge is not "the mere *sum* of the Facts": "The Facts are not only brought together, but seen in a new point of view. A new mental Element is *superinduced*" upon the particulars (153, 139), and this mental element or "*new conception*" is "a principle of connexion and unity, supplied by the mind" (163). In nineteenth-century terms, Thoreau rejected neither poetry nor science, nor did he simply "reconcile" them, collapsing them together. In the "consilience" of Emersonian transcendental wholes with Humboldtian empirical science, he sacrificed neither but attempted to create a way of knowing which combined them both into something new. In today's terminology (Whewell's useful word not having survived, a fact revealing in itself), this integration of knowledge into a new, coherent entity describes the highest level of interdisciplinary studies, where disciplines are not juxtaposed additively but integrated into a new synthesis. The ultimate goal is something even more comprehensive: what Erich Jantsch, in an influential formulation, calls "transdisciplinarity," signifying "the interconnectedness of all aspects of reality, transcending the dynamics of a dialectical synthesis to grasp the total dynamics of reality as a whole."[17] Here is a twentieth-century return to that nineteenth-century hope, poised as it was on the verge of the disciplinary fragmentation that made it seem an unrealizable romantic dream.

Thoreau was preoccupied with this innovative work from the mid-1840s on, when he was exploring and applying his developing insight in modes that are often experimental and always unstable and interactive. The result could be called a kind of situated knowledge, which celebrates his own locale but also seeks to interlink with each and every locale; to give voice to *all* the agents creating the world he knew, human and nonhuman, present, future, and past; to celebrate his own individuality, not to colonize his readers but to inspire their own parallel celebration; to reach a connective truth through the commonplace particulars of daily life in a place exemplary only in its ordinariness. His act of consilience seeks to give voice to all the participating agents, not by blending them together but by giving each a distinct hearing in a medium of sustained attention. In consiliating literature and science, Thoreau tried to enable and enact both, as real knowledge situated in, not beyond, the world; and since language is the medium of our knowledge for Thoreau, the sustained and daily use of language was the necessary way to knowledge of his world. One final possibility, then: given his persistent concern with the problem of disciplinarity (and his interest in theoretical and actual "transgressions" of discipline), his project may have been less pre- than postdisciplinary, and his experiments may have value to us today, not as antiquarian remnants of time passed, but as models of a postdisciplinary practice. This practice was incipient in his published work, and may have been more fully realized had his early death not left his work unfinished. All the more, then, will rediscovering it help us to interrogate the familiar canonical figure of Thoreau anew, and in the process interrogate ourselves as well.

If nineteenth-century boundaries were fluid and permeable, it is now all the more necessary not to solidify them in retrospect, anachronistically. This conviction has directed the method of this study, which began with a reading of what Thoreau read and what he wrote—together, in chronological sequence—in an attempt to inhabit something of the same field of knowledge as he. One inspiration was Martin Rudwick's call for "empirical studies of science in the making" that would focus on a problem that brought together a group of individuals in "an interacting network of exchange," making "'small facts speak to large issues.'"[18] Hence I hope to make the small facts of Thoreau's curious field studies speak to the larger issues of literature and science in Thoreau's career, and potentially the very large issue of literature and science in nineteenth- and twentieth-century America.

Another inspiration was Robert M. Young's acerbic suggestion: "Instead of burrowing deeper into the minutiae of critical receptions of Victorian novels, it might be worthwhile to pay closer attention to the works and the movements which *evoked* so much Victorian writing"—combined with his caution that "if one works backward from the secondary literature to nineteenth-century documents," the thesis that science, theology, literature, politics, and social and economic theory are part of a single debate "does not even arise." His prescription is "to place oneself in the midst" of the documents of the time "and to discover the highly integrated network of issues in all these spheres."[19] Therefore the study that follows evolved through a roughly chronological reading and interpretation of Thoreau's written environment. I do not pretend to recreate this environment fully—for that a lifetime alone would hardly suffice. The mere catalog of Thoreau's known reading fills a book in its own right,[20] and his *Journal* bulks to fourteen large volumes even in the incomplete 1906 edition. Thus, for my partially recreated "field" I deliberately select those canonical and noncanonical texts which Thoreau is known to have read that bear more or less directly on "scientific" modes of discourse—from Locke and Paley to Emerson, Coleridge, and Carlyle, to scientists such as Lyell, Humboldt, and Darwin. What this leaves out is significant: Greek and Roman classics; literature of travel and exploration; English, Welsh, and Scottish poetry; Hindu scripture; local and colonial history; studies of "American Indians."[21] While Thoreau read exhaustively in all of these areas, they must here be set aside, and of those texts which I admit, I will necessarily emphasize those selected landmark texts which loom largest in Thoreau's imagination, as gauged by the number of references to them and extracts from them in his works.

Finally, it is only proper to acknowledge my own stakes—for as Darwin remarked in a letter, "How odd it is that anyone should not see that all observation must be for or against some view if it is to be of any service." And as Thoreau added, "Every man thus *tracks himself* through life, in all his hearing and reading and observation and travelling. His observations make a chain" (XIII:77; 1/5/60).[22] I am on the trail of that empirical, nonorganicist alternative, which I believe formed a strong competitor to dominant, and ultimately "victorious," organicist views. These views have cast Thoreau as above all else a poet heavily invested in romantic symbolism and organic holism, as evidenced most clearly in his masterwork, *Walden*. This book did indeed, in many ways, climax and conclude Thoreau's career as an Emersonian poetic symbolist, but in other ways it cleared the ground for—and even necessitated—a major new effort which is insufficiently recognized. The irony of marketing Walden Pond as a symbolic artifact was not lost on Thoreau. I would argue that the selling of Walden/*Walden* consolidated his emerging career as an empirical, rather than a rational, holist. In his late essays he celebrates not the crash of metaphysical dualisms but the murmur of multiple voices and actions, not the ecstasy of transcendental disembodiment but embodiment's perilous and bittersweet joys, not the imperial eye and the power of the all-dissolving mind but the immense beauty of, and urgent need for, the vision of all and the wise action of the collective.

In his last decade Thoreau repeatedly invoked chaos and contingency, the generational source, the saving "wild," which alone could redeem an encrusted, static, and alienated civilization. The terms of his late writings suggest something incongruously akin to "postmodernism," an association which turns inside out modernism's own association with organicist wholes, attended by a literary historical periodization which has locked in the very structures Thoreau strove to keep fluid and interactive. The resulting popular image envisions Thoreau as a premodern isolate who turned his back on society to rhapsodize about a pure, untainted, Edenic American nature. This iconic figure reasserts the tragic and sterile dualisms between subject and object, freedom and fate, spirit and matter, the human and the natural, which Thoreau himself inherited from Emerson and Coleridge, which he lived and experienced fully, and which he fought to disrupt and disown in the name of creating a future that might succeed the "evil days" ushered in by romantic alienation in a commodified society.

This seems to me an important story to recover. It appears that the nonorganicist, or "Humboldtian," alternative had liberatory potential which made it at once exciting, dangerous, and vulnerable. If it was indeed suppressed and then written out of our history, we are left with a distorted view of our own past. In recovering it, we may learn of alternatives to present dilemmas, and discover that our attempts to find a "postmodern" ground for literary criticism and social critique are not sui generis but in fact have a history and a tradition from which we may learn, on which we may build.

1

Facts and Truth

Transcendental Science from Cambridge to Concord

Let us not underrate the value of a fact;
it will one day flower in a truth.
—Henry David Thoreau
"Natural History of Massachusetts"

Roughly seven weeks after Thoreau's graduation from Harvard, Emerson's literary prompting of his new young friend (Emerson himself was only thirty-four) resulted in the famous opening of Thoreau's *Journal*, on October 22, 1837: "'What are you doing now,' he asked, 'Do you keep a journal?'—So I make my first entry today" (1:5). A little over seven weeks later, Thoreau formulated his earliest thoughts on natural fact and moral truth into a key statement:

> How indispensable to a correct study of nature is a perception of her true meaning— The fact will one day flower out into a truth. The season will mature and fructify what the understanding had cultivated. Mere accumulators of facts—collectors of materials for the master-workmen, are like those plants growing in dark forests, which "put forth only leaves instead of blossoms." (1:19; 12/16/37)

The mysterious link between fact and truth continued to haunt him. In 1842, when Emerson commissioned what was to be Thoreau's first natural history essay, Thoreau culled the 1837 passage and cast it into the sentence which stands at the head of this chapter. Facts are the seeds of truth, sterile in the hands of the collector, but fertile in the ground of the mind. This formulation, familiar as it is, nevertheless poses some puzzles: how does it happen that Thoreau can conceive of "mere accumulators" as against wiser minds who perceive "true meaning"? Why, on the other hand, should he feel the need to defend the "value" of such evidently "underrated" things as "facts"? And why does he choose to relate the two poles of fact and truth through a vertical image of organic assimilation, growth, and florescence?

Nominalists, Realists, Idealists: Harvard and After, 1837

In the months around his graduation, Thoreau was immersed in a constellation of books which cast fact and truth as polar terms in a larger debate. At Harvard, he had recently studied the mainstays of natural and moral philosophy: John Locke's *Essay Concerning Human Understanding* (1690), Dugald Stewart's *Elements of the Philosophy of the Human Mind* (1792, 1814), and William Paley's *Natural Theology* (1802).[1] That summer and fall, he graduated to the international set of authors at the leading edge of the new philosophy: Cousin, Coleridge, Carlyle, Emerson, Goethe. Gradually and in stages he was moving from Harvard's Unitarian orthodoxy to the radical propositions of romanticism. Thoreau's bookish enthusiasms were deepened by his warming friendship with Emerson. Soon he was accepted into the loose circle of reformers and intellectuals who composed the "Hedge Club," a group notable enough to attract public attention and a name: the "Transcendentalists." Thoreau's allegiance to the new philosophy was immediate and total, and the ideas he first encountered in 1837 shaped his thought for over ten years. Through it all, Emerson acted as motivator, mentor, conduit, and catalyst for exciting ideas that promised liberation—indeed, that promised no less than intellectual and social revolution.

Liberation from what, exactly? From something that was condemned as "mechanism," or as "atomism," or even "the corpuscular theory." This oppressive enemy was associated with Locke and his sensation-based, skeptical philosophy, but "Locke" and "mechanism" were often little more than code words for discontent with the past and the inherited ills of the present. This broad condemnation was focused on the church, philosophy, science, the commercial and industrial economy, social disintegration. Behind all these aspects arose still the memory of the French Revolution, the specter of atomistic mechanism's terrible potential.

Yet the texts Thoreau was reading do not stake out simplistic positions in a debate between eighteenth-century "empirical" or "inductive" reasoning, and the "idealist," a priori rationalism of the new philosophy. Terms and positions turn out to be more complicated. For example, the "empiricists" do not after all advocate the "mere accumulation of facts," but value theory and hypothesis as much as the "idealists"; while the "idealists" are just as anxious to claim validation of their theories with inductive, "empirical" evidence. Transcendentalism repudiated Locke, Stewart, and Paley, while demonstrating close ties with all three; while in Cousin and Coleridge, romantic liberation assumes some problematic and oddly authoritarian implications, which in Coleridge become an explicit social program. Some of the resulting stresses between "dry" particulars and the transcendent whole are exemplified in Goethe, and eventually find their way into Thoreau's 1842 essay "Natural History of Massachusetts," which plays off much of his reading in philosophy. That is, given the universal agreement, across this spectrum, on the need to employ *both* fact and theory, how exactly were particular facts of the natural world to relate to "truth"?

The dichotomy creates the problem: How is the gulf between the two halves to be bridged? Or can the two finally be married into one complete whole? Locke's *Essay* marked one extreme: human understanding was like a "dark room," "a closet wholly shut from light" with some small openings to let in ideas (II.xi.17). Locke's philosophy, "empiricism" or "sensationalism," dominated American academic and clerical institutions. Acceptance of Locke defined what Emerson called the "pale negations" of the Unitarian old guard, as much as rejection of Locke and all he stood for came to identify the transcendental radicals.[2] The totality and sheer exuberance of transcendental opposition are suggested in Emerson's notorious passage from *Nature* (1836): "I become a transparent eye-ball; I am nothing; I see all; the currents of the Universal Being circulate through me; I am part or particle of God" (CW 1:10). From a dark, nearly impenetrable closet, to a raw eyeball bathed in divine light: the battle lines were, it would seem, clear.

But of course transcendentalism also built on Locke. In their rational, "natural" questioning of tradition, both Locke and the transcendentalists meant to clear the ground for a new start. As Emerson asked, "Why should not we also enjoy an original relation to the universe?" (CW 1:7). Locke's *Essay* is permeated by a heady anti-authoritarianism and a demand that men think for *themselves*: "The floating of other men's opinions in our brains, makes us not one jot the more knowing, though they happen to be true" (I.iii.24). Truth, then, comes not from tradition, but from experience with particular things, which is all there really is; "*general* and *universal* belong not to the real existence of things; but are the inventions and creatures of the understanding, made by it for its own use" (III.iii.11). Things therefore have no "real essence," but acquire only a "nominal" one, as we sort them out and name them according to our own interests.

However, Locke's claim that our ideas are abstracted from things has two consequences which transcendentalism sought to circumvent. First, since we lack control over both the objects of our sensation and the operation of our minds, our understanding is fundamentally passive. Hence transcendentalism proposed a higher faculty, Coleridgean "Reason," to overcome the limitations of our sense-based understanding. But this bypasses a second interesting consequence: since even our simplest ideas are not "given" to us but actively produced by our perception, our complex ideas are not passively received but actively framed by the mind (II.xii.1). Against the absolutist imperatives of Coleridgean "Reason," Thoreau would develop this idea at length, calling it the "intentionality of the eye." In so doing, he forwarded Locke's suggestion that our minds participate actively in creating knowledge, a concept that, by the 1850s, will link Thoreau with the vanguard of scientific theorists, including Herschel, Whewell, and Humboldt. But sacrificing "Reason" meant sacrificing certainty: to these theorists, even scientific knowledge is at best probabilistic. In such a universe, lacking in boundaries, certainties, and absolutes, all things are interconnected in ways which we cannot yet perceive. The idea traces to Locke: "things, however absolute and entire they seem in themselves, are but retainers to other parts of nature," and the most "complete and perfect a part that we know of nature" owes its being and its excellences "to its neighbors" (IV.vi.11). One might almost call such a vision proto-ecological.

Thoreau always remained sympathetic to Locke's most central idea: knowledge would be renewed amidst falsehoods, and truth discovered, by going back to things in themselves as they exist, letting experience teach where Reason cannot.[3] "What is more common than error?" Thoreau asked, in an 1837 class essay; the confidence of the "embryo philosopher"

> in the infallibility of reason is shaken,—his very existence becomes problematical. He has been sadly deceived, and experience has taught him to doubt, to question even the most palpable truths. He feels that he is not secure till he has gone back to their primitive elements, and taken a fresh and unprejudiced view of things. He builds for himself, in fact, a *new world.* (EE 103; emphasis added)

It is not the last time Thoreau will discover a "new world" through a return to those "primitive elements" of truth. The caricature of the empirical drudge, accumulating mere facts, may operate as a rhetorical counterweight to the image of the poet's fanciful flights—but the poet may also be confined to the old world of preconceptions, while the "drudge" emerges from the patient study of primitive particulars to spread forth for our consideration a vision of a remade universe. In one form this theme becomes *Walden*'s Artist of Kouroo; in another, the naturalist who builds for himself a new world of Concord.

Locke's philosophy was first introduced to Harvard students through Dugald Stewart, whose position as a contributor to the new philosophy was, like Locke's, problematical. Transcendentalists worried that the Scottish philosophy, by establishing a common inner moral sense, only meliorated what should be rejected outright; yet the acceptance of a common moral sense did provide a sort of "stepping stone" to a more radical formulation.[4] Stewart's definition of "philosophy" as that which mediates between fact and truth—"the analytical investigation of general laws from the observed phenomena" (II:482)—would be recurred to by Thoreau. Along another pathway, this notion would surface in Darwin's oft-quoted statement "that science consists in grouping facts so that general laws or conclusions may be drawn from them."[5] According to Stewart, it was absurd to contrast theory and experience as if they were opposites. A general idea and its supporting facts needed each other: without theory, "experience is a blind and useless guide; while, on the other hand, a legitimate theory . . . necessarily presupposes a knowledge of connected and well ascertained facts, more comprehensive, by far, than any mere empiric is likely to possess" (II:444). Most discoveries had actually started with a theory, which was "generally the best guide to the knowledge of connected and of useful facts" (II:406). The *real* danger, for Stewart, lies in bad theories and vicious theorists who twist "facts" around, "amassing a chaos of insulated particulars" to support their own systems, all the while declaiming against the "fallacy" of systems. Whom might Stewart have had in mind here? Without naming names, he gives a single, telling example: to refute the theory that slave labor is actually less productive than that of freemen, certain dubious "evidence" has been offered from "the other side of the Atlantic" (that is, America)—while *accurate* examination has shown "how wide of the truth" such calculations based on "experience" really are (II:449).[6] In other words, statistics can lie, and "facts" be manipulated for political ends. To correct this danger, Stewart calls for "philosophical studies" with a proper distrust of human ability to embrace complex detail, and "a religious attention to the designs of Nature," as displayed in general laws (II:451).

One will not find, in these Harvard-assigned empiricists, blind collectors of facts. Indeed, Dugald Stewart explicitly supports one of Emerson's central transcendental doctrines, the theory of correspondence. The material and moral worlds clearly form "parts of one and the same plan," a conclusion confirmed by "all the discoveries of genuine science" (II:398).

> The presumption unquestionably is, that there is one great *moral system*, corresponding to the *material system*; and that the connections which we at present trace so distinctly among the sensible objects composing the one, are

> exhibited as so many intimations of some vast scheme, comprehending all the intelligent beings who compose the other. (II:399; emphasis in original)

Furthermore, the evidence is precisely of the same sort as Newton's; it is merely "accidental circumstance" that Newton could verify his conjectures by appealing to "sensible facts." Emerson found in Swedenborg confirmation of what he had already learned from an older, and far less fashionable, source; and Thoreau, reading *Nature* in the summer of 1837, met in the concept of correspondence an old and already familiar friend.

In the debate that is taking shape here, what Locke and Stewart share is their position as "nominalists" who argue that categories of human thought do not really exist in nature, but are mental constructions; words are not things but "mere signs of ideas," names we have given to things according to our own interests and concerns.[7] Their opponents were termed (confusingly, for us) "realists," for they held that the categories of human thought do "really" exist in nature, that they name a prevailing order that is preexisting and determined by God. Thus, for instance, William Paley's order is as absolute, bounded, and certain as Locke's was probabilistic. Yet across this divide, Paley shares significant common ground with Locke and Stewart: he, too, insists that theories and facts generate, and depend upon, each other. This interdependency is demonstrated exhaustively by his classic work, *Natural Theology; or, Evidences of the Existence and Attributes of the Deity, collected from the Appearances of Nature*, a required text for decades of Harvard students presenting a single vast theory: God created a harmonious universe. This theory is explicated through the facts or "evidences" that both give rise to it and support it. As a "realist," Paley defies Locke, maintaining that categories are not human constructions but divinely ordained. The construction, or "contrivance," is there; however, it is not ours, but God's: "There cannot be design without a designer; contrivance, without a contriver; order, without choice. . . . Arrangement, disposition of parts, subserviency of means to an end, relation of instruments to a use, imply the presence of intelligence and mind" (8). Not that such a "presence" was denied by Stewart or Locke, but with Paley we are on the other side of the gulf of knowability: material and moral worlds aren't just *analogous*, they are on some level *the same.* God's mind is commensurate with ours, his works of nature with our mechanical devices, and his world is fully knowable by our minds.

What makes Paley so attractive is his sheer wonder at the ingenuity of God's various contrivances, and his delight in his ability to solve the puzzle God so gamesomely sets before us. *Natural Theology* is an exuberant celebration of the unity of so vast a plan, built out of such an astonishing wealth of detail: "How many things must go right for us to be an hour at ease! how many more for us to be vigorous and active!" (90). "The universe itself is a system," interdependent and interconnected (312). What is more, the design is beneficial, for the Deity has superadded pleasure beyond what was necessary.

> It is a happy world after all. The air, the earth, the water, teem with delighted existence. In a spring noon, or a summer evening, on whichever side I turn my eyes, myriads of happy beings crowd upon my view. . . . If we look to what the *waters* produce, shoals of the fry of fish frequent the margins of rivers, of lakes, and of the sea itself. These are so happy, that they know not what to do with themselves. . . . (317–18)

In this genial appreciation of sensuous delight, Paley and Emerson might have sat down together, expanding mutually in the air "like corn and pumpkins" —and in Emerson's insistence on "this general grace diffused over nature" (CW 1:14) one hears Paley too, especially in Emerson's theory of compensation. For Paley, everything is part of God's plan. Even pain has its benefits, and "The horror of death proves the value of life" (348). For Emerson, "Even the corpse has its own beauty" (CW 1:14). For Paley, "*Money* is the sweetener of human toil; the substitute for coercion; the reconciler of labour with liberty" (353); for Emerson, it is, "in its effects and laws, as beautiful as roses" (CW 3:136). In both, the same assumption—that the universe is a rational, harmonious unity—leads to a Leibnitzian optimism: all that is, is right. What are, for twentieth-century readers, some of Emerson's most indigestible claims flow out of this certainty, which is reiterated constantly in this literature and which forms the bedrock of the doctrine of natural theology. And though the optimism may be tempered, the unity of plan on which Paley rests his argument also grounds Goethe's contemporaneous search for the lawful, unified plan of nature. The romantics did not disavow natural theology, but remade it according to new specifications.

For by 1830, the positions of nominalist and realist did not completely describe the field. A new contender had entered the debate, and it too claimed science for support of a theological system. "Idealism" reached America in a number of forms, from England, Germany, and France. Among the most interesting of these were the works of Victor Cousin, whose philosophy of "eclecticism"—yet another attempt to unify all under a single plan—enjoyed a vogue in America in the 1830s and 1840s.[8] In Cousin's startling and fascinating book, *Introduction to the History of Philosophy*, Paleyan optimism takes an even more extreme form. The universe is in constant harmony, and in universal life nothing perishes, all is balanced, even in history: in Cousin, all that *was* is right, too. War, for instance, is useful, because it causes both the "defeat of the people that has served its time," and the victory of "the people which is to serve its time in turn, and which is called to empire"

(275). "In reality, not a single great battle has taken a turn detrimental to civilization," Cousin declares (281). Success and victory are self-justifying: "the vanquished is always in the wrong" (327). Similarly, every vicious act will bring its own punishment, and "nations as well as individuals" are sure to "meet the fate which they deserve" (284).

Cousin explicitly repudiates Lockean nominalism, since thought is prior to words. But realism stops short too, since ideas are not just things but conceptions of "universal, absolute, and infallible reason" (124, 128). In Cousin's deep idealism, ideas—the useful, the just, the beautiful—are inherent in the mind of man, and only fitfully revealed by a nature which "veils and obscures" (13). The ideal alone is pure; "All that is actual, is mingled and imperfect" (13–14). Man's task on earth is to reform defective nature into art; his goal, "the entire absorption of nature into humanity" (9). Since the real is imperfect, thought becomes the "real" reality: ideas are neither mere words, nor beings, but "conceptions of human reason" and therefore of "absolute reason"; "they are nothing else than modes of the existence of eternal reason" (129).

In the course of the book, Cousin develops an elaborate theory of history without reference to a single historical event, person, object, or place. As he says: "We cannot apply to facts: for what information can we obtain from facts? From them, we can learn only their existence; but neither its reason, nor its necessity. We must, therefore, according to our ordinary method, apply to thought" (214). Yet in spite of his dismissal of "particularities" as "evanescent" and "nothing but illusions" (182), Cousin nevertheless insists his method "is, in its ultimate analysis, no other than the method of observation and induction," and rests on nothing but the "terra firma of facts"—that is, "the manifestation of reason within the narrow but luminous sphere of individual consciousness" (228). History, according to Cousin, is human consciousness writ large, and the "three integral elements of reason" form the three and only three elements of history: first, "the infinite, unity, God" (218); second, the finite, multiplicity, variety; third, their combination, or generation. The result is "history modeled upon on human nature. This is no abstract system,—as some may suppose; it is a system founded upon realities; for it rests upon the very centre of all our real thought" (229). Cousin affirms that his system is, indeed, the only moral, beautiful, and "scientific" one:

> In fact that which constitutes science is the suppression of every thing anomalous; it is that which substitutes order for what is arbitrary, reality for appearance, and reason for sensation and imagination;—it is that which recalls and raises particular phenomena to a strict accordance with their general laws. (233–34)

Law is not "investigated" from phenomena or "drawn" from facts, but used to marshal the unruly into strict order. Hence Cousin could offer an ordered history of nations and peoples, pure of any contaminating or anomalous actual human life, but defended as resting on facts, observation, and inductive reasoning.[9]

In the "new philosophy" which was being introduced to Thoreau through Cousin, Emerson, and Coleridge, objects of nature had no independent existence. They were the thoughts of God. Cousin, for example, stressed that every fact had its significance: "The world of ideas is hidden within the world of facts," which in themselves are insignificant, "but impregnated by reason they manifest the idea which they envelope, they become reasonable and intelligible" (235). Sense data, or Lockean experience, and ideas in our own minds are correspondent and so become potentially interchangeable. Indeed, ideas like "Unity" and "Variety" are far more reliable than the sense data from which they are derived, as they perfect the impure and mingled state of nature.

It might seem a peculiar form of arrogance to solve the universe by a priori deduction, treating empirical data with such contempt—and yet paradoxically, the empirical study of nature was not treated with contempt, but honored and prescribed. Coleridge, Carlyle, and Emerson are all in the habit of pointing, like Paley, for evidence to the data of the senses. Idealism does not deny "facts," but on the contrary insists that it rests on "facts." But it does so by redefining the fact, the material object, as a thought of God. Therefore the reverse is also true: thoughts are as much "facts" as material objects are facts. Indeed, more so. As Cousin said, thoughts are not merely "things like other things," but better, for thoughts are linked directly to the divine mind, whereas objects are intermediary and hence unreliable. So, if thoughts are facts, deduction a priori becomes the purest form of science. If facts are thoughts of God, then the collection and collocation of facts becomes good science too, a secondary but still essential method of establishing and celebrating the nature of God's universe. This is the basic reason why the "new philosophy" embraced by Emerson and the transcendentalists conceived itself as not opposed to, but allied with, "science," defined as the ordering of particular phenomena "to a strict accordance with their general law," a return of the many to the One.

The notion that facts are "thoughts of God," present in both realist and idealist positions, creates a curiously dovetailed pair of results. First, if thought or ideas are primary, real-world facts are secondary—worse, are unreliable or even threatening allies of ideas. For without the coherence of a stable, ideal system they threaten to dissolve into "isolated particulars," or fly apart into the chaos of atomization. This theme is implicit in Paley and

Cousin, and developed obsessively by Coleridge, Carlyle, and Emerson. Second, one effective way to compensate for the apparent unpredictability of facts is to insist on their absolute, given, and divinely significant status, in a universe still larger than our understanding of it. To gather and exhibit facts, whether in cabinets and museums or taxonomies and treatises, exhibits one's control of the *truth.*[10]

The new philosophers longed to establish a unity that would combat incoherence. Cousin, for example, stressed that "eclecticism" would combine and harmonize the dangerous and, as he saw it, unnecessary fragmentation created by the proliferation of philosophical systems. By his book's end he has worked through all of history and is ready to pronounce the wave of the future: "Idealism is as true and was just as necessary as empiricism" (392), he intones, but both have been taken to their limit, and no third is possible. The solution is to combine and reconcile both "by regarding them in a point of view which, being more comprehensive than that of either the one or the other system, may be capable of including, and thus of explaining and completing them both." Thus emerges his philosophy, which unites the opposites of Kant's "subjective idealism" and "the empiricism and sensualism" of Locke "by centering them both in a vast and powerful eclecticism" (414), a system that reckons with, interrogates, and judges all other systems. This is already happening: Cousin points to current philosophy, in which the reign of exclusive systems is past and "idealism" and "sensualism" are attempting to meet and mingle (427–28). Eclecticism, he proclaims, is "necessarily the philosophy of the present century" (440). Cousin's grand unification theory exemplifies the urge to all-embracing coherence which characterized natural theologians, from the Anglican William Paley to Cousin, Coleridge, and the German *Naturphilosophen.* But Cousin, who made room for all, was more ecumenical than his more familiar contemporaries, who were militant on the question of what must be recovered and empowered, and what must be jettisoned or destroyed. Emerson, Coleridge, and Carlyle were haunted by a sense of crisis, anxious to combat the threat of the "past" they sought to supersede, and which they identified with Locke, materialism, empiricism, and the dangers of social revolution.

Romantic Theologies

Emerson's most immediate break was with the "dead forms" of the Unitarian church, which he resigned in 1832, to pursue his ministry in the less doctrinal forums of the lecture and the essay. The decisive moment was his 1838 address delivered (by the students' invitation) to the graduating class of the Harvard Divinity School. In the aftermath Emerson was banned from

Harvard. In "The Divinity School Address," Emerson claims that revelation seems "somewhat long ago given and done, as if God were dead," and that the institution—*their* institution—has become "an uncertain and inarticulate voice" preaching sermons that "clatter and echo unchallenged" (CW 1:84, 86). The resulting "loss of worship" precipitates the nation into calamity:

> Then all things go to decay. Genius leaves the temple, to haunt the senate, or the market. Literature becomes frivolous. Science is cold. The eye of youth is not lighted by the hope of other worlds, and age is without honor. Society lives to trifles, and when men die, we do not mention them. (CW 1:89)

Emerson envisions nothing less than a society in total disintegration. His conviction led to a pitched battle with orthodoxy in the name of his proposed solution: "dare to love God without mediator or veil" (CW 1:90). That one bold move is at the root of Emerson's proposed revolution: jettison the dead shell of the past, and breathe life into the forms of the present by opening your soul "to the influx of the all-knowing Spirit" which annihilates our petty distinctions. Emerson closes the address by looking toward the Teacher who shall follow "those shining laws" full circle, "shall see the world to be the mirror of the soul; shall see the identity of the law of gravitation with purity of heart; and shall show that the Ought, that Duty, is one thing with Science, with Beauty, and with Joy" (CW 1:93). Notice, in the midst of this climactic hyperbole, the references to the law of gravitation and to science: Duty, Science, Beauty, and Spirit were thoroughly embedded in each other, in this peculiar mix we have come to call romanticism.

Coleridge ended in the arms of the church, as Emerson began there. *Aids to Reflection* (1825), a book of the first importance to Emerson, explicitly declares that its aim is to lead us back to a contemplation of "the Supreme Being in his personal attributes," in the doctrines of the Trinity, Incarnation, and redemption (347)—breathing, as Emerson desired, new life into old forms. That the doctrines are so apparently different is in this context unimportant, for they achieve similar ends. Coleridge is concerned, out of horror for pantheism, to recover the real and substantial "Jehovah" from Wordsworth's indefinite "sense sublime" (347). Emerson rejects God as a person ("the soul knows no persons" [CW 1:82]) precisely to enable a personal God, whom each of us can know directly.

What horrifies Coleridge at least as much as the vagueness of the "sense sublime" is the equation of God with his "physical attributes," the laws and objects of nature. He writes, "I more than fear the prevailing taste for books of natural theology. . . . Evidences of Christianity! I am weary of the word" (348). For to point to any *external* evidence is to fall into a trap:

> The last fruit of the Mechanico-corpuscular philosophy . . . is the habit of attaching all our conceptions and feelings, and of applying all the words and phrases expressing reality, to the objects of the senses: more accurately speaking, to the images and sensations by which their presence is made known to us. (348–49)

The person who exemplifies the worst dangers of this habit is none other than the late Dr. Paley, for whose fame and literary grace Coleridge can profess nothing but the highest envy. Nevertheless, he is "bound in conscience to throw the whole force of my intellect in the way of this triumphal car." For Paley is "plausible and popular" because he is "feeble," favors "mental indolence," and "flatters the reader" with his easy and "rational" Christianity (350–51). The effect of the harmful split which makes faith dependent upon reason (when it should be the other way around) is indicated by the story he tells about the reader of *The Friend* "who dropped the book and exclaimed, 'Thank God! I can still believe in the Gospel—I can yet be a Christian!'" (*Letters* I:26).

What Coleridge condemns here is "the source, the spring-head" of all he is opposing: "Materialism," now gone underground as an avowed dogma, but present everywhere in the popularly accepted division between body and soul which attributes "reality" only to the material. By this logic, "mind" becomes an absurd ghost and men lose touch with the truly real. As is typical in the rhetoric of Coleridge, Carlyle, and Emerson, declaration of the enemy is followed immediately by affirmation of the truth. In this instance, Coleridge affirms that the "dogmatism of the Corpuscular School" has been dealt "a mortal blow from the increasingly dynamic spirit of the physical sciences now highest in public estimation" (*Aids to Reflection* 340–41). Time and again in this literature the "dead past" is finally being buried by the forces of the new science, which brings not "materialism" but the very opposite, the Law that underlies and circulates through mere matter and so unites all in the all-encompassing Spirit. Thus the break that is postulated is not between old mechanistic science and new dynamic philosophy, but between old, mechanistic "natural philosophy" and the new dynamism which is ratified, even led, by "science." Poetic cognition was not fighting a desperate rearguard battle with modern science; it was to be coextensive with science. Romantic rationalism found in the most "modern" science of its day its source and energy.

The new philosophy characterized the old orthodoxy, both religious and scientific, by a variety of descriptive and philosophical terms. The most fundamental is "dualism," designating both the Cartesian divide between mind and body, and Locke's division between a passive understanding and the objects of sensation. Coleridge credited the first introduction of the "Mechanic or Corpuscular scheme" to Descartes, with the qualification that it might have been useful as a "fiction of science." However, Descartes "propounded it as a truth of fact," and instead of a world filled with productive forces,

> left a lifeless machine whirled about by the dust of its own grinding: as if death could come from the living fountain of life; nothingness and phantom from the plenitude of reality, the absoluteness of creative will!
>
> Holy! Holy! Holy! let me be deemed mad by all men, if such be thy ordinance: but, O! from such madness save and preserve me, my God! (345)

The ambivalence of Locke's legacy has already been noted. Conservative Protestants in America had found ways through the Scottish "common sense" compromise to bridge that dangerous gulf between mind and body,[11] but Emerson was still unwilling to accept the implied passivity of the mind before matter, and the dualism which the church had stitched so precariously together. He discovered his solution in 1830, in Coleridgean Neoplatonic romanticism, which bound the universe firmly through the agency of the mind as an active and shaping power grounded in the divine.[12] As Coleridge wrote in *The Friend*, all speculations must begin with "Postulates" supplied by the conscience, and the "chasm" between things that are seen and the "One Invisible" can be filled only by spirit and religion (*Letters* I:30).

Failing that, the division or chasm could grow to threaten a universal atomism, what Carlyle, in *Sartor Resartus*, called a "general solution into aimless Discontinuity" (121). One of the formulaic phrases of Protestant literature, the "fortuitous concourse of atoms," was intended to indicate the absurdity of a material world left to its own devices, uncontrolled by divine purpose.[13] Coleridge was equally clear on the absurdity of imagining that matter could organize *itself*: "To say that life is the result of organisation, is to say that the builders of a house are the results" (*Letters* I:233). That atomized matter could ever cohere in a meaningful way was clearly a logical contradiction. Worse yet was the emotional burden of such a philosophy, especially when compared with its alternative, the "ideal theory." According to Emerson, the latter

> beholds the whole circle of persons and things, of actions and events, of country and religion, not as painfully accumulated, atom after atom, act after act, in an aged creeping Past, but as one vast picture, which God paints on the instant eternity, for the contemplation of the soul. Therefore the soul holds itself off from a too trivial and microscopic study of the universal tablet. It respects the end too much, to immerse itself in the means. (CW 1:36)

Matter without spirit, as materialism proposed, could be no more than an aggregate of little pieces, and to *study* it in that fashion was to reduce oneself to the "trivial." During his early immersion in empirical science, and soon after writing "Kubla Khan," Coleridge protests in a letter to Thelwall that he can at times feel "the beauties" but more often "all *things* appear little," and all knowledge, the universe itself, "what but an immense heap of *little* things? —I can contemplate nothing but parts, & parts are all *little*—! —My mind feels as if it ached to behold & know something *great*— something *one* & *indivisible*—."[14]

The fear that man might fall to the level of brute creation is implicit in the distinction Coleridge makes here between "sense" and "mind" or "spirit," a distinction only superficially symmetrical. As Carlyle protests, "Soul is *not* synonymous with Stomach" (122). The real fear is that, in a universe where the same law of cause and effect controls both the physical universe and the moral world, "Stomach" will rule. A metaphysics that subjects our entire being to the laws of nature denies the existence of anything which is *not* nature—in other words, in Coleridge's terms, anything *super*natural, or spiritual. "What," asks James Marsh, "is the amount of the difference thus supposed between this being and the brute?" Both are bound to the law of their nature "as by an adamantine chain," and all that is left is "to give man over to an irresponsible nature as a better sort of animal, and resolve the will of the Supreme Reason into a blind and irrational fate" (35–36). We are little better—only a little better off—"than our dogs and horses" (44). Marsh, writing in the widely read "Preliminary Essay" to the first American edition of *Aids to Reflection* (1829), finds it nothing short of astonishing that a Christian community has adhered so long and tenaciously "to philosophical principles, so subversive of their faith in every thing distinctively spiritual," while principles so much closer to "the truly spiritual in the Christian system . . . were looked upon with suspicion and jealousy, as unintelligible or dangerous metaphysics" (50–51). Thus was Coleridge introduced to America.

Marsh's was a powerful argument, but there were other, more immediate reasons why Emerson took up and advanced the Coleridgean principles which Marsh advocated. Above the historical apparition of the French Revolution loomed contemporary forces that threatened the disruption or even the disintegration of society. One of the groups included in Carlyle's sweeping satire of mechanism was "Mechanical Profit-and-Loss Philosophies" (124), pointing toward the commercial nexus underlying the ills of the spirit. To equate soul with stomach was also to put one's own material self-interest ahead of the common good. The evidence was under Emerson's patrician nose: Boston was devolving from a self-sufficient and well-integrated community into a city of competing commercial and political interests, dominated by unchurched newcomers and "sick with the disease of competitive factionalism." Against this background, Emerson struggled to adapt the Federalism he inherited "to new times and new circumstances."[15] Coleridge developed his own metaphysics in response to similar but much intensified pressures. As David Bloor summarizes: "Anything to avoid following the empiricists and the French. Empiricism leads to pantheism, pantheism to atheism, and atheism to the Terror."[16]

It remains ironic that in all these insistent declarations of unity, we are aware of nothing so much as a cascade of dualisms. "Nothing that he sees but has more than a common meaning, but has two meanings," as Carlyle writes of his hero "Diogenes Teufelsdröckh"—"God-born Devil's-dung." The two meanings are meant, of course, to be finally one: "For matter, were it never so despicable, is Spirit, the manifestation of Spirit" (49). Matter and spirit are divided to reassert their ultimate identity: objects and thoughts are finally the same. As Emerson says, "Nature is made to conspire with spirit to emancipate us" (CW 1:30). Thus much does "unity" protest in the face of the fundamental and irreconcilable difference inscribed into its foundational principles. Nature is spirit only until man is properly apprized of his *true* place, and thereafter instrumental nature, the channel of power, is again dead matter. And so the pairs will multiply, behind the twin standards of Reason and Understanding: spirit/matter, man/nature, mind/sense, subject/object, human/animal, ME/NOT ME. Yet while nature appears on the mechanical side of the pairing, *science*, the search for the unifying law, is an aspect of mind or intellect, and is therefore firmly enlisted on the side of Reason. Thus it is that this literature is permeated with the latest scientific metaphors, with science rationalizing truth in the realm of matter. Behind this effort to synthesize what Emerson also called "beauty" with "truth" lay the axiom that science was the intimate ally of religion. Coleridge insists that the mistakes of scientific men had never injured Christianity, "while every new truth discovered by them has either added to its evidence, or prepared the mind for its reception" (*Aids to Reflection* 231). Emerson picked up the refrain: "All science has one aim, namely, to find a theory of nature." For, as he adds, "nature [is] ever the ally of Religion: lends all her pomp and riches to the religious sentiment" through the moral law that "lies at the centre of nature and radiates to the circumference" (CW 1:8, 26).

Thus a pantheon of unifying ideal systems, ranging from Anglican to high romantic and legitimized by science as moral law, was erected against the multifarious threats posed by social and physical atomization. While romantic radicals read nature and man together in the higher Law, conservative Protestant theologians built Baconian, inductive science into a bulwark protecting doctrinal purity and security, their own deliberate counterthrust against the eighteenth-century effort to portray science as indifferent, or even hostile, to traditional Christianity. John Herschel, then the very image of a major modern scientist, reiterated that science is nothing less than the contemplation of God's world. The conviction that science was an aid to biblical piety extended to popular science and the popular press. It became a truism to assert that the alleged conflict between science and religion was illusory, and to insist that "science was a confirmation of the existence and wisdom of God," resulting in a popular science "organized around the notion of the natural world as an emblem of divinity."[17]

Hence the complexities of this situation cannot be resolved simply by labeling science in this period as empirical and "Baconian." Customarily this establishes a neat division between the old-fashioned or conservative "Baconians," practicing the "normal" science of their day, and their romantic (and doomed) challengers. Baconian science supposedly rejected speculation and hypotheses, in the naive belief that the mere accumulation and classification of isolated fact would produce scientific truth. Eventually the burden of accumulated fact became so overwhelming that the system collapsed, clearing the way for a more modern and theoretical approach.[18] This would define and dispose of the situation nicely except for two difficulties, one theoretical and one historical. First, information can be gathered and classified only within a framework of theory; facts are not facts until a hypothesis gives them significance, as Locke and even Bacon himself recognized. Indeed, the second, historical difficulty lies in finding *anyone* who openly advocated the collecting of isolated facts. On the contrary, all sides mocked such blind-as-a-Dr.-Bat behavior, a tradition which James Fenimore Cooper (in *The Prairie*) engages with zest, as does Dugald Stewart when he defends hypothesis against "the professed followers of Bacon," and ridicules those who collect "a chaos of insulated particulars" (II:402, 442). According to Herschel, Bacon's real teaching was that natural philosophy consists of generalizing from particulars, and reasoning from generals back to particulars; thus we reason back and forth. As Bacon himself said, "our road does not lie on a level, but ascends and descends; first ascending to axioms, then descending to works."[19] In 1830, Herschel formalized this method as the feedback loop of deductive and inductive reasoning, combining both experiment and theory to form "an engine of discovery infinitely more powerful than either taken separately" (104, 181). Cousin similarly recommended that the "good physical philosopher" use a balanced approach: "Begin with the method a priori, and give to it by way of counterpoise the method a posteriori. I consider the identity of these two methods as the only torch by whose light we may find our way in the labyrinth of history" (103).

Perhaps the fad for collecting and displaying natural history specimens in cabinets accounts for some of the "Baconian" stereotyping, but cabinet collecting was not an inductive science but a fashionable demonstration of the unity of God's universe (which was, in any case, an initiating hypothesis).[20] If objects are thoughts of God, any object, no matter how trivial, is important, as important as any other. The resulting fascination with the "contrivances" of God's great puzzle connects Paleyan theism with Jeffersonian deism, and both schools held the same basic premise as the believers in Coleridgean Reason: mind and body were the same, either wholly material (as Jefferson held) or wholly spiritual, in a universe created as a harmonious, designed whole. Out of confidence in this whole one could suppress (like Cousin) or ignore those particular pieces judged to be anomalous or imperfect, while collecting, displaying, and celebrating those that verify God's unity of design.[21] Of all of these groups, the "new philosophers" would have seemed the most liberating to spiritual seekers like Thoreau because they alone allowed for intuition, that spiritual leap for which Emerson is so famous.

The evidence suggests that "Baconians" were a nineteenth-century straw man given life by later historians who assumed that, since everyone took turns deriding Baconian science, such a thing must have existed. Susan Faye Cannon discusses, and dismisses, the label of "Baconian," for as she says, Bacon was taught in the nineteenth century as an advocate of "a method of analysis, generalization, and deduction: a rather modern-sounding combination." While there were some people who thought it called for "facts and more facts," these "were not always scientists; some of them were clergymen or review writers," and a scientist found approving of Bacon is not necessarily "Baconian."[22] Nineteenth-century writers themselves tended to draw the boundary not between fact and theory or science and religion, but between ways of *connecting* the two: between "empiricists," who are often religious conservatives and perhaps also scientists, and "idealists," who often are cast as radicals (though they can be thoroughly doctrinal, as Coleridge shows) and who may also be self-declared scientists. The "empiricists" are fond of casting snide remarks in the direction of the "speculators" or "unintelligible" metaphysicians (as Marsh's testy defense of Coleridge indicates); sometimes the simple adjective "German" is sufficient. For their part, the idealists (who entered the twentieth-century literary canon in far greater numbers) are equally fond of sneering at the plodding gaze of the empiricists. Emerson supplies a more temperate instance of the genre: "Empirical science is apt to cloud the sight, and, by the very knowledge of functions and processes, to bereave the student of the manly contemplation of the whole" (CW 1:39). But each side invokes the name of Francis Bacon against the other. Conservative Christians, in adopting Bacon as their standard, also used him as a stick with which to beat the idealists (and eventually, Darwin): they were insufficiently Baconian, too speculative, wildly theoretical. At the other extreme, Reverend Marsh attempted to seize the standard for *his* cause, appealing to none other than "Lord Bacon" as the authority for Coleridge's principles, asserting that "the fundamental principles of [Bacon's] philosophy are the same with those taught in this work," even to the crucial distinction between reason and understanding, which Bacon had called "wisdom" and "reason" (50). Emerson agreed, and claimed Bacon for the transcendentalists and the Platonists—though troubled by a certain weakness in "my Lord's" moral character.[23]

Given this lineup, natural theologians who wished to sling mud found their material ready to hand. All factions appear to have enlisted the science compatible with their ideologies, to have distanced themselves rhetorically from anything that resembled skepticism or materialism of the eighteenth century, and to have allied themselves with the authority and achievements of the seventeenth century. (Interestingly enough, the actual scientific workers in the field were encountering problems which could not be solved by natural design, nor by idealism, nor yet by doxological Baconianism—but that is a story that must keep for Chapter 5.)[24]

In sum, natural theology, including its "Baconian" forms, and idealism's various all-embracing systems all relied on the foundational premise that objects in the real world manifest the thoughts of God. This key assumption underwrote both the concern for collection and accurate assemblage of facts, and the urge to dissolve the irritating anomalies and imperfections of the actual into the unifying realm of spirit. Curiously, these two paired complexes have come down to us as opposites: Baconianism versus idealism. Far from being opposites, they are mutually dependent and closely related. Each expresses the fear and the hope of the other. The so-called "Baconian" fears dissolution into airy speculation unattached to social and religious convention, and hopes to repair this danger by assembling the given certainties represented by facts into the stable and harmonious order assured by God's beneficence. On the other hand, the idealist fears dissolution into a chaos of isolated particulars, and hopes to repair this danger by assuring a strong organizing system which will shepherd and control wayward particularities. Yet the "Baconians" were just as contemptuous of isolated particulars as the idealists, just as anxious to immure them into God's beneficent system; while the idealists feared their speculations lacked substance and were anxious to ratify them through the agency of empirical fact. Thus both were centrally engaged by the search for harmonious order, both named this search "science," and both regarded it as their surest ally; both invoked inductive reasoning as their true method and the name of Francis Bacon as their standard.

By designating only one half of this complex, the term "Baconianism" mythologizes the mid-nineteenth-century field into "Baconian" "scientists" and "antiscientific" "idealists." At one level, this has helped to obscure the deep involvement with science of the British and American romantics: Wordsworth, Coleridge, Carlyle, the Shelleys, Emerson, Thoreau, Poe. On a smaller scale, this has helped make Thoreau's career into a "problem," precisely because it cannot be made to fit comfortably into either of these two constructed categories. Indeed, Thoreau himself came to regard the nature of his career as a problem, for the rhetorical polarization outlined here was crystallizing around him, and one of the reasons it has stayed so long in place is that we have so long accepted as definitive the rhetorical positions of factions that, across common ground, defined themselves against each other.

In the midst of these debates, it was the idealist position that Thoreau encountered most fully during his "Transcendental apprenticeship." As should be evident by now, for all its declarations it was less a departure from, than an extension of, the natural theology he had been taught at Harvard; and the science that was institutionalized at Harvard in the 1850s, and that Thoreau would come to reject, consisted largely of a mix of both. For Coleridge's romantic science, far from dwindling into obscurity, came to dominate British mainstream natural science through the idealist synthesis of Carlyle's friend and the leader of the "Coleridgean intelligentsia," Richard Owen.[25] Louis Agassiz brought a related metaphysics to Harvard, where it settled in companionably with both the old-fashioned natural theology and the newfangled idealism of one of Boston's leading citizens, who would soon be Agassiz's close friend—Ralph Waldo Emerson.

Yet for all his enthusiasm, the transcendentalist's idealism brought in its train implications that could not have sat well with the young Thoreau. He found much of Cousin, for instance, echoed—even to the vocabulary—in Emerson's *Nature*, which he read immediately afterward. Yet how must some of Cousin's assertions have sounded to him? Cousin completely nullifies the individual, whom he regards as utterly unimportant in the scope of the great "idea" manifested by his nation and by the great sweep of history. Ordinary men only make good soldiers; true individuals, those who are "original," represent nothing but themselves, hence are worthless; but the great man, the *representative* man, is both the people and himself (300–301). How could Thoreau entirely nullify the individual—was he himself, if not "representative," therefore "worthless"? Could he so blithely disdain the "mingled and imperfect" particulars of his world? Yet if he were inclined to value the sensual, individual, and particular, the way back to Locke was clearly forbidden by the idealists to whom he was so intensely drawn. In the triad of positions sketched in this chapter, nominalist skepticism had been eliminated as a live option by the combined attractions of realism and idealism. "Experience" had been liberating to Locke; now Locke's experience *was* the authority, and idealism the liberating force. "Transcendentalism and sensationalism!—these were the poles of the philosophy of mind, and among the elect of the new movement to call a man a sensationalist was a polite way of informing him that he was an intellectual and spiritual dullard."[26] Thoreau called himself a transcendentalist to the end, but it took him years to find, in Humboldt's return to experience, an alternative yet respectably romantic path to the unique and "original"—an alternative wherein the debt to Locke was so deeply inscribed as to have been forgotten.

If Thoreau indeed felt a tension between the rich, sensual particulars of experience and the intoxication of the All, he found a similar tension in Goethe, specifically in *Die Italiänische Reise*, which he also read, in the original German, at this time. Goethe's descriptions of his travels are embedded in the specifics of natural history. The opening paragraphs are typical:

> I find I can quickly get a topographical idea of a region by looking at even the smallest stream and noting in which direction it flows and which drainage basin it belongs to. Even in a region which one cannot overlook as a whole, one can obtain in this way a mental picture of the relation between the mountains and the valleys. (5)

The particulars are used to reveal the whole, while the whole is composed of meaningful particulars: Goethe's vision works back and forth between both extremes. He insists on the importance of exact, sympathetic observation, but his particulars are not just pieces in a preformed puzzle. They have their own wonder:

> Today I watched the amusing behaviour of the mussels, limpets, and crabs. What an amazing thing a living organism is! How adaptable! How there, and how itself! How useful my knowledge of natural history, scrappy though it is, has been to me, and how I look forward to increasing it!

He continues to describe the sea wall, the tide, and the creatures, especially the crabs hunting, and failing to capture, limpets: "I did not see a single crab succeed, although I watched for hours" (84–85). Indeed, the book is arguably *about* how to "watch for hours," how to observe and by observing, *perceive*. Goethe studies drawing in Rome, and though the results are amateurish, he does not despair: "The few lines I draw on the paper, often too hasty and seldom exact, help me to a better comprehension of physical objects. The more closely and precisely one observes particulars, the sooner one arrives at a perception of the whole" (161). Drawing, for Goethe, serves as an effective means of observation. One cultivates it less to create fine objects than to learn how to see.

But *what* Goethe wishes to see is precisely what cannot *be* seen: "the whole." One of the book's themes is his search for the Ur-plant: "There certainly must be one. Otherwise, how could I recognize that this or that form *was* a plant if all were not built on the same basic model?" (251). He wants contact with "the things themselves" (343), but he wants to find in and through them a general or higher law, which, once he knows it, he can apply everywhere. Accordingly, his Ur-plant becomes a model for all actual plants, and his search for it carries him to works of art or to works of nature as if they are fundamentally equivalent. In Rome he concludes: "These masterpieces of man were brought forth in obedience to the same laws as the masterpieces of Nature. Before them, all that is arbitrary and imaginary collapses: *there* is Necessity, *there* is God" (383). The discourses of Truth and Beauty have the same root, in a lawful universe that embraces the opposites of Nature and Art, even as he himself reconciles objective vision and subjective feeling, the part and the whole, individual things and higher law. Although Goethe's obsession with unity does have an empirical edge to it, yet the imperial goal is never lost.

Natural History before Walden

Goethe's practice comes closest to Thoreau's own as it developed through the 1840s. One of his earliest *Journal* entries praises Goethe's precise attention to the facts of his experience:

> He is generally satisfied with giving an exact description of objects as they appear to him, and his genius is exhibited in the points he seizes upon and illustrates. His description of Venice . . . is that of an unconcerned spectator, whose object is faithfully to describe what he sees. . . . It is this trait which is chiefly to be prized in the book—even the reflections of the author do not interfere with his descriptions.
>
> It would thus be possible for inferior minds to produce invaluable books. (1:16; 12/8/37)

Genius consists in the selection of facts from the totality of the field, and faithful, "unconcerned" description. The author need not speak, need not—*should* not—reflect, but should allow the facts to speak for themselves. This version of authorship demands the ideal of transparency, the poet as conduit for, not maker of, the truth without to the witness of the page and the reader. This curiously scientific assumption of objectivity—reporting just the uncontaminated facts—recalls Emerson's figure of "The Poet," who does not create but transmits the truth. All Thoreau should have to do, accordingly, is open his senses wide and let the world write itself through his hands; the spirit that circulates through and connects all things will take shape on the page.

> We should not endeavor coolly to analyze our thoughts, but keeping the pen even and parallel with the current, make an accurate transcript of them. . . . The nearer we approach to a complete but simple transcript of our thought—the more tolerable will be the piece, for we can endure to consider ourselves in a state of passivity—or involuntary action; but rarely our efforts, and least of all our rare efforts. (1:35; 3/7/38)

Of course in practice this is impossible. The world won't write itself, facts won't select themselves. To think this way is to flirt with the "chaos of insulated particulars." The way out that was shown to Thoreau was to reach through the particulars to the single law that organized them into a harmonious whole—and suddenly the writer of truth is precipitated from a passive recorder to an active seeker, of whom is demanded a heroic act. Facts cannot be trusted to flower all by themselves, as Goethe shows in his strenuous search for the unifying whole. Yet how can one be sure that one has drawn, out of the massed particulars, the real truth, not imposed upon them an arbitrary one? One cannot, said Locke. Cousin and Emerson pointed to a more satisfying answer: the hero leads because he is "representative" of all. Truth does not stand alone. If what you have found *is* truth, everything will rush to verify it, to "harmonize" with it.

Thus, though the brave soul may have "unhinged" the social universe, unsettling governments and conventions, nature itself will follow him: "All woods and walls echo back his own spirit, and the hostile territory is then preoccupied for him. He is no longer insulated but infinitely related and familiar. The roll-call musters for him all the forces of nature" (1:95–96; 12/39). Bravery becomes a form of health, a kind of marching to the drumbeat of the universe:

> To the sensitive soul, The universe has its own fixed measure, which is its measure also, and as a regular pulse is inseparable from a healthy body, so is its healthiness dependent on the regularity of its rythm [*sic*]. . . . When the body marches to the measure of the soul, then is true courage and invincible strength. (1:96; 12/39)

March alone to the drumbeat of truth, and the universe will march with you.

In effect, truth guarantees itself—the heroic act will put one in a harmonious relationship with the natural, if not the social, universe (and give the lie to society). It is a difficult path to choose, and Thoreau gives over much of his early journalizing to exhortations about bravery and heroism, at a time when Emerson was being excoriated for the "Divinity School Address," and Thoreau himself had been fired from his teaching job for refusing to cane his students, according to the conventions of classroom discipline. In such a world, nature becomes more than a refuge—it becomes the seat of health and the source of creative energy. Align the axis of the soul with the axis of the universe, and one will acquire "a perfect sphericity" (RP 6); or in a musical analogy, "The human soul is a silent harp in God's quire whose strings need only to be swept by the divine breath, to chime in with the harmonies of creation" (1:50; 8/10/38).

The most central man, the most "spheral" man, or the most harmonious man, will of course be the poet. This figure of "the poet" moves toward a solution to the dilemma posed by Goethe's impersonal faithfulness to sheer fact:

> He must be something more than natural—even supernatural. Nature will not speak through but along with him. His voice will not proceed from her midst, but breathing on her, will make her the expression of his thought. He then poetizes, when he takes a fact out of nature into spirit —— He speaks without reference to time or place. His thought is one world, her's another. He is another nature—Nature's brother. (1:69; 3/3/39)

The poet will breath into nature the life of spirit, and cooperatively they shall speak together—"Each publishes the other's truth," he concludes. Thoreau's poet, in the best romantic tradition, will not compete with but embrace science, generalizing science to a wider, a moral, sphere:

> Facts must be learned directly and personally—but principles may be deduced from information. The collector of facts possesses a perfect physical organization—the philosopher a perfect intellectual one. One can walk—the other sit—one acts, the other thinks. But the poet in some degree does both and uses and generalizes the results of both—he generalizes the widest deductions of philosophy. (2:53; 1842–44)

In "Natural History of Massachusetts," Thoreau first publicly practices this transcendental act of taking up facts into the spirit, on the way to generalizing "the widest deductions of philosophy." Emerson prompted Thoreau to write the essay, as a review of a series of government-sponsored reports surveying the plants and animals of the state. He published it in the *Dial* for April 1842, making this one of Thoreau's earliest published essays.

The volumes Thoreau takes up are the purest of "Baconian" surveys, written, as Thoreau admits at the end, to serve the state's need for "complete catalogues of its natural riches, with such additional facts merely as would be directly useful" (NHE 28). This would not seem to provide the apprentice poet with a very wide or promising field, and yet Thoreau, taking the "license" of a preacher selecting his text (27), uses the texts as a springboard for his own reading of the text of nature. If that reading was not yet wide nor deep, still it provided him with a useful opportunity to point through the scattered facts of his experience to higher truth, in words full of echoes of his reading, and culled from nearly five years of his *Journal* writing.

"Books of natural history make the most cheerful winter reading," he opens, bringing pleasant summer images into the snowbound parlor. But all society is a kind of winter: "The merely political aspect of the land is never very cheering; men are degraded when considered as the members of a political organization. On this side all lands present only the symptoms of decay" (2). Nature is more than a compensatory diversion, it is a prophylactic against a diseased society:[27]

> I would keep some book of natural history always by me as a sort of elixir, the reading of which should restore the tone of the system. . . . To him who contemplates a trait of natural beauty no harm nor disappointment can come. The doctrines of despair, of spiritual or political tyranny or servitude, were never taught by such as shared the serenity of nature. (NHE 3)

How can nature do so much, guarantee so much? Thoreau turns to Paley's *Natural Theology*: "Surely joy is the condition of life. Think of the young fry that leap in ponds . . ." (4). And how does one attain access to this joy? Through science, by which one takes the private joys of nature into the public sphere, making it "admirable training" for "the more active warfare of life! Indeed, the unchallenged bravery which these studies imply, is far more impressive than the trumpeted valor of the warrior" (4). Thoreau trumpets the valor of the scientist—of Linnaeus himself, who systematized the universe and created the project for global knowledge of which the Massachusetts reports were a minor, local tributary, and who surveys the artillery of knowledge:

> with as much complacency as Bonaparte a park of artillery for the Russian campaign. The quiet bravery of the man is admirable. His eye is to take in fish, flower, and bird, quadruped and biped. Science is always brave; for to know is to know good; doubt and danger quail before her eye. (NHE 5)

The eye of science sees a world of certainty and security, "breaking ground like a pioneer for the array of arts that follow in her train," literally making the world safe for civilization.[28]

Thoreau then turns from the figure of the Linnaean scientist/explorer as Napoleonic hero, to the "promised topic" of his essay. He takes in turn his own private survey of insects, birds, "quadrupeds," fish, and reptiles, giving some species barely a nominating phrase from Buffon by way of William Smellie, while giving others—those with whom he has made personal contact—the benefit of several paragraphs. These are the animals that have taken hold of his imagination and show the specificity of actual experience: The solitary loon leads him on a sly game of hide and seek (12). The fox leaves tracks that Thoreau treads "as if I were on the trail of the Spirit itself which resides in the wood, and expected soon to catch it in its lair. I am curious to know what has determined its graceful curvature, and how surely they were coincident with the fluctuations of some mind" (15)—not all Mind, but a *fox* mind, something that becomes, as Thoreau continues, altogether more mysterious and strange. He wants the "sympathy" and fellowship of the "minnow in the brook"—but fellow feeling is not enough: "I would know even the number of their fin-rays, and how many scales compose the lateral line" (16).

But what has not yet emerged from this dispersed survey of Massachusetts animal life is the necessary law that will hold it together. For that, Thoreau turns not to the reports (which as he notes are held together rather by a concern for resource inventory), but to Goethe and his search for the Ur-plant. Winter would seem to give the botanist little material, but the idealist doesn't need actual plants anyway; in winter he may turn to "a new department of vegetable physiology, what may be called crystalline botany," and study the hoar frost (24). The winter of 1837 (with his eyes just sharpened by Goethe) was a good year for ice foliage, and the resemblance gives Thoreau the uniting generality he seeks to bring together winter and summer, death and growth, ice and life, fact and truth:

> It struck me that these ghost leaves, and the green ones whose forms they assume, were the creatures of but one law; that in obedience to the same law the vegetable juices swell gradually into the perfect leaf, on the one hand, and the crystalline particles troop to their standard in the same order, on the other. As if the material were indifferent, but the law one and invariable, and every plant in the spring but pushed up into and filled a permanent and eternal mould, which, summer and winter forever, is waiting to be filled. (25)

Thoreau generalizes further: "The same independence of law on matter is observable in many other instances, as in the natural rhymes" or resemblances in nature, implying an "internal melody, independent of any particular sense." Where the law of spirit leads, matter will follow. He speculates still further: since the law of growth is more obvious in crystals than in vegetation, "would it not be as philosophical as convenient to consider all growth, all filling up within the limits of nature, but a crystallization more or less rapid?" (26).

The evidence of his own senses confirms his reading, revealing harmonies everywhere, encouraging Thoreau to read Law back into objects—even if it means conceiving growth as a "filling up" of the predetermined "limits" of nature. This line of thought flows directly from the assumptions of divine design, which see the end implied in the beginning, such that the narrative of growth and development is the unfolding of what is given to its necessary end.[29] Thus Thoreau constructs for himself the dilemma that haunted Emerson: "Each spirit makes its house; but afterwards the house confines the spirit" ("Fate" 946). The only way out will be to question the organicist assumptions that construct such a polarity. Meanwhile, Thoreau brings his essay to a close by returning again to the "promised topics," disposing quickly of the state's reports, whose labor and research, though valuable, nevertheless "imply more labor than enthusiasm." The volumes of "measurements and minute descriptions" are like "those plants growing in dark forests, which bear only leaves without blossoms." Yet even this is a start, needed in frontier country; and it leads him to his key statement:

> But the ground was comparatively unbroken, and we will not complain of the pioneer, if he raises no flowers with his first crop. Let us not underrate the value of a fact; it will one day flower in a truth. (28)

Facts here are *vital.* Metaphorically they are seeds, mysterious in their potency and their unpredictability: who knows what sort of flower they will bring forth? Yet even as ends are built into origins, the "truth" is intrinsic, contained ready in the fact like a preformed embryo coiled in the mother's egg, in the germ theory of reproduction. Finally, it falls to the philosopher to cultivate the pregnant fact, toward its eventual unfolding or birth into truth. Thoreau's role is not intermediary, but mediating—a midwife, a curiously feminine position, nurturant and participatory rather than controlling or masterful, and riddled with implications for his future—labor. In the oddly androgynous position of the naturalist, the Napoleonic will to conquer through violence transmutes to the desire to possess through love and "sympathy," a desire haunted by its potent shadow, and comforted by the knowledge that the object reciprocates in "sympathy," like the "minnow in the brook" whose fin rays and scales Thoreau wants to count.[30]

Thoreau continues: "Men are knowing enough after their fashion"—that is, their utilitarian fashion; but, continuing with a Thoreauvian pun (of which more later), "it is much easier to discover than to see when the cover is off" (28–29). He concludes the essay with a key summation of his views on science:

> We must look a long time before we can see. Slow are the beginnings of philosophy. He has something demoniacal in him, who can discern a law or couple two facts. . . . The true man of science will know nature better by his finer organization; he will smell, taste, see, hear, feel, better than other men. His will be a deeper and finer experience. We do not learn by inference and deduction and the application of mathematics to philosophy, but by direct intercourse and sympathy. It is with science as with ethics,—we cannot know truth by contrivance and method; the Baconian is as false as any other, and with all the helps of machinery and the arts, the most scientific will still be the healthiest and friendliest man, and possess a more perfect Indian wisdom. (29)

Thoreau's themes here will resonate for the next twenty years of his work: there is something a little suspect, "demoniacal," about the search for truth behind appearances. The "true" scientist will take in those appearances with senses more acute and fine-tuned than the ordinary; in spirit he is a poet, perhaps *the* poet of the natural world; and he will learn not by any method, not even the best scientific method, nor after all by conquering nature like a Napoleon, but through "intercourse and sympathy." "Inference," "deduction," the application of law, do not produce a *scientia* that is a true knowing. True knowledge comes only from long familiarity, even friendship, the experience of long association, as long and wise and intimate as that of those wild men, the Indians.

Thoreau wrote the first version of this passage in October 1840, at a time when he was reading works from Greek and Roman antiquity, Bacon's original *Essays*, Cudworth's *True Intellectual System of the Universe*, Sir Walter Raleigh, and ancient Oriental scripture. His 1840 *Journal* shows him eager to link local Concord with all times and all places: "Within the hour" he can go from Minot's barn to the "στοα at Athens" (1:172; 8/14/40); one revolution of his body makes historical epochs and geographical boundaries disappear. As he reads, he relishes his connection through the "solid turf-clad soil" to Goethe's Italy (1:30; 2/27/38), and through the "shady mountains, and resounding seas" to Homer's Greece (1:33; 3/5/38). The familiar "'pine, larch, spruce, and silver fir'" gave him a "foothold" even on the far "Himmaleh range" (1:177; 8/28/40). Yet as he exulted in the constancy of these earthly particulars, he nevertheless reassured himself that "go where we will we discover infinite change in particulars only—not in generals" (1:150; 7/5/40). His chosen particulars will serve transcendental ends, whether he celebrates them, or imagines their dissolution.

It was in this context, in 1840, that Thoreau read his first work of "contemporary hard science," Charles Lyell's three-volume *Principles of Geology* (1830–33).[31] Lyell's work, which was widely read and enthusiastically reviewed, was not a compilation of geological knowledge but a vigorous, eloquent argument for a central principle of scientific reasoning. The book's subtitle spelled it out clearly: "being an attempt to explain the former changes of the earth's surface, by reference to causes now in operation." Lyell opens his argument, which he wages against both scriptural geologists and catastrophists, by reviewing the history of the concept of geology (much as Humboldt would shortly do with the concept of "cosmos"). Lyell bypasses the Bible and Mosaic tradition entirely by recovering and allying himself with a much older tradition, dating back to Thoreau's beloved Orientals—indeed, to Thoreau's touchstone text, the "Institutes of Menu."

Thoreau opened his first work of serious contemporary science, then, first to be assured that "researches into the history of nature" would bring to light "astonishing and unexpected . . . connexions" with past civilizations (I:2), and then to encounter his favorite Hindu scripture. Though only a week before Thoreau had written of the "demoniacal" power of the man who can "discern a law, or couple two facts" (1:187; 10/11/40), he found Lyell's myriads of empirical facts coupled to good purpose: the establishment and verification of the ancient idea of an ageless and cyclical nature, whose daily occurrences speak to the deepest past, and in which progressive development has no place. As Stephen Jay Gould writes, Lyell's great empirical work is no "impassive compendium of accepted information" but "a brief for a world view—time's stately cycle as the incarnation of rationality."[32] The contending balance of the two great "antagonist forces," the igneous and the aqueous, creates the shape of the earth; what fire builds up, water levels down (I:167). Geological strata, according to Lyell, record phases of renewal and destruction "in eternal succession," supplying evidence for the ancient idea of the "great year" (I:116). Lyell applies the global researches and theories of Alexander von Humboldt to the evidence of past changes in the earth's climate, showing how periodic variations in the proportion and distribution of land and sea might bring about climatic change: the earth in geological time experiences "winter" when high land masses are collected in the polar regions, "summer" when they gather around the equator. Our current era of approaching cold or "winter" will be followed by a cosmic spring and summer, during which time "The huge iguanodon might reappear in the woods . . . while the pterodactyle might flit again through umbrageous groves of tree-ferns."[33] Yet these momentous revolutions are so gradual that even "a few thousand years" would produce no sensible change (I:115–23).

So this grand notion of the endless cycle of time, the "great year," having been traced to impeccable sources in ancient India, Egypt, and Greece, had as well the empirical support of the most modern, Humboldtian science, and further, embraced a central principle of scientific explanation, the attribution of past changes to causes observable in the present. As Thoreau writes, "We discover the causes of all past change in the present invariable order of the universe" (1:191, 411–12; 10/19/40). Thoreau was quick to transcendentalize this idea, which he worked up into a long passage, rewrote in 1842, and spliced into *A Week on the Concord and Merrimack Rivers*:

> As in geology, so in social institutions, we may discover the causes of all past change in the present invariable order of society. The greatest appreciable physical revolutions are the work of the light-footed air, the stealthy-paced water, and the subterranean fire. . . . The hero then will know how to wait, as well as to make haste. (WK 128)

Thoreau's fundamental insight—the collective power of many small strokes—would have powerful repercussions in his work to come, as he turned Lyell's antiprogressivist agents into the forces of revolution. Meanwhile, to his transcendental imagination the Hindu sources of Lyell's geology acquired their own geological dignity, as ancient facts turned treasure: Hindu sentences—the Laws of Menu—"are clean and dry as fossil truths, which have been exposed to the elements for thousands of years, so impersonally and scientifically true that they are the ornament of the parlor and the cabinet" (WK 149). This passage transcendentalizes facts in the old rational holist way, removing them from their environment and context into the perfect array of the display case. Thoreau also found confirmation in Lyell for his transcendental belief in the errancy of the particular as against the constancy of the general. In another passage loosely datable to this period and also revised into *A Week*, he writes, "The eyes of the oldest fossil remains, they tell us, indicate that the same laws of light prevailed then as now. Always the laws of light are the same, but the modes and degrees of seeing vary" (WK 157; 1:418).

What would most impress the nineteenth-century reader of Lyell was his imagination of deep time. The ancient scriptural texts that opened his trilogy marked but the first notch in a scale that extended back beyond reckoning. Thoreau read the lesson back into his own imaginings of the Concord forest, gathering from it the same mystical assurance of agelessness:

> I sit now on a stump whose rings number centuries of growth. If I look around I see that the soil is composed of the remains of just such stumps, ancestors to this. . . . I thrust this stick many æons deep into its surface, and with my heel make a deeper furrow than the elements have ploughed here for a thousand years. . . . The newest is but the oldest made visible to our senses. (WK 153–54)

Lyellian deep time was not a way of introducing history into nature, but a way of eliminating history entirely, dissolving it in the grand cyclicity of the eternally returning Great Year. Thoreau had been trying to dissolve the barrier between past and present by identifying with the constancy of earthly details across space and time. Science proved itself a strong ally, engaging the same cosmic questions that most concerned Thoreau, and providing answers with poetic reach and empirical power. In terms of this chapter's opening question, Lyellian science showed that many small facts might combine to one great truth.

The other science that proved compatible with transcendental imaginings was astronomy, which expanded the field in space as well as time. Thoreau was excited by those "revelations of the Real," such as the amazing concept that Venus was not just a bright spark but "*another world* in itself," whose phases were predicted by human reason a century in advance of the telescope (WK 385). In 1851 both Emerson and Thoreau interested themselves in Perez Blood's 85-power telescope, going together to look through it at the skies. Two days later Thoreau made the trip to Cambridge to examine Harvard's new observatory, to check out some books for Blood, and to talk with Harvard's astronomer, William Cranch Bond.[34] Thoreau made frequent references to the astronomer's calling, especially as the human "eye" that could redeem the necessary instrumentation (4:249–50; 1/12/52)—though he was also glad to hear from Bond that the naked eye could still contribute useful observations (3:296–97; 7/9/51). In the journal he kept at Walden he indicated the reasons for his affinity: "Astronomy is that department of physics which answers to Prophesy the Seer's or Poets calling It is a mild a patient deliberate and contemplative science. . . . This world is not a place for him who does not discover its laws" (2:359; after 12/2/46). While geology enhanced the understanding of *this* world, astronomy implied we could "obtain some accurate information concerning that OTHER WORLD which the instinct of mankind has so long predicted. Indeed, all that we call science, as well as all that we call poetry, is a particle of such information . . ." (WK 385). Nevertheless, Thoreau did not dedicate himself to astronomy, maintaining at most a casual and metaphoric interest. Neither the deep past nor deep space could hold his imagination for long. He was interested in the one as it was manifested in present particulars, and he came to complain of the other that it was so abstract as to be barren, like "eternal law": the unchangeableness of the heavens could impress him as "poverty," in contrast to the "unhandselled wilderness" of the forest; he saw the heavens "pierced with visual rays from a thousand observatories" (IV:469; 1/21/53).

As contemplative sciences, geology and astronomy proved congenial to the poetic or transcendental imagination in a way that the experimental sciences of physics and chemistry were not, although Thoreau did read Justus von Liebig's *Animal Chemistry* (1842) at Walden, alluding in *Walden* to Liebig's concept of animal heat (WA 13). One who knew Thoreau by his later reputation alone might expect him to have engaged in the "contemplation" of natural history as well, but at this stage botany and zoology interested him only peripherally. Their association with cosmic origins and pure, absolute Law was equivocal at best (crystalline botany only begged the question), and the problem of naming the myriad particulars of the natural world, thereby relating them systematically, did not yet interest Thoreau. Years later he recalled using a botany twenty years before (probably while teaching science to his students in the Concord Academy); he remembered learning a few names but soon forgetting them for lack of any "system" (IX:157; 12/4/56). What he lacked was not only a botanical system, but any reason to take notice of individual plants and animals; he was after "generals," not the infinitely changeable individuals that composed his immediate world. Though the journal of these years is full of nature notes, he pushes quickly from the object to his meaning, and is impatient with the cumbersome tools that would retard his poetic ascent. In March 1840, for example, he announces smugly, "I learned today that my ornithology had done me no service—"; the birds he heard, "which fortunately did not come within the scope of my science," sang as freshly as the morning of creation, with all the untrodden wilderness "of the soul" for background (1:115; 3/4/40).

The vague reference to "birds," as if birds were pretty much interchangeable, is not unusual. Yet there are moments of intense observation too, as when, two weeks before the bird passage, Thoreau notices that the flooding Concord River has driven out the muskrats to be hunted by sportsmen. He admits himself "affected" by "the sight of their cabins of mud and grass" as "of the pyramids, or the barrows of Asia" (1:111; 2/22/40). The move is again to dispose of the particular in the cosmic, but he does take notice. This same month he writes out the passage which flares so unaccountably amidst the dutiful prose of "Natural History of Massachusetts," a passage without precedent: "A very meagre natural history suffices to make me a child—only their names and genealogy make me love fishes. I would know even the number of their fin rays—and how many scales compose the lateral line." He is the wiser for knowing "that there is a minnow in the brook— Methink I have need even of his sympathy—and to be his fellow in a degree—" (1:109;

2/14/40). Knowing as an activity of sympathy is here explicitly invoked for the first time: counting fin rays is conceived as an act of affection and fellowship, not tedium, part of specifying that it is a *minnow* in the brook and not a bream or a pike.

Curiously, Thoreau had the capacity to be "touched" by fish far earlier than any other "fellow."[35] Sometime between August 1844 and March 1845 he drafted the catalog of his "finny contemporaries" in the Concord River, included in the opening chapter of *A Week*, and complete even to their scientific names (2:104–16)—though it was not until he first really *looked* at a chivin, during the Walden years, that he actually recorded the number of fin rays in his *Journal*: "Single Dorsal fin 10 rays about as high as wide" (2:250; 6/20/46). That is, Thoreau had to *learn* to look, not beyond but at things—"We must look a long time before we can see"—before he could teach himself to draw his looking into words. To look at them would be the first step in recognizing that they too looked back at him. As he says in the minnow passage, forerunner of so many more: to truly call the minnow his "fellow," he cannot just give it sympathy—he needs the minnow's own sympathy back.

It is hardly surprising that Thoreau's interest in plants and animals was still so casual. He had intended to become a schoolteacher and a literary artist, although the first had proved abortive by April 1841, when he and his brother John had to close the Concord Academy due to John's ill health. He spent the decade of the 1840s developing his literary reputation, under Emerson's constant encouragement. His first literary projects were, in fact, directed and suggested by an Emerson who held out immense hope for this promising Yankee scholar. Thoreau contributed to the *Dial*, both on and behind the scenes, living with the Emersons as a handyman in the meantime. Thus it was that "Natural History of Massachusetts" was prompted less by his own genius than by Emerson's need for copy. He was working hard on an anthology of English poets, again at Emerson's prompting.[36] He was finding this last a dreary job, too, if one can judge from a *Journal* entry written on a collecting trip to Cambridge: "When looking over the dry and dusty volumes of the English poets, I cannot believe that those fresh and fair creations I had imagined are contained in them. . . . Poetry cannot breath [*sic*] in the scholar's atmosphere." Perhaps, he wonders, it would be better to step at once from the library into the field or wood; as he runs over book titles he is "oppressed by an inevitable sadness" (1:337–38; 11/30/41). Not surprisingly, he abandoned this pet project of Emerson's when he moved back in with his own family, during Christmas 1843. This put an end to Emerson's other project for Henry, an association with the literary lights of New York, underwritten by a tutoring position at the home of Ralph's brother William Emerson on Staten Island.

Thoreau spent 1844 out from under Emerson, working not as a literary hopeful—apparently not even keeping his journal—but as an engineer and carpenter. He redesigned and perfected the "John Thoreau and Son" lead pencils, which after years of improvement were just then judged to be the finest made in America, and as fine as any made abroad.[37] His family was now prospering from the pencil business, and he helped build them a new, larger house of their own, the "Texas House." Yet his mind was not wholly given to production and construction: in the midst of this activity Thoreau was conceiving the first major literary project that would be his own and not Emerson's. In March 1845 he began building his house on Walden Pond (on Emerson's land), in which to retire from the family factory, and write.

His output at Walden included an essay on Carlyle, his first lectures, and the first manuscript of *Walden*, but the major literary project he took with him to the pond was, of course, *A Week on the Concord and Merrimack Rivers*, planned as his tribute to his brother John. Their two-week trip in September 1839 had taken on mythic dimensions as Thoreau's youth slipped into the past and, more especially, after John's shocking death of lockjaw on January 11, 1842. It is worth remembering, as part of the story of this book and its aftermath, that Henry was so stricken that he too came down with a sympathetic case of lockjaw, and the family actually feared for his life. The *Journal* for this period is poignant. As John falls ill, the stoic Henry writes, "Am I so like thee my brother that the cadence of two notes affects us alike?" (1:362; 1/8/42). After January 9, Henry falls silent for six weeks. On the twentieth of February he writes again: "When two approach to meet they incur no petty dangers, but they run terrible risks" (1:364; 2/20/42). The next day he sounds, for the first time, a theme that threads through the *Journal* until his death:

> I must confess there is nothing so strange to me as my own body— I love any other piece of nature, almost, better.
>
> I was always conscious of sounds in nature which my ears could never hear—that I caught but the prelude to a strain— She always retreats as I advance— Away behind and behind is she and her meaning— Will not this faith and expectation make to itself ears at length. (1:364; 2/21/42)

A nature that retreats from his "advances," eludes him even as he seeks in "her" the spiritual equivalent for his lost brother: this longing and its romantic defeat echoed across the banks of Walden Pond. On the death of his second closest friendship, with Emerson, Thoreau's sense of defeat and loss become so urgent as to propel the orbit of his imaginings to a new realm.

But meanwhile Emerson was all support as Thoreau revived his literary career, embarking on the book that was to gather up his life's reading and writing, link it to the adventure of his lifetime, and establish his literary reputation once and for all. The day after he moved to Walden Thoreau also opened a new notebook, in which he would record his life at the pond. This, and a companion volume which served as a literary workbook, formed the germ of *Walden*, originally planned as a sequel to *A Week*. Thus the life experience that would shape *Walden* was recorded as it was lived. But since Thoreau had taken no notes on his journey with John, he had to rely on memory, compiling material and sorting it out chapter by chapter, evolving a structure through accretion of like with like (2:452–3). The result reads rather as if he were organizing an anthology of his own writing rather than that of others; the reader of the early *Journal* volumes will find sentences written across the span of a decade spliced into single, continuous meditations. The thematic clusters are hung on the narrative frame of the actual trip, resulting in a strongly split-level text that shifts, with the occasional audible wrench, from the lived reality of boats and rivers to the detached plane of contemplation. From on high, the disembodied narrative voice discourses on such topics as religion, Greek poetry, and friendship, in "reflections" periodically interrupted by some irruption of the outer world.

Nor is this binary construction limited to the exigencies of the book's creation—journal reflections projected onto a summer's outing, recollected in bittersweet nostalgia. Thoreau's writing is permeated with Emersonian dualisms, which jar uncomfortably as Thoreau's attention shifts now to one side, now to the other, unable or unwilling to resolve the doubleness into unity: action or thought; real or ideal. Thoreau's riverine site allows him to work the fundamental dualism into a series of productive metaphors: land and water, water and air, "reflections" and reality, with the boat as the "amphibious animal" (WK 16) conducting the brothers now to one realm, now to the other, comfortable cohabitants of both.[38] Out of his sense of double placement Thoreau observes that his vision too is double:

> We noticed that it required a separate intention of the eye, a more free and abstracted vision, to see the reflected trees and the sky, than to see the river bottom merely; and so are there manifold visions in the direction of every object, and even the most opaque reflect the heavens from their surface. Some men have their eyes naturally intended to the one, and some to the other object. (WK 48)

His transcendental training, finding its object in the river's surface, induced him to read intentionality as an either/or phenomenon: either one saw ethereal reflections, or one saw bottom mud. As with Emerson, splitting the field into two allowed Thoreau to subordinate one side to the other, as implied by the metaphor. "It is more proper," he wrote in 1841 (after reading Coleridge), "for a spiritual fact to have suggested an analogous natural one, than for the natural fact to have preceded the spiritual in our minds" (1:231; 1/24/41). Later, he observed that out of the same field he could choose to evoke either poetry or science. Later still, when the field was no longer double but multiple, the intentionality of the eye became a powerful and dangerous tool, capable of creating by sheer force of desire, or slaying whatever it chose not to see.

Yet Thoreau refuses to be pinned down even to this seemingly secure assurance, the superiority of sky to mud. By habit of mind he would invert any certainty he recognized in himself to explore its opposite; thus, inverting the sentiment quoted above, "we reason from our hands to our head" (4:46; 9/5/51). For "There are two sides to every sentence," he writes, one contiguous to himself and the other facing the gods (1:220; 1/13/41). In this Janus vision, both sides were right: "Today You may write a chapter on the advantages of travelling— & to-morrow you may write another chapter on the advantages of not travelling," he notes (4:177; 11/11/51).[39] Emerson found Thoreau's taste for paradox an irritation. "The trick of his rhetoric is soon learned," he decided in 1843; "It consists in substituting for the obvious word and thought its diametrical antagonist." But the habit ran deeper in Thoreau than mere mannerism, for paradox allowed him to declare a simultaneous doubleness and keep both sides in suspension: not either/or, but both/and. It was good to travel; it was also good to stay home, though not in the same way. One had to calculate tradeoffs: traveling spoiled the horizon's mystery, but it gave one another horizon farther off (4:177, 436; IV:31).

From the text of *A Week*, Thoreau can be quoted to sound like the perfect transcendentalist, the mystic of the empyrean, or like a rock-bound realist. The text oscillates from one position to the other. At one moment nature is "elsewhere," eternally unreachable in some separate and ethereal realm; then again, nature is here, solid, heaven under our feet. Each position produces the other, in a rhythm that becomes predictable: "Our actual Friends are but distant relations" of our *real* friends, those higher ones "to whom we are pledged" (265); but in our narratives, "We can never safely exceed the actual facts," for "A true account of the actual is the rarest poetry" (325). All material nature is in flux and flows, and "Let us wander where we will, the universe is built round about us, and we are central still" (331); yet it pleases him to find that "The landscape is indeed something real, and solid, and sincere, and I have not put my foot through it yet" (350). He has "no respect for facts," and laws depend but little on "the number of facts observed" (363–64); but we can accumulate knowledge only through our own life's "experience" (365). "Here or nowhere is our heaven," he proclaims (380); but "we live on the verge of another and purer realm" whence sweeter fragrances waft to us (381). "We need pray for no higher heaven than the pure senses can furnish, a *purely* sensuous life" (382); but we need only sound senses "to teach us that there is a nature behind the ordinary" (383), and we are "provided" with those senses "to penetrate the material universe" to the "immaterial starry system," of which the material "is but the outward and visible type" (386).

Recalling Emerson in *Nature*, the longer Thoreau disavows and dissolves the merely material world in favor of another higher realm, the more intensely does he turn to reclaim and rehabilitate his earthly mother. The movement is amusingly recapitulated in a sequence of *Journal* entries in winter 1841. On February 7, he ventures out into the snow without a greatcoat, and proudly asserts that rather than defend himself against nature, he will make her "his constant nurse and friend—as do plants and quadrupeds" (1:255). A week later he confides that he is confined to the house by bronchitis, and peevishly asserts that now he has found his chest is not of tempered steel, nor his heart of adamant, "I bid good bye to these and look out a new nature. I will be liable to no accidents" (1:265–66; 2/14/41). Thoreau was particularly stung when he felt betrayed by his body, because it was perennially the senses of the body that acted as his channel to the realm beyond the mundane: "The least sensual life is that experienced through pure senses" (1:469; 9/28/43). It was *pure* sensual experience that led him to transcendence, through this world to a better one: "I believe that there is an ideal or real nature, infinitely more perfect than the actual as there is an ideal life of man," he wrote; how else—in a deduction reminiscent of Goethe's Ur-plant—could he have imagined those "glorious summers"? (1:481; 11/2/43).

Not mind but pure sense, in other words, conducted him to the realm of Coleridgean Reason, where moral law stood revealed; dull understanding grubbed in the fallen world of "science": "To the indifferent and casual observer the laws of nature are science— To the enlightened and spiritual they are morality—or modes of divine life" (2:78; 1842–44). The difference obtained in the seer's own inner state of mind, which was ever in danger of being corrupted or polluted if proper discipline was not maintained. This demand invests the seer with sole responsibility not only for the recognition, but for the *generation*, of divine life, and he becomes culpable for any failure to connect with transcendent reality.

> Sometimes we would fain see events as merely material—wooden—rigid— dead—but again we are reminded that we actually inform them with little [*sic*] life by which they live. That they are the slaves and creatures of our conduct. When dull and sensual I believe these are cornstalks good for

> cattle no more nor less—The laws of nature are science but in an enlightened moment they are morality and modes of divine life. In a medium intellectual state they are aesthetics. (1:468; 9/28/43)

Aesthetics—a sense of beauty—show an awakened sensual state, which then can mature to a fully trans-sensual or divine state. The romantic subject stands in splendid isolation, God's lonely agent on earth, his objects but "slaves and creatures" of his actions—which is to say, he is enslaved to himself, to his own bodily or creaturely frailties. In this system, defeat is foreordained, for the subject is guaranteed the lonely doom of Coleridgean despair to which Thoreau alludes: "we receive but what we give, / And in our life alone does Nature live."[40] In this mode nature will always, necessarily, be "elsewhere," like Poe's ideal of unreachable beauty. Thoreau, had he maintained this posture, would have assured himself disappointment at Walden Pond and contracted with the reader not for a chanticleer brag but for yet another "ode to dejection" after all.

How, then, did Thoreau succeed in escaping from the romantic gulf of alienation? One answer might be, by temperamental defiance of the tenets of rational holism. He was instinctively at odds with the rational willingness to dispose of material particulars, and shows an early relish for the way particulars could set any rational system at naught:

> The infinite bustle of nature of a summer's noon, or her infinite silence of a summer's night—gives utterance to no dogma. They do not say to us even with a seer's assurance, that this or that law is immutable—and so ever and only can the universe exist. But they are the indifferent occasion for all things—and the annulment of all laws. (1:122; 4/19/40)

Facts become the "indifferent occasion" for our laws, as if they might just as easily choose *not* to obey our misconceived dictates. We are always, Thoreau asserts, a step behind: "The universe will not wait to be explained. Whoever seriously attempts a theory of it is already behind his age. His yea has reserved no nay for the morrow . . ." (1:122; 4/20/40). Here is a clue to Thoreau's principle of paradox: always reserve a nay for the morrow. There is a deliciously subversive strain to this vision: "Think of the Universal History—and then tell me—when did burdock and plantain sprout first?" (1:131; 6/18/40). The trickster can be a bird: "The wood thrush is more modern than Plato and Aristotle. They are a dogma, but he preaches the doctrine of this hour" (1:159; after 7/27/40). Or a quadruped, like "Jean Lapin," who sat trembling before his cabin door, "a poor wee thing lean & bony—with ragged ears," as if "the earth stood on its last legs"—until

Thoreau takes "two steps—and lo away he scud with elastic spring over the snowy crust in to the bushes a free creature of the forest—still wild & fleet—and such then was his nature . . ." (2:225; 2/22/46).

F. O. Matthiessen established an important distinction between Thoreau and Emerson when he observed how much more concrete and loyal to the object Thoreau was. Where Emerson outlined an organic society in which men would take their appointed place, a young Thoreau defiantly countered in his journal that "properly it will read—Society was made for man" (1:35; 3/7/38). Things must be ends in themselves, not instruments for another's use; hence the tension between earth's objects and "higher laws" which tugs back and forth in *A Week*. Nature as instrument serves, like a slave or creature, the symbolic ends decreed by man, God's agent on earth. But Thoreau refused to locate "reality" on some other, unearthly plane. However abstract this conflict seems to us, Thoreau found it compelling:

> The frontiers are not east or west, north or south, but wherever a man *fronts* a fact, though that fact be his neighbor, there is an unsettled wilderness between him and Canada, between him and the setting sun, or, further still, between him and *it*. Let him build himself a log-house with the bark on where he is, *fronting* IT, and wage there an Old French war for seven or seventy years, with Indians and Rangers, or whatever else may come between him and the reality, and save his scalp if he can. (WK 304)

The "frontier" became metaphorically not a place but a relationship, a religious war waged for the salvation of one's very soul from the forces that would distract and disperse it. Thoreau imagines himself alone on the edge of the civilized world, fronting that other, without mediation, battling to the death for the right to realize it in its absolute existence—a kind of transcendence that would be not beyond, but *within*, the object. What he wanted, finally, was to declare that burdock and wood thrushes had an absolute reality of their own, beyond symbolic construction. "May we not *see* God? Are we to be put off and amused in this life, as it were with a mere allegory? Is not Nature, rightly read, that of which she is commonly taken to be the symbol merely?" (WK 382).

2

The Empire of Thought and the Republic of Particulars

> The exercise of the Will or the lesson of power is taught in every event. From the child's successive possession of his several senses up to the hour when he saith, "Thy will be done!" he is learning the secret, that he can reduce under his will, not only particular events, but great classes, nay the whole series of events, and so conform all facts to his character. Nature is thoroughly mediate. It is made to serve. It receives the dominion of man as meekly as the ass on which the Savior rode.
>
> —Ralph Waldo Emerson, *Nature*

> But nature is the domain of liberty . . .
>
> —Alexander von Humboldt, *Cosmos*

The terms of the cultural field in which Thoreau moved, from the late 1830s through the late 1840s, were largely set by Emersonian transcendentalism and its European associates. It is commonly observed that Thoreau "moved away" from Emerson thereafter, for good psychological reasons. But Thoreau moved away from more than just a father figure or a friendship—which in any case was resumed, if under strain. The demand transcendental idealism imposed, that Thoreau value facts not for themselves but for the higher truths they encoded, was already problematical by 1842. How can the conquering and lawgiving Napoleon also unite with his subjects in friendly sympathy or in "Indian wisdom"? The implicit incompatibility of these stances will become more evident to Thoreau when he begins to explore nature not in literature but on foot. While this conflict may have been exacerbated by Thoreau's own temperament, the question here is how that "temperament" was cast into the terms made available by his culture.

Thoreau identified himself as a transcendentalist to the end, for the name allied him with the quest for a morally significant universe. But to realize that significance while honoring "facts" meant, finally, that he had to reach beyond the single complex described by all the multifarious systems outlined in Chapter 1. That is, orthodox and radical theologies, "Baconians," and idealists were interdependently producing and describing each other's positions. This complex, rational holism, has been taken to be the dominant paradigm of romanticism. But it was not the only paradigm. Like any totality, it embraced the all by locking something out.[1] To open the door to the alternative, it will help to understand the assumptions that grounded rational holism, the metaphors it deployed, and the costs it exacted.

Law as Logos

The fear that the universe might be composed of aggregate matter, without life or spirit yet forming somehow a functioning whole, found its appropriately ambivalent metaphor in the industrial technology of the day: the machine. Carlyle castigated the unregenerate to whom everything was dead and mechanical. Worst of all was "the mechanical Manipulation falsely named Science," which taught the student that his "whole Universe, physical and spiritual, was . . . a Machine!" (79–87). Such teachings led to the pit of existential despair: "To me the Universe was all void of Life, of Purpose, of Volition, even of Hostility: it was one huge, dead, immeasurable Steam-engine, rolling on, in its dead indifference, to grind me limb from limb" (125–26).

The solution was urgent: subdue the senses and unite with spirit through the law of science—*true* science. Carlyle's vehement prose shudders before the horror of the alternative: science indifferent to religion, mechanism content to abandon spirit to "dead forms" of worship, to "the dead Letter of Religion" (87). If religion without science is dead, science without religion is deadly, out "to destroy Wonder, and in its stead substitute Mensuration and Numeration"; "Thought without Reverence is barren, perhaps poisonous" (50–51). Emerson used Carlyle's vision as the emotional trigger for his "Divinity School Address," but more orthodox circles in America were also reacting to "the menace of a 'cold and irrational materialism'" emanating from chemistry and physiology, ominous because of man's tendency to prefer "dead mechanistic operations" over chants of praise for the Creator.[2] The energy behind these denunciations of materialism flowed into the overdetermined distinction between Understanding and Reason.

One effective way to silence your opponents is to take over and redefine their terminology to your own purposes. The authority of Enlightenment "rationality" rather than being opposed was appropriated, and "rationality" redefined.[3] In Coleridge's widely adopted distinction, "Reason" became "the power of universal and necessary convictions, the source and substance of truths above sense, and having their evidence in themselves" (*Aids to Reflection* 211). There can be only one reason, "*one only*, yet *manifold*," for it is the Divine "*Logos*" (212). The old Lockean faculty of aggregative knowledge and accumulated sense impressions became "Understanding." Or in Mary Kupiec Cayton's succinct formula, "Reason focused on the unity of phenomena, Understanding on their differences."[4] The science that was cast under a cloud of suspicion was that outdated, eighteenth-century brand which hid from the light of Reason, tucked into the dark, cold closet of the Understanding. Emerson called it "this half-sight of Science" (CW 1:41), or "the wintry light of the understanding" under which naturalists are apt to freeze their subjects (CW 1:44). Reason was divine, reaching "above sense" to ultimate Mind; understanding, as Coleridge showed, was related to instinct, and therefore animal in its nature, lowering man to the level of the brute (*Aids to Reflection* 232–33). No wonder, then, that Understanding, when shorn of the pure light of Reason, could run to such beastly, such frightening, excess. It is that ancient serpent within us, bound over to animal nature, the cravings and lusts of the senses, and associated, inevitably, with Eve and womanhood (241–43).

Thus, as the clear forces of Reason take up a war of liberation against the clouded half sight of the Understanding, the antinomies fall into place: on the one hand, the old, dead, static past with its materialistic mechanism trapped in the trivial and bound over to the animal, sensual, and feminine. On the other, the new, living, dynamic present with its spiritual organicism breathing significance into a world liberated into the human, rational, and male. As the problems of the past had arisen from a deep and dangerous division, so the watchword for the future was unity: mind and matter, subject and object, spirit and nature. In the new Reason, all the old antinomies would be brought together, the fragmented parts reunited in the whole.

For Coleridge the most persuasive evidence for this unity comes, ironically, from the senses: in the world, we see "everywhere evidences of a unity, which the component parts are so far from explaining, that they necessarily pre-suppose it as the cause and condition of their existing as those parts; or even of their existing at all." As Coleridge continues, this antecedent unity or "all-present power" has been called, from the time of Bacon and Kepler, a *law*. This concept of "law" is central to Coleridge, for in the trinity of its forms, law activates and organizes the universe. When man acts in harmony with this "intercommunion" established by law, it is an all-present power or spirit uniting and becoming one with our human will or spirit; "and by what fitter names can we call this than THE LAW, as empowering; THE WORD, as informing; and THE SPIRIT, as actuating?" (*Aids to Reflection* 106–9). However, this power or spirit or will is unique to man, dividing him from nature. True, all "organized bodies" share life, which is "the one universal soul . . . by virtue of the enlivening Breath, and the informing Word. . . . But, in addition to this, God transfused into man a higher gift, and specially imbreathed:—even a living (that is self-subsisting) soul, a soul having its life in itself," which man does not merely possess but *becomes*. "It was his proper being, his truest self, the man in the man" (70). Therefore nature and spirit are defined against each other as "antitheses." Nature is the aggregated system of all things "representable in the forms of time and space," or comprised in "the chain and mechanism of cause and effect," while spirit is *above* nature, literally "supernatural." Then why should nature even exist? The cause is to be sought for perpetually in something preexisting nature, its antecedent. Through nature, spirit divides from and returns to itself. The very word *nature* means "becoming," in flux, changing, and therefore unable to subsist in itself. In complementary fashion, whatever *can* subsist in itself, "whatever originates its own acts, or in any sense contains in itself the cause of its own state, must be spiritual, and consequently supernatural" (108n, 236). Thus for Coleridge nature consists of dead matter subject to antecedent causes, and it can organize itself into life only when "imbreathed" by spirit. That spirit is life itself, which is *super*natural, for it is self-subsisting and cannot be sought in an antecedent cause; and that spirit is also *beyond* life itself, or the "universal" soul, subsisting in man who alone is graced by the higher gift of a "living" soul.

Hence Coleridge is a vitalist, and his unifying system splits down the middle because it reproduces the Cartesian dualism of mind and matter, while reasserting, in revised form, the classic eighteenth-century solution to the old dilemma: how to animate the bodily machine. Coleridge reinstates his antitheses of nature and spirit in the name of spiritual power. Ultimately he is unconcerned with nature, or with proposing any union between man and nature. What he is intensely concerned with is elevating man above the "chain and mechanism of cause and effect," to find the principle of union in the antecedent "LAW."

In a different metaphor, the actuating power that Coleridge thinks of as "law," Carlyle prefers to call "force." It is this force that renders nature "not an Aggregate but a Whole," and Carlyle, too, arrives at his conviction through the evidence of sense: "O cultivated Reader," he asks, "knowest thou any corner of the world where at least FORCE is not? . . . Thinkest thou there is aught motionless; without Force, and utterly dead?" The thought is absurd; it would imply that something could be a "detached, separated speck." On the contrary:

> Detached, separated! I say there is no such separation: nothing hitherto was ever stranded, cast aside; but all, were it only a withered leaf, works together with all; is borne forward on the bottomless, shoreless flood of Action, and lives through perpetual metamorphoses. The withered leaf is not dead and lost, there are Forces in it and around it, though working in inverse order; else how could it *rot*? (52–53)

Given this, "no meanest object is insignificant," if rightly viewed (53). The universe is saved from unmeaning: "The Universe is not dead and demoniacal, a charnel-house with spectres; but godlike, and my Father's!" (142).

Emerson combines elements of both Coleridge's "law" and Carlyle's "force," but is especially indebted to Coleridge's threefold unity of reason as the law, the word, and the spirit, a trinity which Emerson translates to science, poetry, and action, or truth, beauty, and duty. "Man is conscious of a universal soul within or behind his individual life," which he calls reason. "That which, intellectually considered, we call Reason, considered in relation to nature, we call Spirit. Spirit is the Creator. Spirit hath life in itself." And man embodies spirit in language as "the FATHER" (CW 1:18–19). This is virtual paraphrase of Coleridge. The unity in variety of nature's works impresses Emerson as well: however different, "the result of the expression of them all is similar and single. Nature is a sea of forms radically alike and even unique." Spirit enables us to see the relation of things to virtue, as beauty; reason enables us to seek out the relation of things to thought, "the absolute order of things as they stand in the mind of God, and without the colors of affection" (CW 1:16).

This search for God's "absolute order of things . . . without the colors of affection" is the task of science, which is true so long as it studies not differences but relations, and so attends to the totality of nature. As science aspires to truth, art aspires to beauty, which also subsists in totality. For that very totality, "that perfectness and harmony, is beauty. The standard of beauty is the entire circuit of natural forms. . . . Nothing is quite beautiful alone: nothing but is beautiful in the whole" (CW 1:17). "The plastic power of the artist's eye" can use even a single object to suggest "this universal grace," concentrating "the radiance of the world on one point. . . . Thus is Art, a nature passed through the alembic of man." In the end it is clear that science and art are perfectly symmetrical in their motives, methods, and results: "Truth, and goodness, and beauty, are but different faces of the same All" (CW 1:14, 17). When man at last realizes this truth, he will come into the power that has been rightfully his all along, but masked by his reliance on mere understanding. "This is such a resumption of power, as if a banished king should buy his territories inch by inch, instead of vaulting at once into his throne" (CW 1:43). On the contrary, Emerson asserts, the vision of truth is intuited through the reason, and this vision comes not piece by piece but in a rush like God's vast picture, carrying with it the uttermost certainty and sweeping before it all doubt and confusion.

In short, if materialistic science was atomistic, fragmented, and empirical (and dangerous), romantic science would seize the whole through the agency of law, which would establish the *principle* of relation.[5] Emerson continued to criticize that science which, in his view, failed to strive for the requisite unity, advising naturalists to learn not by "addition or subtraction" but by "untaught sallies of the spirit." Emerson adds,

> I cannot greatly honor minuteness in details, so long as there is no hint to explain the relation between things and thoughts; no ray upon the *metaphysics* of conchology, of botany, of the arts, to show the relation of the forms of flowers, shells, animals, architecture, to the mind, and build science upon ideas. (CW 1:40)

Another name for this ray of relation is "law": the "ambitious soul sits down before each refractory fact," reducing one after another "to their class and their law," and so animating nature by insight (CW 1:54). When Emerson wishes to "build science upon ideas," and seek the "law" of facts, he is making the same equation as Coleridge: "Thus an idea conceived as subsisting in an object becomes a law; and a law contemplated subjectively (in a mind) is an idea" (*Aids to Reflection* 186n). Laws, as antecedent causes, can be understood only by Reason, which is also antecedent and independent, whereas the understanding can only abstract, generalize, and compare.

The relation here between that binding law and those refractory facts and minute details is of peculiar interest. How does one arrive at the law from the facts? One clearly wishes to start with the antecedent and higher faculty, and derive the law from one of those powerful sallies of the spirit. But as Goethe, the "seminal figure" for biological idealism, established, one does not then turn to observation for validation of the law thus arrived at, for the law is a transcendental pattern imposed by the mind, and therefore not dependent on phenomena; instead, it suggests to the observer how phenomena are to be interpreted. Particular phenomena which obeyed the law were not to be interpreted as supporting evidence, but as illustrations of what was already known, a priori. Phenomena which disobeyed the law could be disregarded, because they indicated only the incompleteness of our knowledge.[6] Thus deftly could law be used to dispense with empiricism, when it was *law* one started with.

The idealist who approaches nature in the right frame of mind is assured of finding what he wishes—this, in fact, is the advantage idealism gives us, according to Emerson: "that it presents the world in precisely that view which is most desireable to the mind" (CW 1:36). Rightly interpreted, the data of our senses can point to only one truth. And given the truth, all the data will fall into place. The skeptic might bring the question: if one assumes a harmonious and divinely created universe, will not one see only the data that confirm that assumption? Then adducing those data as evidence does not establish the truth, only that truth is its own tailor.[7] But this is exactly what is wanted: it need not be a very long step from seeing phenomena in accordance with Reason, to seeing phenomena disappear altogether. That, in fact, is the ultimate goal, if one truly wishes to penetrate to the spirit; and what rational transcendentalist does not? In *Sartor Resartus*, Carlyle imagined a mental journey to the point at which nature, the living garment of God, can indeed be seen to dissolve. The "beginning of all wisdom is to look fixedly on Clothes . . . till they become *transparent*," Teufelsdröckh declares (50). The goal of the "Clothes-Philosophy" is "To look through the Shows of things into Things themselves" (155), a state at last reached when he can rend asunder those "mysterious, world-embracing Phantasms, TIME and SPACE," and look "fixedly on Existence, till, one after the other, its earthly hulls and garnitures have all melted away; and now, to his rapt vision, the interior celestial Holy of Holies lies disclosed" (91). Emerson begins *Nature* by himself becoming "a transparent eyeball," but as he writes, the metaphor shifts: in the transcendent moment when the true relation between mind and matter appears to man, "the universe becomes transparent, and the light of higher laws than its own, shines through it" (CW 1:22). Soon the metaphor intensifies: as the mind apprehends the laws of physics, man becomes greater and the universe less, "because Time and Space relations vanish as laws are known" (CW 1:25). As man possesses the light of the law, time and space, the burdensome depth of history and the weight and expanse of material nature, vanish like shadows.

The desire to render nature transparent so that the law can shine through it demands that law precede fact. This is law as "Logos," a concept derived not only from Coleridge but also from the German *Naturphilosophen* and from Protestant natural theology: law is the power of divine will, in effect a "divine edict" handed down from above. Alternatively, scientific law might also be defined as "an expression of the sovereign properties of matter."[8] Yet Coleridge insists that matter is precisely that which has no sovereign, or "self-subsisting," properties. Law can come only from above, and knowledge of it cannot be arrived at through studying the properties of matter. On the other hand, for the "Baconian" Protestants, facts were not expendable but precious, neither more nor less than evidence for the divine structure of will which held them firmly in place, but which was beyond the realm of science to determine. Since the evidence was never complete (a new fact could always turn up), the laws could never be completely known by man; man can fit data into rational schemes only because God wills it so. In effect the orthodox Baconian Protestants, for whom the facts were fixed and the laws incomplete, mirror the radical idealists, for whom the laws were fixed, and the facts, in their essential infinitude and completeness, indifferent. Both were bound by the law as Logos, the will of God in a universe whose objects were before all else the thoughts of God, in whose wake followed the thoughts of man.[9]

Coleridge's binary concept of Reason and Understanding could be so widely accepted because it expressed this subordination of matter and all its train to the rule of law, for Reason stood antecedent to all, determinative of all. In its threefold manifestation, the law, the word, and the spirit, Coleridgean Reason underwrote the mainstream solution to the impending social, political, and intellectual stresses of the early Industrial Revolution, stresses which were interpreted to presage an imminent state of chaos and which were condemned as the evil heritage of a materialistic past. The solution, to dissolve all the old antinomies into the unity of spirit under the law, seems in retrospect to have carried the day, for it has come down to us as the single voice of its age—as, indeed, the emergence of "modernism" itself.

Rational Holism

Emerson's long essay *Nature* presented to its readers a way to reunite that sundering trinity, Nature, God, and Man, through a recovery of the spirit that flowed through all and bound all lawfully into one. This program rests on the belief that the universe is created as a whole, entire and complete and commensurate with our minds. As Emerson says, "Undoubtedly we have no questions to ask which are unanswerable. We must trust the perfection of the creation so far, as to believe that whatever curiosity the order of things has awakened in our minds, the order of things can satisfy" (CW 1:7). This belief and the program of consolidation which rests on it form the crux of that complex I wish to call rational holism, and from this central belief flow a number of consequences:

1. Matter embodies, or "bodies forth," spirit; since spirit is primary, matter is thus a veil or garment.

2. Science seeks the organizing law, which is spirit. Science therefore is an agent of reason and gives moral truth.

3. The organizing law is a process, eternal and immutable, whereas physical nature is always "becoming," in a state of flux and change.

4. The Fall into matter was the originary creation of difference, hence nature is fundamentally polar. This polarity generates the upward movement of nature to man and thence to God.

5. Words are cognate with objects; language names the path of spirit through nature.

6. Man is law on earth.

These six characteristics depend on the fundamental assumption that the universe is a whole formed by an antecedent spirit, and therefore determinate. The whole can be known only through reason's unmediated and intuitive connection with spirit, a vision frequently named "transcendental." This intuitive vision is grounded in the law as Logos, known by and expressed through the faculty of (Coleridgean) reason: hence, "rational holism." The parallel but quite different assumptions which constituted what I call "empirical holism," and which were being worked out simultaneously by another loosely affiliated international network of intellectuals, will be discussed shortly. The following, which is meant to be merely suggestive, sketches in some of the ramifications of the six points listed above.

Coleridge establishes the logical basis for the first characteristic, matter as the embodiment of spirit. According to him, the particles which constitute the body are in perpetual flux, like smoke or a cloud, and only because of the "coarseness of our senses" do objects appear the same "even for a moment." The "combining and constituent" power evolves the body's characteristic shape, and "the material mass itself is acquired by assimilation." The "germinal power" of the plant shows the organic principle at one level; as "the unseen agency weaves its magic eddies, the foliage becomes indifferently the bone and its marrow, the pulpy brain, or the solid ivory." What we *see* is "the Work, or shall I say, the translucence, of the invisible energy" in action (*Aids to Reflection* 343–44). Thus the spirit acts as the organic or organizing principle to assimilate matter to its flow of energy; the organic metaphor becomes the principle for all creation.

Carlyle's more humanized metaphor of matter as "clothing," or "*the living visible Garment of God*," allows him to emphasize its built-in impermanence and play changes on man as a "moving Rag-screen, overheaped with shreds and tatters raked from the Charnel-house of Nature" (41). Matter itself "exists only spiritually, and to represent some Idea, and *body* it forth." Our bodies and lives become emblems or garments for "that divine ME," and language "the Flesh-Garment, the Body, of thought" (54). Emerson exploits this connection when he calls nature "the vehicle of thought": first, God's thought, through natural facts; then *our* thought, through words which are signs of natural facts (CW 1:17). This notion is grounded in the same basic conviction: "There seems to be a necessity in spirit to manifest itself in material forms." Objects "preëxist in necessary Ideas in the mind of God. . . . A Fact is the end or last issue of spirit" (CW 1:22). Each fact is important in itself insofar as it is a thought of God, and once discovered must be subsumed by its rightful place in the system. The true importance of any fact, therefore, subsists not in its own identity at all, but only as it bodies forth a divine thought. Emerson finds another way to say this when he is thinking analogically: "nature is the opposite of the soul, answering to it part for part. One is seal, and one is print." Its beauty, and its laws, are the beauty and the laws "of his own mind" (CW 1:55).

Second, since matter is the embodiment of spirit, science is the intellectual search for embodiment's organizing law, which, being co-determinate with spirit, is inherently moral. On this Emerson is quite clear: "The axioms of physics translate the laws of ethics," he states, and follows with half a page of physical, natural, historical, and proverbial truths that are all ultimately moral (CW 1:21–22). Coleridge's revolutionary plans called for an elect band of natural philosophers, or "'patriot sages,'" who would transfer the "'natural to moral sciences'" and thus be prepared to act as the social engineers of the age.[10]

Third, the laws themselves are eternal and immutable, as Emerson assures his readers: "God never jests with us. . . . Any distrust of the permanence of laws, would paralyze the faculties of man. Their permanence is sacredly respected . . ." (CW 1:29–30). In this sense nature, too, is "permanent"; that is, nature as "process," for "every natural process is a version of a moral sentence. The moral law lies at the centre of nature and radiates to the circumference," and is the "pith and marrow" of every substance, relation, and process (CW 1:26). Law as process is also referred to as "the currents of the Universal Being" which circulate through him, or the "endless circulations of the divine charity" which nourish man (CW 1:10, 11). But while the currents of law are permanent and unchanging, the material through which the currents flow is in perpetual flux. Matter flows through law like ephemeral clouds in a perpetual sky: "The heavens change every moment," while the sky itself is a type of eternal reason (CW 1:14, 18–19). Emerson's grandest statement of this principle occurs, appropriately enough, at the climax of the chapter "Discipline": law binds all things by resemblance. Granite, river, air, light, heat all resemble each other; "Each creature is only a modification of the other," as their "radical law" is the same. "A rule of one art, or a law of one organization, holds true throughout nature. So intimate is this Unity, that, it is easily seen, it lies under the undermost garment of nature, and betrays its source in Universal Spirit" (CW 1:27–28).

Fourth, nature is fundamentally polar. The force that enabled the act of creation was polarity, or the initial differentiation of matter. We have already seen Cousin explain the progress of psychology and history through the polar oppositions of unity and variety, and Lyell the changeless changes of the earth's surface through the polar opposition of igneous and aqueous forces.[11] The notion is used even more explicitly by the *Naturphilosophen*, beginning with Schelling, who defined the fundamental forces of the universe as attraction and repulsion, or in scientific terms the positive and negative poles of a magnet. Lorenz Oken (one of Emerson's favorite scientists),[12] was one of the early researchers to establish this principle, in 1809: "Polarity is the first force which appears in the world. If time is eternal, polarity must also be eternal. There is no world, and in general nothing at all, without polar forces." Additionally, "Galvinism is the principle of life. There is no other vital force than the galvanic polarity" (21, 182). Galvanic polarity became a reigning principle in idealist science. Coleridge used it as a social force as well: "In all subjects of deep and lasting interest, you will detect a struggle between two opposites, two polar forces, both of which are alike necessary to our human well-being, and necessary each to the continued existence of the other" (*Letters* I:53). The same agonistic complementarity found its way into Emerson's structuring opposites—what Myra Jehlen calls his "nonantagonistic dualism."[13] The most important of these he called freedom and fate, or power and necessity, or thought and nature: "History is the action and reaction of these two,—Nature and Thought. . . . The whole world is the flux of matter over the wires of thought to the poles or points where it would build" ("Fate" 964–65).

Physical nature may be ever in flux, but the flux is never random or (at least in the human time scale) strictly cyclical, but always "building," always driven upward by the engine of polar difference. Emerson's circular metaphors would seem to imply a final sameness or an eternal return, but his aesthetic rather describes a spiral, as life force aspires upward from the organic to the spiritual: "within the form of every creature is a force impelling it to ascend to a higher form," nature's "higher end" of "*ascension*" (CW 3:12, 14). The upward ascension of life is even more crucial to Coleridge, who sees in it the will to escape from death, as the creative power strives to rejoin God: "All things strive to ascend, and ascend in their striving" (*Aids to Reflection* 140).

The fifth characteristic concerns the nature of language, which is man's way of piercing the flux of things with names: words are signs of natural facts; natural facts are signs of spiritual facts; words pierce through nature to spirit, using, and then abandoning, the fact as the vehicle of thought (Emerson CW 1:17–18). The poet, famously, is the one who "turns the world to glass, and shows us all things in their right series and procession" (CW 3:12). Language, then, is not merely a human construction, but the necessary articulation of meanings inherent in nature.[14] The power of the name assumes mystical proportions. Carlyle has Teufelsdröckh say: "Not only all common Speech, but Science, Poetry itself is no other, if thou consider it, than a right *Naming*" (66). Names identify the transcendental spirituality common to things and to thoughts, thereby naming the path of spirit through nature. However, as allies of the object, words also share the object world's fatal limitation. As objects dissolve in the face of spirit, so do words: "What are your Axioms, and Categories, and Systems, and Aphorisms? Words, words. . . . Be not the slave of words," advises Carlyle (40). Emerson also associates words with the bane of fragmentation. "Words are finite organs of the infinite mind. They cannot cover the dimensions of what is in truth. They break, chop, and impoverish it" (CW 1:28). A curious thing for a poet to say, but the ambivalence carries through his work: language may aim for the infinite, but in actual use it is doomed to fall short.[15]

Thus, sixth and finally, the worship of heroic human action over words. The quotation above, on the impoverishment of words, continues: "An action is the perfection and publication of thought. A right action seems to fill the eye, and to be related to all nature" (CW 1:28). Words, as objects, are limited to the material world, but action links to process, and thus can become the expression of law *in* the material world. This is man's true role on earth, for man is the agent of spirit, and therefore the law, and stands at the center, organizing all things to the axis of his thought.[16] Thus it is that nature is "thoroughly mediate . . . made to serve" (CW 1:25), fluid, volatile, "obedient" (CW 1:44), and the heroic man is he who has entered most effectively into his proper kingdom of command. "One after another, his victorious thought comes up with and reduces all things, until the world becomes, at last, only a realized will,—the double of the man."[17] The man who thus victoriously humanizes or spiritualizes nature (it amounts to the same thing) is the central man, the "virtuous man" who "is in unison with her works, and makes the central figure of the visible sphere" (CW 1:16). Emerson often calls this man "the poet," he who "conforms things to his thoughts" (CW 1:31). His world-making power is celebrated by Carlyle: "A Hierarch, therefore, and Pontiff of the World will we call him, the Poet and inspired Maker; who, Prometheus-like, can shape new Symbols, and bring new Fire from Heaven to fix it there" (169).

The Prometheus image indicates the link here to Coleridge's clerisy: the artistic genius acts as reason's conductor to "the crowd of lesser men."[18] The course of his victory illumines the pathway for all of us: the transformation of matter into spirit, into man's double, demands *work.* All parts of nature "incessantly work into each other's hands for the profit of man," Emerson writes; but the labor does not stop there: "A man is fed, not that he may be fed, but that he may work" (CW 1:11–12). Carlyle exults in work as the solution to the ills of the age. "Doubt of any sort cannot be removed except by Action," he exhorts us. "*Do the Duty which lies nearest thee!*" "Be no longer a Chaos, but a World, or even Worldkin. Produce! Produce! . . . Work while it is called Today; for the Night cometh, wherein no man can work" (147–49). As Coleridge had it, all life strives upward to put death behind it; Emerson, naming the principle of death the "Beautiful Necessity," also converts it to power: "Fate, then, is a name for facts not yet passed under the fire of thought . . . every jet of Chaos which threatens to exterminate us, is convertible by intellect into wholesome force. Fate is unpenetrated causes" ("Fate" 958). And Emerson ends by chanting, four times, "Let us build altars to the Beautiful Necessity," for the law that rules us *is* us (967–68). This was written late in his career, but the principle is clear in 1836: "That which was unconscious truth, becomes, when interpreted and defined in an object, a part of the domain of knowledge,—a new weapon in the magazine of power" (CW 1:23). It is the appointed role of the twin arbiters, the poet whose end is beauty, and the philosopher/scientist whose end is truth, to postpone "the apparent order and relations of things to the empire of thought" (CW 1:33). No wonder *Nature*, this landmark of American culture, ends with the ascension of man to his proper dominion: "The kingdom of man over nature" (CW 1:45). How could anyone conclude that these men were little more than benign mystics? At the heart of antebellum culture, on the eve of the industrialization of a continent, their call is for the reduction of nature to the empire of thought, the conversion of facts to weapons in the magazine of power.

Facts which refused this conversion or threatened to get in the way of order were pernicious things indeed, and here is one reason why this literature is full of sallies against "isolated facts" and their mongerers. Carlyle supplies a typical instance: "What are your historical Facts; still more your biographical? Wilt thou know a Man, above all a Mankind, by stringing-together beadrolls of what thou namest Facts?" (153). The note of irritation here recalls Coleridge: "parts are all *little*—!" Emerson, in a celebrated passage, shows that he, too, knows what to do with all those data:

> All the facts in natural history taken by themselves, have no value, but are barren, like a single sex. But marry it to human history, and it is full of life. Whole Floras, all Linnæus' and Buffon's volumes, are dry catalogues of facts; but the most trivial of these facts, the habit of a plant, the organs, or work, or noise of an insect, applied to the illustration of a fact in intellectual

> philosophy, or, in any way associated to human nature, affects us in the most lively and agreeable manner. (CW 1:19)

The specter of an isolated fact—"unsponsored, free"—and unmarried to preestablished order or system threatens, if not the integrity of the entire system, at least the dignity and centrality of man.[19] Who could possibly be interested in the habit of a plant or the noise of an insect for its own sake—except the hidebound pedant? As Thoreau said, in a late transcendental mood: "A fact stated barely is dry. It must be the vehicle of some humanity in order to interest us" (XIII:160; 2/23/60). At least it is not "barren," only uninteresting. The ritual of dry-catalog-of-fact bashing must have had a therapeutic function, in reminding the author and his audience that they were *not* naive mechanists, unlike you-know-whom; but on some level it must also have served the purpose of reassurance, that in so complex and heterogeneous a world, all the facts *could* in fact be dried out and stored comfortably and safely in catalogs and cabinets, which then could double as munitions banks in the great war of liberation from the senses.

In such a climate, it is Emerson's saving grace to be inconsistent. At the climax of his chapter "Idealism," just when he has got nature, as he says, "under foot," he interrupts his argument to caution against taking the logic of idealism *too* far: "I have no hostility to nature, but a child's love to it. I expand and live in the warm day like corn and melons. Let us speak her fair. I do not wish to fling stones at my beautiful mother . . ." (CW 1:35–36). A few pages before, he wished to find "the absolute order of things as they stand in the mind of God, and without the colors of affection" (CW 1:16); now he flinches before the chill of the absolute. The same bereft note resurfaces in the next chapter, when he hesitates to recommend a form of idealism which denies "the existence of matter," for "It leaves me in the splendid labyrinth of my perceptions, to wander without end." It "baulks the affections in denying substantive being to men and women," and "makes nature foreign to me." And so he recommends that we "Let it stand then . . . merely as a useful introductory hypothesis" (CW 1:37–38). This is a curious comment from one who takes idealism as the very ground plan and elevation of his philosophy, but it sounds a note early on that will grow louder with time. By 1844, this particular comment will come back deeply tinged with irony: "Let us treat the men and women well: treat them as if they were real: perhaps they are" (CW 3:35). However much Emerson struggles with his "introductory hypothesis," he can never evade its logic—though he qualifies and compromises it in ways that turn out to be enormously productive.[20]

We are assured over and over again in this literature that transcendent unity embraces All in the One, but we see that it embraces the All by throwing off the many. The price of transcendence is the death of the self and the world. Cayton tells us the bargain seemed worthwhile: salvation redeemed the individual by obliterating him, "But in exchange for what it took away, it recompensed a hundredfold the individual who believed."[21] Either the stakes were very high indeed, or the loss did not seem so great. A journal entry Emerson made in 1836 is illuminating: "Science to apprehend Nature in Ideas. The moment an idea is introduced among facts the God takes possession. Until then, facts conquer us. The Beast rules Man" (JMN 5:146). The choice is to conquer or be conquered, to rule the beast of the senses by obliterating it or else to be consumed by it. Through senses man and the world embrace, and the world is neither a trustworthy nor a pleasant place. Emerson describes access to the region of the gods, whether by science, piety, or passion, as renewal; we become "nimble and lightsome," and "life is no longer irksome. . . . No man fears age or misfortune or death, in their serene company, for he is transported out of the district of change. . . . We apprehend the absolute. As it were, for the first time, *we exist*" (CW 1:34–35).

This is an extraordinary paradox: we are to become convinced of our existence through *divorce* from the world and its objects, which are defined as bestial, dangerous—death. The celebration of flux and flow has disappeared in the longing for the absolute, stasis which can never threaten change. To truly live, we are to die out of life. This is an awkward faith for a poet. Vivian Hopkins remarks on the continual fleeing from reality that results in "a disembodied quality" to Emerson's aesthetics. Similarly, in Paul de Man's well-known analysis: "In truth, the spiritualization of the symbol has been carried so far that the moment of material existence by which it was originally defined has now become altogether unimportant; symbol and allegory alike now have a common origin beyond the world of matter." The "ultimate intent" of this rhetoric of "translucence" is indeed synthesis, as de Man states; however, the synthesis is that of the subject not with the object but *with itself.* And the result is indeed a less than "entirely good poetic conscience."[22] All true enough, but Hopkins' disembodiment and de Man's bad poetic conscience sound entirely too amiable. The view they are anatomizing conceives nature and all its train as disposable, to be consumed and discarded. Worse yet is the violence behind it, the fear and even hatred that make Emerson suddenly hesitate and protest that he *does* love his "beautiful mother"—after having wished and manipulated her into inexistence for some thirty pages.

The idealist model of the world divides it into two realms, the transcendent, which is order, certainty, and security, against the worldly, which is chaotic, atomized, sensual. Despite fantasies of dissolution, both worlds are necessary, and "Reason" offers to mediate between higher and lower, allowing them to interact in a dialectic of polar opposites. But the polarity is asymmetrical, and the dialectic is suspended in favor of the higher term. "Reason" will irradiate the lower from on high, both philosophically and socially. In de Man's terms, congruence between the two poles glides away to an ontological priority of one over the other, intersubjectivity over material object.[23] Thus "freedom" is conceived in terms of escape from the world to the transcendent, where one knows the freedom of the obedient subject: Good, Coleridge writes, is "Whatever springs out of *the perfect law of freedom*, which exists only by its unity with the will of God, its inherence in the Word of God," and communion with the Spirit of God (*Aids to Reflection* 268–69).

The "law" of freedom demands purity first of all, for the impure are bound by the petty demands of the senses. Thus Carlyle's first move to spiritual ascension is renunciation, "The Everlasting No" which sheds the whining fear produced by sense and desire through all-consuming defiance which declares, "'*I* am not thine, but Free, and forever hate thee!'" (128). Then follows the indifference to all things that prepares the acolyte for "the first preliminary moral Act, Annihilation of Self" (141), which accomplished, leaves one free to abandon Happiness for Duty: "On the roaring billows of Time, thou art not engulfed, but born aloft into the azure of Eternity. Love not Pleasure; love God. This is the EVERLASTING YEA, wherein all contradiction is solved" (145). Carlyle's conquering hero mocks the idea of *choosing* a sovereign: his Ruler was chosen for him in Heaven: "Neither except in such Obedience to the Heaven-chosen is Freedom so much as conceivable" (187).

It is the absoluteness and purity of heaven which give us the standard by which to measure all things earthly, and nothing merely material can match the power of the unleashed imagination. Carlyle is breathtaking on this point: "Earth's mountains are levelled, and her seas filled up, in our passage: can the Earth, which is but dead and a vision, resist Spirits which have reality and are alive?" (200). (Not until Wallace Stevens' "insolid billowing of the solid" will such romantic rhetoric meet its match!) Emerson, never the dogmatist, is far less arrogant. No sooner does he have his fixed point and standard than he fears for *our* fragility, against the immutability of God's world: our world, and our body, proceed from the same spirit; but unlike our body, our world is "not subjected to the human will. Its serene order is inviolable by us." As the expositor of the divine mind, "It is a fixed point whereby we may measure our departure. As we degenerate, the contrast between us and our house is more evident. We are as much strangers in nature, as we are aliens from God." And that we *are* strangers, everyone knows who has tried to understand a bird or approach a fox (CW 1:39). The monism that began so defiantly by damning the dualism of the past has reinstated a dualism deeper and even more troubling. Bridges, however precarious, *were* built between Locke's subject and object; but this gulf is so deeply inscribed into the structure of its premises as to be bridgeable only by absorbing the object altogether.

In a terrible irony, celebration of flow, showing the dynamic reality below the surface, ends by offering only static principles; a system intended to be liberating ends in tuitions of mastery and repression, mind over matter. Ben Knights details how Coleridge mounted a vision of freedom that was ultimately defensive, against invasion by forces outside the mind's control—"drawing tight the conceptual net over reality" in the name of unified modes of seeing the world.[24] Emerson may have approached the project with greater misgivings, but the same imperative obtains: freedom meant freedom from the ills and temptations of the body and the corruption of change, resulting, as Jehlen shows, in an ideology that preempted the emergence of anything irreconcilable or contradictory. Part of the sadness of this irony is what it cost them as poets. Knights notes how Coleridge gave up the vividness of perception which characterized his best poetry.[25] Critics have long noted the "disembodiment" that weakens so much of Emerson's best poetry. But what could be attributable to peculiar or personal failings has a more general basis, in the ambivalence of rational idealism toward the practicing artist, whose grasp of the All must suffer as he concentrates, for the sake of expression, on a single object.[26] The qualification is directly linked to the suspect association of words with objects: both have the misfortune of being, as Emerson said, finite, as opposed to the potential infinitude of action. In 1849 Thoreau turned the dilemma into a gnomic poem: "My life has been the poem I would have writ, / But I could not both live and utter it" (CP 85).

But what this ambivalence cost art is slight compared with what it has cost science, and ourselves as the inheritors of the Coleridgean tradition. In a sense, it is why this book needs to be written at all. On one level it is incomprehensible that one should need to argue at length for the centrality and vitality of natural science to American literature. Yet here we are. The romantic rationalists could not, try as they might, make the material world disappear, nor could they succeed in killing it off actually. But they could, and did, kill it off metaphorically. And they succeeded in repressing it, together with all that Coleridge contemptuously called "the chain and mechanism," through acts of institutional and cultural mastery that remain part of our cultural hegemony to this day. Subordination of material nature as no more than a passive vehicle for the currents and energies of a life force that is fully home and fully whole only in the human soul has, as one consequence, the relegation of the study of *non*human beings to the technicians of the "understanding." Science is set aside as a specialized province of little interest to those pursuing the higher ideals of literature and art, of "culture."[27] This guarantees that science will remain available as a resource of dazzling power with which to accomplish the ends of a society in an explosive state of expansion, assimilation, and consolidation. But as a mere tool, science itself would be unworthy of—finally not available to—a critique that could have located science within human culture, not exiled from it, whether tucked away in institutions and research laboratories as a resource, or lurking on the margins of the warm human world, the Frankenstein's monster haunting the frontier of our human village.[28]

Emerson wanted to be, in some high-minded sense, a revolutionary. Rational idealism appealed to his need for an organic, hierarchical society; it was formed to the same social fears and stresses, and adapted well to American intellectual fashion. But to *begin* by postulating order pulls you toward orthodoxy and the status quo, even if you are being excoriated as an infidel and are pleased to think of yourself as a radical. To quote Ben Knights: once your advocacy "gathers force as an orthodoxy," you will begin to change from jester to priest:

> In theory you will be in favour of the revivification of reality. But since creative forces are apt to be anarchic, contradiction-riddled, and incomprehensible in the available languages, you are liable not to recognise them when they appear, or even to join the forces opposed to them. (15)

Protection against chaos may seem the ideal way to secure order, and of course it is; but the order you secure has been inoculated against change. Your philosophy has just become politics. In thirty years even Harvard welcomed Emerson back.

The Organic Machine: Making Matter Mind

The need to assimilate the many into the One, to inoculate against change while celebrating growth and flow, found its most useful metaphor deep in the heart of science. Despite the claims of the dissociation of science and society, science is fundamentally social. It grows and extends not by some mysterious inner force but by multiplying social links and associations, becoming ever more technical and ever stronger as the number of linkages increases, until it seems inevitable, invincible.[29] Coleridge labored over decades to assemble his resources and allies in science because he understood how thoroughly science *is* social, and has social consequences.

It is no accident that Coleridge and the romantics, so fond of quoting Bacon—"Knowledge is Power," as Coleridge reminded his readers in the prospectus for *The Watchman*—should have drawn their metaphors from science. Metaphors, as Donna Haraway observes, are technologies. Coleridge found the powerful metaphor which he needed to accomplish his work deep within biological science: the organized being, the organism, organicism, in which the whole is greater than the sum of its parts. The way Coleridge himself used the organic metaphor had little to do with biology, for he remained committed to the machine paradigm. In fact, truly biological organicism developed only when the machine paradigm was fundamentally altered, and this change became possible only at the end of the nineteenth century.[30] But Coleridgean organicism had everything to do with art and society: what organicism gave Coleridge was a technology of unification, and what he applied it to was the state.

Enlightenment thought located life in the "parts," the individuals, who in their association composed the whole; the machinery of association was a deliberately artificial contrivance. For example, the "constitution" of the United States sought to interlink and balance the constituent political elements across levels ranging from autonomous individuals to the state as a federated whole. But the notion of contriving a state artificially, from below, had led to its collapse into atomized pieces and the terrifying excesses of revolution. The solution lay in reversing the assumption and locating life in the whole, returning to essential harmony and interconnectedness through the "dynamical philosophy," which was, as we have seen, the achievement Coleridge credited to the best contemporary science. This move to the true philosophy was essential to the security of the state, for so long as the state was assumed to be an artificial contrivance by men, its stability was imperiled. Hence Coleridge asserted, in a letter: "It is high time, My Lord, that the subjects of Christian Governments should be taught that neither historically nor morally, in fact or by right, have men made the state; but that the state, and that alone makes them men." As he added elsewhere, "Depend upon it, whatever is grand, whatever is truly organic and living, the whole is prior to the parts."[31]

In a state which makes men, and not the reverse, men will have their given parts to play, and education will be a matter of polity. In *On the Constitution of Church and State*, Coleridge accommodates the German idealist philosophy to English cultural conditions by reconstituting the clerisy, that band of enlightened and benevolent intellectuals who would control and coordinate the "body politic" as the head the body human, or the reason the mechanisms of the understanding, forming "a moral unit, an organic whole."[32] Emerson adapted the idea to his own project, though again the differences are as instructive as the similarities. Emerson, too, tried to imagine a social system which would be a harmonious cooperation of parts, for his observations of Boston convinced him that economic individualism hardly correlated with any rise in moral individualism.[33] Emerson's solution was to reject Federalist social organicism for a "natural organicism," which made the health of the social order dependent on the pursuit of *personal* virtue. Of course this was the very climax of antinomianism, and instead of cementing the old social order helped to complete its demise. But this was hardly what the Unitarian church had intended when its leaders embraced social and natural organicism as complements. God was supposed to hold it all together. As Emerson wrote in his journal:

> Thus God had created "an Order," "a System," which was "a harmonious whole, combined & overruled by a sublime Necessity, which embraces in its mighty circle the freedom of the individuals, and without subtracting from any, directs all to their appropriate ends."[34]

Cooperation with God's plan—submission to the "sublime Necessity"—would bring about the social harmony built into the universe. Emerson left the church to cultivate his antinomian idealism in the secular pulpits of the lecture hall and printed essay. Nature, as secured by God and conceived not as a physical environment but as law and process, became "his place to stand amidst the kaleidoscopic shifting of human affairs."[35] Armed with such cosmic organicism, in 1830 he could write in his journal,

> Wrong is particular. Right is universal. (JMN 3:214)

But wait. Isn't organicism, the hallmark of romanticism, a *liberating* metaphor? The romantics themselves opposed it, *and* themselves, to "mechanism." Surely the virtue of a plant or a tree, next to some affair of levers and pulleys or a sooty engine belching smoke, would seem to be self-evident. To wrestle organicism into one suspect manifestation, Coleridge's organic society, and call all organicism static and hierarchical may seem unfair, if not downright counterintuitive.

Organicism is a lovely metaphor, but it conceals a sting in its tail. It makes the initial assumption that the ordering whole is prior to any of its constituent parts, which is why it became the flagship metaphor for the rational holists: the logic was compatible, the ordering force could be identified with spirit, the ordering logic—from the center or polar axis to the farthest circumference—could be applied universally. Best of all, it was beautiful: conceptually elegant, simple, powerful, fertile with visual associations. Emerson, in a classic passage, puts it to use thus: Spirit creates; it does not act upon us from without, in space and time,

> but spiritually, or through ourselves: therefore, that spirit, that is, the Supreme Being, does not build up nature around us, but puts it forth through us, as the life of the tree puts forth new branches and leaves through the pores of the old. As a plant upon the earth, so a man rests upon the bosom of God; he is nourished by unfailing fountains, and draws, at his need, inexhaustible power. Who can set bounds to the possibilities of man? (CW 1:38)

Who can set bounds indeed? When the center is true, cannot the circumference be infinite? Our growth is as limitless as the reservoir we draw from. Life forms and processes shooting outward from within are set against those cruel and restraining forces that act from without, in space and time, as on mere dead mechanical matter. The logic of rational holism reproduces itself in the structure of its key metaphor. One could say it defines itself as alive by defining the world as dead, to animate it with supernatural spirit. Organicism defines itself against mechanism, but produces itself as, finally, mechanism under a sweeter name—a mechanism rendered politically correct.

This paradox is embedded in the history of the concept of organicism. Both mechanism and organicism concern the relationship of parts to whole: in the former, they are assembled from without, extrinsically and by addition; in the latter, parts are connected from within, intrinsically, through growth and assimilation. The mystique of the whole being "greater than the sum of its parts" traces to Plato, who maintained that poems were logically whole, rather like a building or a ship or a living creature; Aristotle added the proviso that no part may be altered or withdrawn without disrupting the whole.[36] But the equation with life itself was not made until Kant proposed that creatures were "organized" wholes, rather like works of art in possessing a formative power from within. Yet Kant rejected the analogy of life with art: it meant locating the natural object's purpose outside itself, in God, or in human purpose, which rendered the natural object no longer natural. The only solution was to regard the natural object as "both cause and effect of itself" (*Critique* 217)—a *self*-organizing being, not beholden to any outside power or agency. So Kant defined a living organism as "*one in which every part is reciprocally purpose* [end] *and means*" (222; emphasis in original). That is, within the Kantian organism, traffic travels both ways. Parts produce the whole, the whole produces the parts.

While rejecting the analogy to works of art, Kant does accept a useful political analogy: to the recent formation of the United States, along democratic principles. Kant finds the word "organization" here particularly apt: "For in such a whole every member should surely be purpose as well as means, and, while all work together toward the possibility of the whole, each should be determined as regards place and function by means of the Idea of the whole" (221 n. 4). But as Coleridge takes over Kant and interprets him for Anglo-America, he collapses the two-way traffic: since matter cannot organize itself, the idea of the whole—an idea that originates with God—alone must determine the place and function of the parts, whether the whole be political, biological, or artistic. Organisms *were*, in fact, works of art—God's art—and therefore they were the model and binding principle for "natural" social organization. Coleridge thus appropriates "organic" to turn it against mechanic, and Kant and Coleridge together were so successful that they changed the language: the two words, hitherto synonymous, emerged from romantic theory as diametrical opposites.[37]

Given such a controversy, we should expect the literature to get technical, as indeed it does. One of Coleridge's most technical, and therefore most difficult *because* most social, texts is the neglected *Theory of Life*—written under the pressure of what was to Coleridge the key controversy at the heart of his social program: how matter was *organized.* In this text Coleridge defines life as "the principle of unity in *multeity*," or "*the principle of individuation*, or the power which unites a given *all* into a *whole* that is presupposed by all its parts" (384–85). This principle enables Coleridge to bring all the material, imaginative, and spiritual worlds into a single, rational, totally comprehensive hierarchical system, in which the highest forms are the most individuated, or those where "the greatest dependence of the parts on the whole is combined with the greatest dependence of the whole on its parts" (387)—his translation of Kant's principle of reciprocity. Life ascends like rungs on a ladder from metals, through crystals and geological formations, to the lowest forms of the animal and vegetable world, and so on to man, in whom individuality is perfected and who begins a new series "beyond the appropriate limits of physiology" (390). Life is thus properly a product of organization from within, the more complex, the higher. Coleridge's organic synthesis is able to absorb mechanism as a special case: "whatever is *organized* from without, is a product of *mechanism*; whatever is *mechanized* from within, is a product of *organization*" (385 n. 2; emphasis added). When life is in a latent condition it is one with the "synthetic" powers of mechanism; but as life manifests itself, it "subordinates and modifies these powers, becoming contra-distinguished from mechanism . . ." (385). Thus life can be explained, but to account for it is quite beyond the understanding: to that question, Coleridge states, "I know no possible answer, but GOD" (379).

The less crucial but more familiar manifestation of Coleridge's organicism is his symbol of the shaping power of the mind, expressed as Imagination: the growing plant. Meyer H. Abrams, in *The Mirror and the Lamp*, codified the five properties of Coleridge's plant: (1) The plant originates as a whole, and the parts are secondary and derived. (2) The plant conveys the process of its growth to the observer. (3) Diverse elements are assimilated into the plant's substance, losing their own identity. (4) The mature form of the plant is directed from within. (5) The parts of the mature form are interdependent, reciprocally means and end, with life interfusing all.[38] One of the best examples occurs in *Aids to Reflection*, when Coleridge educes more evidence for a unity "which the component parts are so far from explaining, that they necessarily pre-suppose it as the cause and condition of their existing as those parts; or even of their existing at all" (106). Coleridge calls to our minds the crocus: "That the root, stem, leaves, petals, &c. cohere to one plant, is owing to an antecedent power or principle in the seed, which existed before a single particle of the matters that constitute" the crocus "had been attracted from the surrounding soil, air, and moisture." The same holds true for the seed, and for all the "countless millions" of seeds, and for all the elements of the system, the world—all demanding their antecedent unity (106–7).

The metaphor of organicism has undergone an extraordinary metamorphosis from ancient Greece, to Kant, to Coleridge and Abrams. Turned inward, Plato's organizing metaphor for works of art has become a model for the shaping, "vegetative" power of imagination and the organizing power of reason. Or turned outward: Plato's unifying artistic metaphor for living creatures has become a defining condition for the organism's existence. The very ease with which Coleridge absorbs mechanistic philosophy suggests how similar mechanism and organicism are. They inhabit each other, and between them totalize the field, cleansing it of any alternative. The theoretical relationship between parts of a successful poem became the necessary relationship between parts of a successful organism, as if an organism were produced to another's purpose, exactly like a poem, or a statue, or a machine. Men thus came to see the conditions for their own works of art as defining characteristics of works of nature—seen, naturally, as the finished works of God's art. The metaphor became a determinant of reality.

The conceptual net, to borrow Knights' image, has indeed been drawn tight. Once the parts are no more than instruments to the whole, the metaphor of organicism—whose promises seemed limitless to Emerson when he was writing *Nature*—circles back to fence in possibility with an iron chain of circumstance. The end is programmed into the beginning. For all the activity Emerson's life manifests, it can only become more itself.[39] The seedling offers the structural formula of the tree; the leaf, whether ice or green, diversity without difference. This is not organicism's paradox, this is the secret of its success: organicism offers growth without change, an evolutionary model that safely contains the revolutionary threat.

Emergent Laws

In brief, rational holism is based on a central organizing law defined as Logos, the Word of God. Matter is in itself passive, even dead, and therefore cannot organize itself or have self-subsisting properties. These are granted by the transcendent principle, the governing trinity-in-one of Law, Spirit, and Word. Matter "organized" by the transcendent principle became alive, "organic," in a term Coleridge adapted from Kant's *Critique of Judgment.* But while Coleridge's Kant emphasized the transcendental principle, Kant himself preferred another possibility. Recall that he stated explicitly that "organic" matter was that which organized *itself,* and a living organism is "*one in which every part is reciprocally purpose* [end] *and means.*" Kant's organic whole is thus *emergent,* parts and whole mutually generating each other in a creative interaction.[40] As we have seen, Coleridge did away with this as errant nonsense, for it meant that parts could interact on their own, without divine agency; or, by extension, that men could create a state on their own, when it was evident that only the state could properly make men.

According to another group of Kant's successors, if matter does somehow combine to make new wholes, it is the business of science to learn how. Metaphysical solutions, such as Coleridge's invocation of God, were bracketed and set aside in a move meant to focus scientific investigation on questions which could conceivably be answered, not relinquished to theology. This move made biology possible. That is, Coleridge's division between matter and spirit (rather than inorganic and organic) meant that *all* matter was "alive"; only some was more alive, more highly organized by spirit, than the rest. Or, *no* matter was alive—only God; so it was finally only to God that the whole question could be referred. Foucault states this with deadpan wit: in the eighteenth century biology was unknown, because "life itself did not exist."[41] Ridiculous! However, not until life, or *self*-organized matter, existed as a category distinct from inorganic matter could the study of self-organizing processes—biology—be imagined. Hence in his survey of the cosmos, Humboldt enables a discussion of life by first distinguishing organic from inorganic matter, than dismissing the question of life's origin: "the empirical domain of objective contemplation, and the delineation of our planet in its present condition, do not include a consideration of the mysterious and insoluble problems of origin and existence" (*Cosmos* I:339–40). This limited science did indeed attempt to operate independently of religion, thereby opening itself to charges of reactionary materialism from idealists like Coleridge, and to charges that it was antireligious from Baconian theologians—though in Humboldt's case, the latter more often accepted the impeccably inductive results of such science and incorporated them within their religious framework.

Another way to imagine the alignments is this: the idealist/orthodox position accepted the idea that parts combined like isolated, dead atoms into organo-mechanic wholes, whose motive force came from without, and which could therefore be understood a priori, or *rationally*. The alternative position held that parts combined interactively to generate individual organisms, whose own interactions generated ever-expanding networks which, since their motive force came from within, could be understood only through experience, or *empirically*. Kant held that linear, mechanical modes of thought were inadequate to deal with the biological realm, since cause and effect there were not mechanical, but reflexive, feeding back into iteration cycles or feedback loops, making each part the "*generative cause* of the other." Thus the organic realm requires a new set of assumptions and strategies. Biology, Kant maintained, could not be deduced from theoretical constructs; given all the laws of physics and chemistry, one could not predict the form, organization, or behavior of a single actual living organism.[42] One must simply go, and look.

Alexander von Humboldt's ideas about the organism as a functioning interdependence, present in his thinking for some years, were brought into focus in the late 1700s by his reading of Kant's *Critique of Judgment*. Humboldt's experiments with muscle tissue and electricity had already shown him that so-called "vital energy" could be excited artificially, shaking his belief in a vital force. Kant's innovative theory marked a new direction in scientific thinking, and Humboldt, in disavowing vitalism, with it disavowed mechanism as well.[43] His goal became the writing of a "physical history of the world." This would be analogous to the writing of "Civil History" in one key respect: both must begin with a description of the actual, in all "its individualities, variabilities, and accidents" (*Cosmos* I:33). "All points relating to the accidental individualities, and the essential variations of the actual . . . do not admit of being based only on a *rational foundation*—that is to say, of being deduced from ideas alone" (I:49–50). Nature, in other words, was not ideal but historical, and therefore could be understood only empirically.

Humboldt was at pains to distinguish his descriptive history from attempts "to reduce all sensible phenomena to a small number of abstract principles, based on reason only." He called his own approach "a rational empiricism," based "upon the results of the facts registered by science, and tested by the operations of the intellect" (I:49). His tone is edged with defensive sarcasm: *his* history "does not pretend to rise to the perilous abstractions of a purely rational science of nature. . . . Devoid of the profoundness of a purely speculative philosophy, my essay on the *Cosmos* treats of the contemplation of the universe"—daring us to belittle such a subject. Humboldt was bringing Kant's lesson home: in complex, living systems, one cannot assume, invent, or rationalize. The parts cannot be deduced from the whole. However, knowledge of the whole can be built up through a comprehensive understanding of its parts. Looking *is* a "partial" activity, and therefore looking is the first requisite of knowing. Transcendent claims to knowledge are false and a betrayal of nature's vital individuality.

Humboldt did remain deeply committed to the search for unity in nature. His prose is full of invocations to the "great natural whole" of the universe (*Cosmos* V:5). But he sought a more adequate way to conceive of that whole, one that would proceed otherwise than by fitting facts into a preconceived whole, or what Kant termed an "architectonic concept," for it "creates the sciences," "wherein the manifold is derived from the whole" (*Geographie* 257). While the human mind does need ordering concepts so knowledge can form a system and not be a mere aggregate, Kant pointedly observed that all our "systems" to date are still really just aggregates, since "we do not have as yet a system of nature" (260). Not for lack of trying—we simply don't know enough for such a system, we have not, as a civilization, acquired enough experience. To fill the yawning gap, Kant proposed "a description of the whole earth," or a complete physical geography, which would be a scientific framework for all other knowledge. Kant's words were Humboldt's calling. His own "physical geography" meant not the dwindled field remaining today, but cosmic description, nothing less than the basis for all knowledge; and it is in attempting to create Kant's new, foundational science that Humboldt can sound at once so plodding and modest, and so heroically ambitious.

In this new form of holism, facts do not fit into, but rather generate, the whole. Order is produced by the interaction of differences, not the law of the same. The process of scientific inquiry, therefore, combines empiricism guided by intuition with theories grounded in empirical evidence. Humboldt called his method "rational empiricism" in order to stress the mutual interaction of empirical "fact" with "intellect," or theory: "The rational experimentalist does not proceed at hazard, but acts under the guidance of hypotheses, founded on a half indistinct and more or less just intuition of the connection existing among natural objects or forces" (*Cosmos* I:74). "Connection" is always the key in Humboldt's science. Convinced of the connectivity of the universe, he sought to turn this central intuition into an entire science of connections which he grounded in geography and extended to the cosmos. Thus what Humboldt wanted was not just taxonomic nature study:

> Observation of individual parts of trees or grass is by no means to be considered plant geography; rather, plant geography traces the connections and relations by which all plants are bound together among themselves, designates in what lands they are found, in what atmospheric conditions they live, and tells of the destruction of rocks and stones by what primitive forms of the most powerful algae, by what roots of trees, and describes the surface of the earth in which humus is prepared.[44]

Or in a more succinct formulation, Humboldt was after "the study of the reciprocal relations between organisms and their environment"—Haeckel's definition of *his* new science, which he proposed in 1885.[45] Humboldt continues: it is this emphasis on connections that "distinguishes geography from nature study, falsely called nature history"; *falsely* because zoology, botany, and geology study only the form and anatomy of the individual. While Humboldt's ambition for "geography"—a "true" natural history, Kant's *Naturgeschichte*—was ultimately both rejected and fragmented into a multitude of specialties, its key idea emerged late in the century as Haeckel's new science, "ecology."

In 1797, Humboldt's electrical experiments and his consequent denial of a vital force drew bitter criticism from his friend Schiller, who wrote in a letter of Humboldt's "poverty of understanding and perception" in using "that sharp, naked reason which is impertinent enough to want to measure and fathom Nature, which must remain inscrutable and awe-inspiring in all its aspects."[46] Humboldt had initially been receptive to the *Naturphilosophie* proposed by Schelling and the Jena/Weimar circle, beginning in 1797. But as Schelling and his colleagues rose to dominate German science in the first decades of the nineteenth century, relations became strained (although Humboldt and Goethe retained their close, lifelong friendship and mutual admiration.) Humboldt's apostasy labeled him, for them, as a narrow, unimaginative materialist. On his part, Humboldt seized opportunities for satirizing the idealist science of the *Naturphilosophen*, as in a passage from *Cosmos* which drips with contempt for "those profound and powerful thinkers who have given new life to speculations which were already familiar to the ancients," and who lead "many of our noble but ill-judging youth into the saturnalia of a purely ideal science of nature," characterized "by a predilection for the formulæ of a scholastic rationalism, more contracted in its views than any known to the Middle Ages" (*Cosmos* I:75–76). So much for the *Naturphilosophen*.[47]

Yet Humboldt was by no means isolated in his beliefs or in his science. On the contrary, his links with scientists in Germany, France, and Britain, from Latin America to Russia and the United States to Australia, made him virtually, if not actually, the founder of international science, and allied him with a number of scientific schools and directions. As his reputation grew, he became for the nineteenth century what Einstein was for the twentieth: not just a man but a Life—the very *symbol* of science.[48] He was part of a major movement in contemporary science, as Timothy Lenoir describes it: Kant's proposed philosophy of biology gave rise to a coherent alternative approach to biological phenomena, "teleomechanism," which was committed to "the notion of special emergent vital forces" within a "holistic conception of life." From this core developed an astonishing range and variety of work, including that of many of the important German scientists of the late eighteenth to mid-nineteenth century: Johann Friedrich Blumenbach (whose ideas influenced Kant), Carl Friedrich Kielmeyer (a possible source for Coleridge), Karl von Baer's foundational work in embryology, Johannes Müller's in physiology, and Justus Liebig's in chemistry. Lenoir identifies Humboldt as a member of the teleomechanical circle and of the German expatriate community in Paris, both united in "their opposition to *romantische Naturphilosophie*." In rejecting *Naturphilosophie* they were rejecting the search for prior law, considering their task to be "the construction of general descriptive laws" in their various areas of research, and the synthesis of their research "into a unified system. Only once the systematic interdependence of these laws could be demonstrated in terms of a single unifying principle did they think that a causal explanation could be attempted: as Humboldt expressed it, only then would *Naturbeschreibung* give way to *Naturgeschichte*."[49]

The search for causal explanation was not, therefore, to be deferred forever, only until enough was known so that unifying principles could emerge. Humboldt reminded his readers that to understand, for example, the relationship of plants to each other, the soil, or the air "which they inhale and modify" meant applying oneself to every variety of available knowledge. "The progress of the geography of plants depends in a great measure on that of descriptive botany; and it would be injurious to the advancement of science, to attempt rising to general ideas, whilst neglecting the knowledge of particular facts" (*Personal Narrative* [1852] I:x). It is in this spirit that Humboldt added that his own search had not been so much for the collection of *new* facts or species, but for "the connection of facts which have been long observed. . . ." For connections cannot be made until facts *are* "observed." As Thoreau would pun, it is "easier to discover than to see when the cover is off. . . . We must look a long time before we can see" (NHE 29). Factual knowledge of the kind Humboldt advocated did not lead to the static "filling up" of a preordained system or grid, nor did he naively suppose that facts would mysteriously "add up" to a principle or an organizing structure. On the contrary, the kind of reflexive, feedback-dependent approach Humboldt and his colleagues advocated demanded a good deal of creative imagination. The power of this kind of fact gathering and theory building was realized, for instance, in the work of Charles Darwin, who followed Humboldt to South America in the 1830s, and cast his work directly and explicitly in the tradition of Humboldt. In 1832, immersed for the first time in the tropics, Darwin exulted in his journal: "The mind is a chaos of delight, out of which a world of future and more quiet pleasure will arise. I am at present fit only to read Humboldt; he like another sun illumines everything I behold."[50]

Historians of science have generally accepted the notion that Humboldt initiated, or at least participated in, an alternative tradition that has been too long ignored. However, Humboldtian science tends to be collapsed back into one side or the other of the supposed polarity of inductive "Baconianism" versus idealism or "philosophical" science, according to which Humboldt either is read as an impoverished "Baconian" who analyzed the earth's surface through statistics, or, downplaying the "Baconian" elements, is absorbed back into the idealist or *naturphilosophische* tradition.[51] It is easier to assert that Humboldt was no idealist than it is to clarify why his science is not, as one historian has it, "a peculiarly nineteenth-century Baconianism," meaning a demographic form of natural history which takes a census in hopes that numbers alone may reveal something, and which puts off universal truth to some future Newton.[52] Humboldt did indeed pioneer the use of statistics as a powerful tool, inventing, for instance, "botanical arithmetic" as a way to study the distribution of plants: statistics proved an effective way to convert raw data into a meaningful pattern. Though Humboldt's theoretical and moral aims were obscured in the hands of certain of his followers, who perpetrated "a flurry of calculation for its own sake,"[53] it is a mistake to write off all of this material. Some of it provided crucial evidence to Darwin, and general ideas certainly can be based on carefully collated information.[54] Of more immediate interest is the possible relationship between the fashion of tabulating data, as one outgrowth of Humboldtian science, and the 650 separate tables and charts of data which Thoreau assembled in the last years of his life.

Thoreau's charts have, predictably, been condemned as symptomatic of a breakdown into pointless activity, even as the fashion of counting and tabulating statistics seems "pointless" to many of Humboldt's commentators. Labeling Humboldt's philosophy "Baconian" proved an effective way of dismissing it, by associating it both with pointless numeration and with eighteenth-century radicalism.[55] Humboldt, who proudly called himself "a man of 1789," used every opportunity to spread his radical reformist ideas, even after they were suppressed and he was forced to silence his writings on social problems. Humboldt's preferred tool was indeed the old-fashioned Baconian one of using empirical data both to forward intuition and to unmask social oppression, employing statistics to expose awkward and unpleasant social realities which many of his supporters (and more of his enemies) would have preferred kept an open secret.[56]

The most forceful argument for a distinctive "Humboldtian" science has been made by Susan Faye Cannon, who points toward "a complex of interests for which there has never been a completely satisfactory phrase," a complex "which includes astronomy and the physics of the earth and the biology of the earth all viewed from a geographical standpoint, with the goal of discovering quantitative mathematical connections and interrelationships—'laws,' if you prefer, although they may be charts or graphs." This science she names after Humboldt, not because he invented all of its parts, but because he elevated "the whole complex into the major concern of professional science for some forty years or so," from about 1800 to 1840. Humboldtian science she defines as "the accurate, measured study of widespread but interconnected real phenomena in order to find a definite law and a dynamical cause." And it was, for nineteenth-century Americans, a way to participate "in the latest wave of international scientific activity," to join, in fact, "the avant-garde."[57]

Nor was the Humboldtian complex restricted to science, given the range of Humboldt's concerns. He shared with Goethe the belief that man's wonder at the beauty and boundlessness of nature could inspire poetry as well as science. From this belief he hoped to arrive, as Margarita Bowen shows, at a synthesis of poetry, philosophy, and science,

> although he realized that it involved a different interpretation of all three elements—a poetry more meaningful and less sentimental, a philosophy not entirely divorced from scientific inquiry, and an empirical science less narrowly conceived and more productive of generalizations. All of these lead to comprehension of the world; it was to the sciences and the problems of empirical method that Humboldt directed his energies.[58]

Humboldt's universal science, in embracing so much, reorganized itself into a way of reunifying mind and matter that is still challenging today—what Bowen calls "the ecology of knowledge," in which "all knowledge occurs within the functioning ecosystem and itself forms an integral part of that system, at least while mankind survives." The preexisting universe awakens wonder in us, the emotion that stimulates a creative search for order in nature. Thus begins empirical science, which

> is the means by which the contact of man's mind with the external world can lead to intellectual development: the ordering of experience contributes to the control and increase of ideas; thought and its products are incorporated in the world totality. In Humboldt's view the two spheres of the one cosmos are in dynamic interaction.[59]

As Humboldt writes in *Cosmos*, we can if we like oppose nature to intellect, pretending that the latter is not comprised within the former, or we can oppose nature to art, by defining the latter as "human"—but these contrasts "must not lead us to separate the sphere of nature from that of mind, since such a separation would reduce the physical science of the world to a mere aggregation of empirical specialities" (I:76). Humboldt thus traces the fragmentation of science to a false dualism between mind and nature. As he continues, "Science is the labor of mind applied to nature," but nature, the external world, exists for us only through "the medium of the senses." As intelligence and speech, thought and words, "are united by secret and indissoluble links, so does the external world blend almost unconsciously to ourselves with our ideas and feelings"; quoting Hegel, "'External phenomena are in some degree translated in our inner representations'" (I:76).

In this way Humboldt dissolves the dualism between mind and nature. He imagines each actively creating the other, in a process that parallels the way in which the reciprocal interaction of parts, elements, and forces generates the organism, the living whole, or life itself. The principle works at all levels, from chemical to cosmic. In this environment, neither fact nor truth is an absolute to be discovered. Both are social accomplishments, an insight that links Humboldt to current theories asserting the social construction of knowledge.[60] For Humboldt, scientific truth advances as concepts and instruments are tested and accepted by societies, and cast into the symbolic representations—words, pictures—which allow them to be shared across space and time. This is why Humboldt used the entire second volume of *Cosmos* to follow the historical development of "the idea of the Cosmos as a natural whole" (iv), from its earliest germ in the intuition of natural fact, to its extension in the invention of the telescope and the discovery of electricity, together with its manifestations in literature, art, and gardening. *Cosmos* was subtitled "a Sketch of a Physical Description of the Universe"; for Humboldt that "Sketch" entailed a book-length interpretive history of an idea.

Given all this, it is hard not to be haunted by the fact that Humboldt's innovative science was never incorporated into any tradition, for reasons that will be broached in the following chapter. Meanwhile, although Humboldt's belief that perception influences knowledge, that mind and nature create each other, and that science involves imagination may not have found validation in the social mainstream, it did continue "underground" and in fractured form within both the traditions which, after Humboldt's death in 1859, were alleged to be "at war": poetry and science. To identify such a fractured tradition, it will help to name it, and to suggest an anatomy.

Empirical Holism

Both rational and empirical holisms share one foundational assumption: the universe is a unified whole. If romanticism looks to a unified living nature that is the totality of all things, then Humboldt's central statement is fundamentally romantic:

> Nature considered *rationally*, that is so say, submitted to the process of thought, is a unity in diversity of phenomena, a harmony, blending together all created things, however dissimilar in form and attributes; one great whole . . . animated by the breath of life. (*Cosmos* I:24)

This oft-quoted passage sounds distinctly Coleridgean. The difference, though, is that Humboldt's "great whole" is "moved and animated by *internal* forces" (I:vii, emphasis added), not from without. Nature is made—rather, nature makes itself—whole through "a chain of connection, by which all natural forces are linked together, and made mutually dependent upon each other" (I:23). The rational holist view uses "the whole" as an architectonic idea which, as Kant said, precedes the parts and derives the manifold from its own identity. Parts are isolated and lack identity until they can be subsumed into the whole, and if they cannot be, they are discarded as inconsequential. As Coleridge's disciple J. B. Stallo summarizes in his 1848 introduction to German "Nature Philosophy," "The manifold must arise from the One, not the One be a composition of the Manifold" (17).

By contrast, empirical holism knows no way of beginning other than by considering the parts, the elements of a material universe which is conceived as preexisting and hence, in some irreducible fashion, ungraspable by human ideas. The parts, not the whole, are antecedent, and the only way to know the whole is through them. Humboldt asserts that his science, "one of the most beautiful fields of human knowledge,"

> can only progress by individual study, and by the bringing together of all the phenomena and creations which the surface of the earth has to offer. In this great sequence of cause and effect, nothing can be considered in isolation. The general equilibrium, which reigns amongst disturbances and apparent turmoil, is the result of an infinity of mechanical forces and chemical attractions balancing each other out.[61]

Nearly forty years later, Humboldt wrote of the links among beauty, wonder, and science: "The powerful effect exercised by nature springs . . . from the connection and unity of the impressions and emotions produced; and we can only trace their different sources by analyzing the individuality of objects and the diversity of forces" (*Cosmos* I:27). This process characterizes aesthetic experience and expression as well:

> Thus, in the sphere of natural investigation, as in poetry and painting, the delineation of that which appeals most strongly to the imagination, derives its collective interest from the vivid truthfulness with which the individual features are portrayed. (I:34)

As all combine to generate the whole, inconvenient parts cannot be discarded, for they may indicate gaps in our knowledge, rather than "accidents" in design or interpretation. It is not for the whole to stubbornly subdue recalcitrant facts; it is for the stubborn facts to defy those wholes whose existence depends on their exclusion or silencing.

The rationally determined whole is, in its preexistence, closed and determinate, even though it may be infinite in extent; as William James points out, "only in the form of totality" is the whole conceived to be "fully divine."[62] The empirical whole is, since reflexive, necessarily indeterminate and open-ended; no "totality" is possible. We are thus in no danger of exhausting it or reaching the "end" of knowledge or of nature. Humboldt anticipates those who mistake this point:

> Weak minds complacently believe that in their own age humanity has reached the culminating point of intellectual progress, forgetting that by the internal connection existing among all natural phenomena, in proportion as we advance, the field to be traversed acquires additional extension, and that it is bounded by a horizon which incessantly recedes before the eyes of the inquirer. (*Cosmos* II:294)

Whether this is cause for exhilaration or for terror is perhaps a matter of temperament.

These, then, are the fundamental assumptions of the empirical holist: the universe is a whole of antecedent parts, open-ended and radically indeterminate, hence knowable only through empirical contact. From these gathered assumptions flow a set of consequences which I have twinned with those outlined for rational holism (see page 60–61):

1. Matter is. Appearance counts, for dualism is illusory.

2. Science seeks laws, which are the emergent properties of matter and in themselves amoral.

3. These laws are constant and inherent in the structure of matter, while nature is always in flux.

4. The movement of nature lacks destination and is generated by the interaction of random differences with and within living organisms.

5. Nature is an expressive artist with its own voice, and all elements of nature speak.

6. Moral law is a communal accomplishment, and in that community man is a partner or associate.

Full exploration of these characteristics is beyond the scope of this study—clearly, beyond the scope of a career. The concepts broached here are touchstones for some of the most active (not to say virulent) debates in critical theory today, across any number of disciplines—an observation which makes their nineteenth-century derivations all the more curious, and invites questions (some of which will be suggested in the final chapter) regarding the accustomed periodization of literary history. But some aspects of this complex with particular bearing on this project can be mentioned here.

First, matter does not embody spirit; it *is*, in itself. Thus it cannot be a "veil" or "garment" for something else, and the attendant mind/matter dualism is dissolved. As Thoreau speculated, "Is not Nature, rightly read, that of which she is commonly taken to be the symbol merely?" (WK 382). It never occurred to Humboldt that matter was a shroud for something higher or transcendent. On the contrary, matter "embodies" spirit in a very literal way: exalted powers and sublime inspiration arise from an intercommunion of man with material nature, and Humboldt is capable of irony in referring to the "speculative philosophy" which no doubt finds such aims insufficient and would demand something "still more exalted" (*Cosmos* I:75). Humboldt is content to find the sublime *in* nature, not beyond it in the realm of "perilous abstraction" (I:49).

Facts, since they are not "thoughts of God," are no longer objective givens which can be uncovered, uncontaminated by inference or subjectivity. They are, as indicated in the Latin root *facere*, things *made* by the interaction of mind with external nature, and hence not things at all, but *events*. As an event, an "isolated" fact is an oxymoron, an illusion dispelled by knowledge: "On being first examined, all phenomena appear to be isolated, and it is only by the result of a multiplicity of observations, combined by reason, that we are able to trace the mutual relations existing between them" (*Cosmos* I:48–49).

Kant had divided art from science, and intuition from observation, to avoid the danger of allowing human purpose and linear reasoning to impose itself on the nonhuman realm. One outcome of this caution was the division of human mind from nonhuman nature with the resultant dualism that characterized Coleridgean thought. Goethe, and after him the *Naturphilosophen*, sought to reunify the division by tracing the laws of beauty and of nature to the same source. For Goethe this meant that he could study the appearances of plants and from them derive the determinative natural law that controlled all individual manifestations. Humboldt's solution was similar: the fact of nature, in its harmonious unity, is both beautiful and meaningful, and it inspires both the artist in his inward goal of ennoblement of the intellect, and the scientist who seeks knowledge of the laws and principles of unity. Both work with different aims toward the same truth. Indeed, it was Humboldt who, when he joined the Weimar circle in 1794, inspired Goethe to turn to the study of comparative anatomy, and they traded thoughts on science until Goethe's death in 1832.[63]

However, the distinctiveness of Humboldt's aims emerges when one compares his study of plants with Goethe's. Whereas Goethe sought the One, the Ur-plant that gave the law of transformation to the many, Humboldt studied the composition of the many, what he called the "physiognomy," or "face," of plant communities, to understand how it is that their sheer mass stamps a "peculiar character" on the "total impression" a landscape produces. To do this he grouped plants by their forms, habits, and appearance, and finally by their geographical distribution and ecological roles (*Aspects* 236–37). The "types" of vegetation he arrived at were not prescriptive or normative essences like Goethe's Ur-plant, but descriptive categories in the first stage of building to a comprehension of the laws that regulate the global distribution of species. Plants had no "essence" obscured by the actual form; the "essence" *was* the form, and it could be known not through disembodied reasoning but only through the senses. If the focus is not on what nature symbolizes, or on ideal types which nature imperfectly manifests, but on what nature *is*, then the information that the senses bring us becomes our point of contact with the world, and the senses become of central importance. Appearance, then, the outward diversity of phenomena, is not a veil of truth but truth itself.

Second, communion with nature moves us emotionally and awakens an intuition "of the order and harmony pervading the whole universe" (*Cosmos* I:25). Sensation having thus revealed to us, by inspiration, the existence of laws, science then seeks "to establish the unity and harmony of this stupendous mass of force and matter" (I:24). But Humboldt does not believe that man can ever, "by the operation of thought . . . hope to reduce the immense diversity of phenomena comprised by the Cosmos to the unity of a principle . . ." (I:73). Laws there are, descriptive of the observed properties of matter, but a single governing principle either does not exist or will never be found by us: "Experimental sciences, based on the observation of the external world, can not aspire to completeness; the nature of things, and the imperfection of our organs, are alike opposed to it" (I:73). But in the meantime, science, like a voyage of discovery, advances in showing us progressively how more and more phenomena connect; every step in this increase and dissemination of knowledge is upward.

As for the morality of science, although we may intuitively sense an "occult connection" between forces "inherent in matter" and those governing the moral world, such views go beyond the range of science, which

> is limited to the explanation of the phenomena of the material world by the properties of matter. The ultimate object of the experimental sciences is, therefore, to discover laws, and to trace their progressive generalization. All that exceeds this goes beyond the province of the physical description of the universe, and appertains to a range of higher speculative views. (*Cosmos* I:50)

This is the limitation of science: it describes only what it observes, and cannot prescribe moral truth. Moral law does not belong to the province of science. Yet although science was not in service of moral truth, neither was it independent and unrelated to morality. In Humboldt we meet the third position excluded by the polar dialectic which demanded either the absolute identity of scientific and moral law, or their total severance: the linkage of the physical and moral universes, such that each contributes to the understanding of the other. For Humboldt, the greatest lesson offered by nature was freedom: "But nature is the domain of liberty," he declared in the opening of *Cosmos*.[64] Nature's harmony through diversity evidenced a great political and moral truth, that peace and free self-determination alone would bring all humankind together under conditions allowing the perfect development of each individual: a republic of particulars.

The third aspect flows from the belief in law as that which is constant, with science an attempt "to trace the *stable* amid the vacillating, ever-recurring alternation of physical metamorphoses" (*Cosmos* I:xii). But this attempt to "trace" stability yields laws that are not immutable, but tentative, for what is known is ever changing as knowledge, or connectivity, increases. Humboldt admits rather sadly that his own work is inevitably doomed to become "antiquated" and thus "consigned to oblivion as unreadable" (I:xii). Laws as properties of matter are certainly constant, but our understanding of them continues, in this open-ended universe, to change as it builds on its own past. This change-in-constancy is further complicated by nature's constancy-in-change, for an attempt to describe the constancy of what exists *now* is continually embedded in its own past. Nature passes through a succession of stages, and we can never entirely separate present from past. The past is revealed in the present, and description of the present "is intimately connected with its history," so that "we behold the present and the past reciprocally incorporated, as it were, with one another. . . ." In this, "the domain of nature is like that of languages, in which etymological research reveals a successive development. . . ." Basalt cones and pumice deposits, for instance, inscribe their own past: they act on the imagination "like traditional records of an earlier world. Their form is their history" (I:71–72). The flux of nature is not continually dissolving and reassembling, like clouds in the blue eternal sky; it stubbornly leaves traces of all its past, and presents to us in the present a face imbricated with its own history.

Fourth, Humboldt is confident that nature presents a series of stages, but does not insist that each stage is progressively superior. The flux of nature is not drawn upward by a teleology given to it through design, but driven outward by the emergent properties of matter. For instance, Humboldt denies that cultivated Europe is superior to the tropics. On the contrary, Humboldt tends to measure value by diversity of sense impression, and on this scale Europe is rather impoverished. To the "upward striving" of life that raises some races of mankind above others, Humboldt is adamantly opposed. Progress is a historical achievement, not a preordained law. Mind does develop and progress, but humanity is one species, and all races have the same capacity of mind, differently influenced by local environment and historical experience. Speaking in concert with his brother Wilhelm, he writes, "we . . . repel the depressing assumption of superior and inferior races of men." All races are alike noble, and "All are in like degree designed for freedom"; and as Wilhelm adds, the one idea that testifies to the perfectibility of the "whole human race" is that "of establishing our common humanity," of striving "to treat all mankind, without reference to religion, nation, or color, as one fraternity, one great community . . ." (*Cosmos* I:358). Humboldt was always compelled by the idea of community and connection, of "secret analogy" and "mysterious communion." He was unimpressed by the "Fall into difference" and theories of polar forces which were the supposed engine of progress. He speaks with contempt, for example, of Aristotle's misguided attempts to explain all phenomena of the universe by this one principle:

> All things were reduced to the ever-recurring contrasts of heat and cold, moisture and dryness, primary density and rarefaction—even to an evolution of alterations in the organic world by a species of inner division (antiperistasis) which reminds us of the modern hypothesis of opposite polarities and the contrasts presented by + and —. The so-called solutions of the problems only reproduce the same facts in a disguised form. . . . (III:13–14)

To conceive polarity as the key to the universe is, for Humboldt, not only the most errant reductionism, but an ancient and still-vicious circle.

The fifth aspect concerns another central result of the binary mode of thought which, by giving language to man, reduces nature to silence or to a transparent symbol system. Basalt cones and pumice deposits in effect "speak" their past, to the observer who stops to "listen." Their record is written in their very form, to be read by the attentive observer. Humboldt's entire process of knowledge is initiated by sensation, which communicates at the point of contact between observer and observed: hearing, sight, touch, taste, smell. This initiating stance of the observer must be receptive, for it is only to the "pious and susceptible spirit" that nature reveals her "many voices" (*Views* 201; *Aspects* 214). Only in a listening stance can nature "impress" us, or affect us emotionally and inspire our response. Humboldt believes in an aesthetic link between nature and man, as the beauty of nature moves us first to emotion, then to thought, then to action. Part of this chain of thought inhabits Humboldt's metaphors: nature presents to us a "face," a "physiognomy" whose various parts combine into a distinctive expression. But this communication does not travel in just one direction. As our hearts are moved by simple contact with nature's beauty or sublimity, that wave of emotion leads us to seek for pattern and meaning. New knowledge adds to our enjoyment; new perceptions are awakened in us, and we enter "into a more intimate communion with the external world" (*Cosmos* I:36), in a spiral of pleasure and knowledge. In this reciprocal interaction of mind and nature, each flows into the other, until they create each other like thought and language:

> The external world only exists for us so far as we receive it within ourselves, and as it shapes itself within us into the form of a contemplation of nature. As intelligence and language, thought and the signs of thought, are united by secret and indissoluble links, so in like manner, and almost without our being conscious of it, the external world and our ideas and feelings melt into each other. (I:64)

Our emotional response to natural beauty helps us to compose it, but saying this does not exclude nature's own part in initiating and cooperating with our response: both sides are contributors. That we will never fully grasp this process creates "a sense of longing" which binds "still faster" the links connecting the material and the ideal worlds, and "animates the mysterious relation existing between that which the mind receives from without, and that which it reflects from its own depths to the external world" (I:80).

The literary artist who describes nature should strive to recreate this process by keeping to a plain, pure diction, allowing art to emerge from the details, poetry from the truth:

> The poetic element must emanate from the intuitive perception of the connection between the sensuous and the intellectual, and of the universality and reciprocal limitation and unity of all the vital forces of nature. The more elevated the subject, the more carefully should all external adornments of diction be avoided. The true effect of a picture of nature depends on its composition; every attempt at an artificial appeal from the author must therefore necessarily exert a disturbing influence. He who . . . knows how to represent with the simplicity of individualizing truth that which he has received from his own contemplation, will not fail in producing the impression he seeks to convey; for, in describing the boundlessness of nature, and not the limited circuit of his own mind, he is enabled to leave to others unfettered freedom of feeling. (*Cosmos* II:81)

Words are truer, the closer they are to nature. "Speech is enriched and animated by everything that tends to and promotes truth to nature" (*Aspects* 206). Nature "animates" speech, and this suggests a corresponding alliance: language also "animates" thought. If language can interpret thought, and "paint with vivid truthfulness the objects of the external world, it reacts at the same time upon thought, and animates it, as it were, with the breath of life. It is this mutual reaction which makes words more than mere signs and forms of thought . . ." (*Cosmos* I:56). Humboldt suggests a "communion" between nature and language, words and thought. Language expresses nature, but nature reacts upon language to animate it; words express thought, but words also react upon thought and animate it. Nature and words are allied here in a way that mirrors rational holism, but there both were allied in limitation; here they alike "animate" with "the breath of life" and move beyond passive expression to that mutual interchange which sets imagination free from the bonds of mind. What enslaves is not words, but mind; what liberates is not transcendence, but contact.

Sixth and finally, this mutuality of man and nature is possible only because man is part of nature, quite literally—the chemical elements and forces of nature form our very bodies and minds. It is this literal kinship that makes it possible for nature to produce "impressions," or to "im-press" us with its beauty and a dawning intuition of the interconnectedness and unity of the whole. This communion also makes possible the naturalist's key task, to give "*ein Gemälde*" of nature, "a superior and harmonious synthesis of physical nature with moral nature"—or "an historically emerged holism."[65] Man also affects nature more directly, as his restless activities "gradually despoil the face of the earth" (*Aspects* 232). And nature affects man directly: the character of nature in different regions "is intimately connected with the history of man, and of his civilization" (235). In the introduction to the

Personal Narrative (1852), Humboldt describes the "general results" that were his goal as comprising

> in one view the climate and its influence on organized beings, the aspect of the country, varied according to the nature of the soil and its vegetable covering, the direction of the mountains and rivers which separate races of men as well as tribes of plants; and finally, the modifications observable in the condition of people living in different latitudes, and in circumstances more or less favourable to the development of their faculties. (I:xiv)

At every level, Humboldt sees man as embedded in nature, physical and moral nature as mutually interactive. He cannot study nature independently of the human social and political systems which inhabit and influence it, or the art and literature which generate the very *idea* of "nature" in our minds and in our cultures. Even science is not disembodied knowledge but a social tool. In this vast community that is the cosmos, man acts as a partner in the creation of knowledge, nature as an associate in the creation of poetry. Man cannot act as lawgiver to nature, only describe the laws whose operation he discerns. Moral law receives no automatic authority from science. Instead it is generated by the community of free individuals, and the individual's first moral obligation is to work toward abolishing the archaic social barriers to freedom.

From this point of view, man is not the consummation of creation. Nature is vastly larger than he, and within its domain, man has been able to trace the empire of a few natural laws. Nature's scale of time is beyond that of entire civilizations; the "traces of nations fade and disappear," yet "an ever new life springs forth from the bosom of the earth" (*Aspects* 186). Far from reaching its end, nature "presents itself to the human intellect as a problem which cannot be grasped" (*Cosmos* I:80). This, to Humboldt, makes Schiller's charges against him absurd. The notion that nature's inscrutable mystery was diminished by knowledge seemed to him "narrow-minded" and "morbid sentimentality" (I:38). The observations of the careful student of nature can lead only to an enhanced sense of nature's majesty and sublimity as the horizon recedes and the view expands, and to "unveil her secrets" is to produce an impression still more "imposing and more worthy of the majesty of creation" (I:40).

The last few pages have offered an interpretive codification of a coherent philosophy which Humboldt himself never gathered into a single document or declaration. The very nature of its root assumption, the reciprocal interaction of all with all, means every point of his philosophy touches upon every other, and thus it is difficult and in some degree false to analyze and resolve it into separate elements. Yet the network of ideas I have attempted to sketch here runs through all of Humboldt's writing, and, as his personal and professional influence grew, aspects of the philosophy I attribute to him reached widening circles of readers. In the late 1840s, a surge of publications, translations, and reviews would begin to make Humboldt's work widely available, for the first time, to a popular, English-speaking audience. The timing was significant. As empirical science fed imperial ambitions, Americans acclaimed Humboldt as their second Columbus, helping America discover itself. And reading Humboldt soon after his life at Walden Pond, Thoreau began to visualize a way to banish the tragic dualisms that set man apart from nature, and to affirm that knowledge could be a form of love, and experience the truest form of knowledge.

3

Seeing New Worlds

Thoreau and Humboldtian Science

How novel and original must be each new mans view
of the universe—for though the world is so old
—& so many books have been written—
each object appears wholly undescribed to our experience
—each field of thought wholly unexplored—
The whole world is an America—a *New World.*
—Henry David Thoreau, *Journal*

The polar terms of rational holism set the classic dilemma of romantic alienation from nature, a gulf between mind and nature which only the heroic, central man can overcome, through moral and physical purity which strips his own soul clean for the ascent through nature into the transcendent, whence he can speak with the voice of prophecy. James McIntosh's influential study, *Thoreau as Romantic Naturalist* (1974), can be taken as a paradigmatic analysis of Thoreau from this viewpoint: Thoreau wants involvement with nature, yet feels apart from it. His sense of separation and desire for involvement work together in a dialectic, "opposed attitudes vibrating against each other in the crucible of an essay, a poem, or a day's journal." Disappointed in his hopes for mutual sympathy, he eventually moves away to "a soberer conception" in which he realizes nature will offer him nothing but his own perceptions in return, even though "he never relinquishes the idea of a generous interchange altogether. . . ." Thus he arrives at a "shifting stance" toward nature, a temporary stay against "the possible breakdown of his romantic thinking," a "skeptical faithfulness."[1] For once the fundamental dualism between mind and nature is granted, no contact can be anything but temporary, fleeting, provisional, a promissory note for a state of integration rendered impossible by its initial premise.

This standard interpretation builds from the renunciation of material nature for which I have tagged Coleridge as standard-bearer, with its resulting queasiness about the experience of the senses, unsponsored empirical data, the hands-on *techne* of the artist, and the labor of the natural scientist. Framed in this way, Thoreau's growing absorption in sense experience and his methodical exposure to field data can be interpreted as at best only an intensification of the romantic dilemma, at worst a falling off from the unifying wholeness of transcendental Reason into the half-sight of the materialistic Understanding. Thoreau was, indeed, on a cusp. From one angle he could reproduce the Coleridgean squeamishness about raw matter, and berate himself for falling off from the "higher law." But from another angle he could joyfully embrace "raw matter" and exult in a methodology of contact. This alternative approach to the problem of material nature and its link with higher law took shape through a combination of temperament, immersion in nature, and a course of reading which familiarized him with a very different philosophical system from that of Coleridge and Emerson, a system whose central spokesman and representative in America was Alexander von Humboldt. Thoreau had already encountered intimations of this other mode of approaching nature in 1840, in Charles Lyell's *Principles of Geology*. Points of contact multiplied through the 1840s, until by late 1848 or 1849 he finally had access to the original writings of Humboldt, newly available in America.

Alexander von Humboldt, "The Napoleon of Science"

Alexander von Humboldt and Napoleon both turned twenty in 1789, the year of the French Revolution and the year with which Humboldt identified for the rest of his days.[2] He had just joined his brother Wilhelm as a student at Göttingen, effecting his own brief declaration of independence from his mother's insistence that he pursue a safe and stable career in commerce. Instead, at Göttingen he met Georg Forster, the explorer and naturalist who in 1772 had sailed, at eighteen years of age, with his father Reinhold on Captain James Cook's second voyage around the world, and whose account of the journey had since established his fame. Thereafter Humboldt was set on pursuing exploration and the sciences. Already he was developing the habit of going and looking: in 1790 he left Göttingen behind to travel with Forster down the Rhine and to London, studying the culture of plants and of people. Bastille Day, 1790, found them in Paris to join the celebration. While

Forster joined the Revolution (dying tragically in disgrace and misery four years later), Humboldt returned to his studies, reaching a temporary compromise: he would join science, exploration, and commerce together in a career as a geologist and a mining engineer, studying with Abraham Gottlob Werner at the Mining Academy at Freiberg and joining the Prussian civil service.

Humboldt had shown early talent in art, and while growing up with Wilhelm on the family estate near Berlin, Alexander learned to render plants and landscapes in careful detail—a skill he put to good use while exploring South America. He also studied botany with his friend and mentor Karl Ludwig Willdenow, who had earned his fame as one of the first naturalists to consider the "history of plants," or plant distribution and diversity, as a separate field. Thus Humboldt was introduced to botany as the study of interrelationships, an idea which seized hold of his imagination; one of his first books discussed the growth of plants in dark underground caverns.[3] As a member of the Weimar coterie that included Goethe and brother Wilhelm, he was earning a reputation as a man of letters. Schiller asked him to contribute to his new philosophical journal, *Die Horen*, and in 1795 printed Humboldt's old-fashioned allegorical essay "The Rhodian Genius," which showed how the vital force alone kept the elements of living matter from embracing in the ecstatic dance of putrescence—a theory Humboldt disavowed soon afterward. When the death of his mother in 1796 brought him freedom and his inheritance, he did not apply it to literature. Instead, Humboldt quit his mining job and used his fortune to bankroll the ambition of his life: the exploration of unknown lands. As he made clear in a letter, his plans had a very particular, Kantian-inspired purpose:

> I shall collect plants and fossils, and with the best of instruments make astronomic observations. Yet this is not the main purpose of my journey. I shall endeavor to find out how nature's forces act upon one another, and in what manner the geographic environment exerts its influence on animals and plants. *In short, I must find out about the harmony in nature.*[4]

Time and again his travel plans—to the Far East, to Africa—were thwarted by wartime blockades, but at last, having sweet-talked the Spanish crown into granting them permission to range freely in the Spanish colonies (in itself an astonishing feat), he and his companion, the botanist Aimé Bonpland, sailed to South America in June of 1799.[5] They were the first white European explorers to venture beyond the coasts and port cities, penetrating deep inland, mapping and collecting: first in Venezuela, where they traced the Orinoco waterway, then up to Cuba in 1800, and in 1801–2 to Colombia, Ecuador, and Peru, sailing in March 1803 to Acapulco and traveling overland through Mexico City to Vera Cruz. From there they sailed to Havana and on to Philadelphia, in March through May of 1804, braving shipwreck and the British naval blockade of the U.S. coast to see the young American democracy, and allow Humboldt to pay his respects to Thomas Jefferson. After six weeks in the United States, Humboldt and Bonpland returned to France, arriving in August 1804 to international wonder and acclaim.

Humboldt soon settled in Paris, by then the center of European science, and began the mammoth task of publishing the scientific accounts of the expedition, which he completed in thirty-two years and thirty volumes, draining the last of his share of the family fortune. He interspersed a dozen other works among the scientific tomes: on plant geography, on Mexico and on Cuba, on preconquest Native American cultures, on climatology, on geology, on historical accounts of voyages to America—in which he was the first to trace the name "America" to its source, in the first map of the New World, by the German professor Waltzeemüller.[6] An uplifting lecture series in Berlin during the grim winter of 1806–7 resulted in his most popular book, *Ansichten der Natur* (1808), a series of "views" or portraits of separate landform communities: steppes and deserts, the Orinoco jungle, the forms of vegetation generally. A second edition appeared in 1826, and a third, much enlarged, in 1849; this one was immediately translated twice into English.[7] It took Humboldt ten years to overcome his reluctance and accede to the popular demand for travel narrative. The unfinished *Personal Narrative of Travels to the Equinoctial Regions* (1815–26) also appeared in two English translations, though the second was published in 1852–53, long after the first was out of print. The insistent radicalism of his social writings (notably the *Political Essay on the Island of Cuba*, the last volume of *Personal Narrative*) was brought to a halt in 1827, when his patron, King Frederick William III, effectively revoked Humboldt's leave of absence and forced his return from Paris to Berlin. Humboldt marked his return by another series of lectures, on the cosmos, and after their phenomenal popularity (they had to be repeated in a larger hall—no trivial task, since he had delivered them extemporaneously), he began revising them into his crowning masterwork, *Kosmos*. This ambitious and, as he put it, "perhaps too boldly imagined" work was an intellectual portrait of the cosmos as man had come to understand it—"a physical description of the universe, embracing all created things in the regions of space and in the earth" (I:viii). Writing it occupied the last years of Humboldt's long life. After filling five volumes (1845–62), it was still unfinished at his death in 1859.[8] Meanwhile, the first edition of Volume I sold out in two months (80,000 copies had sold by 1851), and it was translated immediately into virtually all the European languages. Reviewed widely and enthusiastically, *Kosmos* consolidated Humboldt's reputation as one of the greatest minds of the age, and, after Napoleon, the most famous.

He was famed as the courageous romantic explorer of South America and Central Asia, who in scaling Mt. Chimborazo had climbed the highest of any man in history; as the prolific contributor to scientific fields ranging from botany to physical geography, geophysics to meteorology; as the organizer and promoter of international science and the patron saint of young scientists and artists; finally as the figurehead of universal wisdom, aging gracefully in his Berlin study where he entertained visitors with his cosmopolitan wit and sat for innumerable portraits.[9] After all this, what he has continued to be known for is his South American expedition and the methodology which he there developed and applied.

Humboldt's field method consisted of four principal commandments: explore, collect, measure, connect. His typical tasks during the expedition were to observe the forms and habits of the vegetation, take and calculate geophysical measurements, take exact geographical bearings, and collect plants, animals, and minerals.[10] It was a program that meant, first of all, *explore*: to study nature, one must go, not to the study, nor the cabinet, nor the laboratory, not the museum or the botanical garden—but into the field, to nature in its own state, unappropriated by man. Or, more simply, into the *wild*. For ideas, no matter how beautiful, cannot tell us what is actually on the ground and what it might be doing. What one finds there will, Humboldt assures us, always surpass expectation. The increase in "variety and grace of form, mixture of colours, and . . . energy and vigour" of life in the tropics can be denied only by "those who have never quitted Europe"; "Individual plants languishing in our hot-houses can give but a very faint idea of the majestic vegetation of the tropical zone" (*Aspects* 232, 246).

Once on the scene, one must *collect* whatever is detachable: plant and animal specimens, rocks and minerals, ethnographic artifacts, extensive field notes. These collections will help provide the masses of objects necessary for new patterns to emerge, as they are classified and categorized; classification itself is an active process of mental organization, of asserting affinities and distinctions which will help one to grasp connections. The other way to detect patterns which may lead to an understanding of connections is to *measure*: in Cannon's words, "Humboldt preached that all matters which vary with geographical position should be measured, in their variations." Above all else he demanded the utmost precision in these measurements, such that Humboldt "measured accurately what explorers had reported inaccurately," and his own expedition is a record of erroneous and misleading maps put right.[11] The value of measurement lies in the clues it gives to the constancy of laws within the flux of nature:

> The interest derived from measurements of this kind, which I made a special subject of inquiry in the western hemisphere, is increased by the consider-

> ation, that the objects to be measured vary in magnitude at different points. A philosophical study of nature seeks, in considering the changes of phenomena, to connect the present with the past. (*Views* 362)

Humboldt reminds us that in order to investigate any "periodic recurrence" or "the laws of the progressive changes in nature," it is essential to obtain "certain fixed points" as a basis for comparisons (363). Only through specific comparisons can connections be made, not through some absolute, abstract standard.

The purpose of collection and measurement, then, is ultimately for comparison, for only by comparing can one *connect*. Humboldt distinguishes his own science from both the abstractions of physical science, which consider only general properties of bodies, and the enumerations of the features of the globe (which belong to old-style geography) or the pompously titled "catalogues of organized beings" which are but "pretended systems of nature." He wants a science of "mutual connection": "It is by subjecting isolated observations to the process of thought, and by combining and comparing them, that we are enabled to discover" the relations which impart a "nobler character" to the science of the cosmos (I:60–61). Indeed, the "deeply-rooted prejudice" that science must "chill the feelings" has been created and fostered by the accumulation of "unconnected observations" —while those who "cherish such erroneous views" fail to appreciate that it is the detail of isolated facts which can lead us on "to general results," where each step opens new paths and new excitement (I:40–41).

It is true that the scientist or the worker in the field might not feel this excitement, and, restricted by the demand for accuracy, feel oppressed by the weight of so many facts which might indeed appear isolated. While he labors, he might not feel his imagination sufficiently at liberty. Yet the work must be done, and well, with its true purpose kept in mind:

> The astronomer who . . . measures patiently, year after year, the meridian altitude and the relative distances of stars, or who seeks a telescopic comet in a group of nebulæ, does not feel his imagination more excited—and this is the very guarantee of the precision of his labors—than the botanist who counts the divisions of the calyx, or the number of stamens in a flower. . . Yet the multiplied angular measurements on the one hand, and the detail of organic relations on the other, alike aid in preparing the way for the attainment of higher views of the laws of the universe. (*Cosmos* I:39)

Finally, only when one has tramped the ground, smelled the air, listened attentively to whatever voices one can hear, measured the blue of the sky and the temperature of the water, considered the rocks and handled the plants—then one is ready to speculate, to make the intuitive leap to pattern and meaning. At the end of an essay on volcanoes, Humboldt permits himself to add to the facts he has collected some "uncertain and hypothetical conjectures," for

> The philosophical study of nature rises above the requirements of mere delineation, and does not consist in the sterile accumulation of isolated facts. The active and inquiring spirit of man may therefore be occasionally permitted to escape from the present into the domain of the past, to conjecture that which cannot yet be clearly determined, and thus to revel amid the ancient and ever-recurring myths of geology. (*Views* 375)

Once the individual scientist had made his own connections between facts and conjectures, the next step was to bring facts and scientists together into union through scientific associations. Humboldt was an enthusiastic member of Lorenz Oken's "Gesellschaft deutschen Naturforscher und Ärzte," founded in 1822 and the model for both the British and the American Associations for the Advancement of Science. In his 1828 presidential speech, Humboldt spoke of the value of the "voluntary association of men in search of truth," in increasing "the mass of facts and opinions which are here brought into one common and useful union."[12] Isolated connections are "sterile" too; only if brought together in a human community do they become significant and useful.

Hence the hallmarks of Humboldtian science involve translating and communicating information: measurement and visual display. Cannon claims that "Humboldtians are easy to spot, because of the two technical characteristics of Humboldtian science." One was "a finicky fiddling with instruments," in their demand for ever greater accuracy. It was, after all, the improvement of known instruments, like the barometer, thermometer, and sextant, which made "a science of the whole world possible." The demand for accuracy of data and a fanatical concern for accurate instruments were inherited from the eighteenth century but systematically addressed by Humboldt and those influenced by his methods. The opening pages of the *Personal Narrative* enumerate the instruments he considered necessary for a description of natural phenomena. The annotated list of the instruments he collected, field-tested, and took with him to South America goes on for six pages; the final item is "a great number of small tools necessary for travellers to repair such instruments as might be deranged from the frequent falls of the beasts of burden."[13] We are reminded of the conditions under which Humboldt worked—and the amount of practical advice he could give to aspiring expedition leaders.

The other characteristic is a conceptual tool ubiquitous now, but novel then: the visual display of numerical results. Cannon concludes, "In short, if you find a 19th-century scientist mapping or graphing his data, chances are good that you have found a Humboldtian."[14] Presenting data visually became a powerful tool for clarifying connections and patterns otherwise invisible, as in Humboldt's famous and innovative isothermal lines, ubiquitous in today's weather maps. Humboldt invented visual ways of organizing information because he understood that the eye can recognize patterns more quickly than the mind. Thus it is that he developed theories of the physiognomy of plants based on the patterns of growth they exhibited and the "total impression produced" by a region's characteristic "vegetable covering" (*Aspects* 236–37). Analyzing these groups would give new ways of understanding the dynamics of plant communities. He also designed innovative methods of illustration, such as the mountain cross-section showing plant associations correlated with various kinds of data and arranged according to altitude, casting his "zones of habitation" into a memorable image of ascension from lush tropical jungles to high alpine snows, mirroring the climatic bands that ribboned the earth from equator to poles.

Nor was Humboldt restricted to "scientific" forms. His own South American paintings were respectable, but not such as to draw fame on their own, and once he returned to Europe he had little time for the paintbrush. But his interest in the power of images to impress the mind with grand ideas led to a long discussion in *Cosmos* of landscape painting. His enthusiasm for the tropics, where "an inexhaustible treasure remains still unopened by the landscape painter" (II:93), together with the scope of his theories, inspired Frederic Church to become his disciple, the model Humboldtian explorer-artist. In 1853 Church traveled to Quito, Ecuador, to live in the same house Humboldt had occupied fifty years before. Any one of Church's paintings offers a visual representation of Humboldt's fusion of detail into the cosmic sublime.[15]

Public lectures were another effective means of making connections vivid before the public eye, and Humboldt exploited the form brilliantly in separate months-long series given in both Paris and Berlin, speaking extempore (in French and German, respectively), without notes and without charge to audiences that were, by all accounts, gratifyingly enthusiastic. As he says in the preface to *Cosmos*, "public lectures seemed to me to present an easy and efficient means of testing the more or less successful manner of connecting together the detached branches of any one science" (I:ix).

Encouraged by their reception, he went on to present in writing the connections he had surmised. Humboldt's technical or scientific volumes cannot concern us here, although they established his reputation among the scientific community and generated new areas of research, even new scientific fields. Humboldt himself realized that his popular writing presented different problems, and he was uncertain whether he had adequately overcome the difficulties. What he wanted to do was present an all-inclusive network of specific and general observations and thoughts, together with the events that prompted both. He struggled to find or invent a literary form capable of doing so much, but no linear prose line could absorb such spatial and temporal burdens. Yet his failures are instructive, for Humboldt struggled with difficulties similar to those Thoreau faced and surmounted.

In *Cosmos*, Volumes I and II, Humboldt confined himself to a kind of intellectual portrait painting, and here he was confident of some success: science could be made clear to the layman if one remembered not to confuse the one who *collects* details and studies relations with the one to whom the results are presented; for the latter audience (extending the painterly metaphor), details must be "suppressed" to allow the "great masses" to be better seen. This approach results in some loss of distinctness, but the viewer's, auditor's, or reader's intellect could still be enriched and imagination vivified (I:46–48). Furthermore, it is important to attempt this kind of generalization, because it makes scientific knowledge the common property of "all classes of society" (51). Nevertheless, the remaining three volumes of *Cosmos*, intended to present the very latest scientific knowledge, sag to unreadability under the weight of detail, and the project collapsed under its own ambition. Scientific results were coming in faster than even Humboldt could assemble them. *Cosmos* was not only unfinished, it was unfinishable.

When it came to the descriptive and narrative writing in his other popular books, Humboldt was acutely aware that the fortuitous union of lifelike truthfulness, accurate detail, general results, and literary grace usually eluded him. In retrospect, he muses: "I scarcely venture to hope . . . that I have myself avoided the shoals and breakers which I have known how to indicate to others" (*Cosmos* I:ix). The *Personal Narrative* (1852) was an attempt to split off those aspects of his scientific journey which would be of general interest to "every enlightened mind": climate and its influences on "organized beings," the aspect of the country, the lives of the people (xiv). In addition, Humboldt understands that the narrative should be unified by the narrator's own presence, for it is he whom we "desire to see in contact with the objects which surround him." However, his own travel was undertaken with so many different purposes that even this simple unity becomes "scarcely possible," and he realizes he has again encumbered his narrative with too many detailed observations. Humboldt's uneasy compromise is to follow a narrative of phenomena "in the order in which they appeared," interspersed with chapters considering them "in the whole of their individual relations" (xx).

He adopts a similarly two-tiered solution in *Ansichten der Natur*, but neither is this fully satisfactory. Here he relegates the "details" to long sections of "Illustrations and Additions" following each essay. These assemblages of miscellaneous notes are so much longer than the original essays (over one hundred pages, in one case) that they engulf what they are intended to supplement. The advantage, though, is seen in the essays themselves, which having been cleared of the footnotes, are gems of kinetic description within an open-ended "accordion" form. Imagine Thoreau or Whitman dutifully footnoting every single item in their polished presentation pieces to all the supporting source material they had amassed in their vast notebooks! Humboldt felt it was his public responsibility to publish it—all of it—in a single work. Even so, his explorations of narrative form had value for those of his readers with similar ambitions (and lighter consciences.)

The story of the fortunes of Humboldt's flawed but popular and influential writings, and the wider story of his global reputation, cannot concern us here, though it is a story which has not been, and needs to be, told for a fuller understanding of nineteenth-century intellectual culture. One remarkable chapter in it is the extraordinary importance Humboldt assumed in antebellum American culture, followed by an even more extraordinary eclipse and erasure of his memory from both popular culture and intellectual history—though interestingly enough, memory of Humboldt in Latin America has been and remains quite alive. This erasure has made it the more difficult, and the more important, to recapture the centrality of Humboldt's presence in American culture during Thoreau's lifetime, particularly in the years between 1845 and 1860, when Humboldt's popularity in the United States reached a peak.[16]

Humboldt's fame in the United States followed several pathways, for his long career intersected American social and scientific history at several points. His own visit to America in 1804, though brief, established important and lifelong friendships with Jefferson, Albert Gallatin (secretary of state), and James Madison, among others; Humboldt was already a world-famous scientist and explorer, and his advice was of great strategic importance to the government of a country just embarking on its own era of exploration and expansion. Humboldt served as the unofficial advisor, enthusiastic promotor, and active and powerful mentor for a number of American scientists and explorers: Captain Charles Wilkes, Lt. William H. Emory, Lt. John Charles Frémont, the geographer Arnold Guyot, Louis Agassiz. Written correspondence between Humboldt and a number of prominent Americans was enhanced by personal visits paid him, over the decades, by the likes of Washington Irving, Bayard Taylor, George Bancroft, Benjamin Silliman, and George Ticknor. Yet these contacts alone did not result in his national fame, for his role in the "'great reconnaissance' of the American West" was indirect, however crucial;[17] and though the visits and letters turned up as worldly gossip in fashionable American periodicals, this did not establish, but only signaled, his reputation. It was not until *Kosmos* began appearing in 1845 to widespread and enthusiastic acclaim that Humboldt achieved prominence as a cultural figure—as Henry T. Tuckerman dubbed him (in *Godey's Lady's Book*) in September 1850, "the Napoleon of science." Oliver Wendell Holmes affirmed the pairing in the poem "Bonaparte—Humboldt," written for the 1869 Boston centennial of Humboldt's birth (organized by Louis Agassiz): "Hero of knowledge, be our tribute thine!"[18] A strong and steady stream of books by and about Humboldt coursed through the American market from the mid-1840s through the definitive biography of 1873. In the wake of the books followed flotillas of notes, articles, reviews, and reprints of speeches and lectures. In this same period, no less than thirty-seven place-names were bestowed in the United States in Humboldt's honor, more than in any other region of the world.[19] Emerson's journals are full of references to Humboldt, from 1821 to 1870, reiterating his status as a Universal Man; Emerson contributed an honorary address to the 1869 Boston centennial. Edgar Allan Poe dedicated *Eureka* (1848) to Humboldt, "With very profound respect"; Thoreau, as I will soon argue, steeped himself in Humboldt's works to great effect.

Humboldt's reputation rode high on the crest of what Theodore Parker called the "German craze," but he was also personalized as "the friend of Young America," as George Bancroft put it.[20] Humboldt's old-fashioned radical materialism made him peculiarly available to a number of ideological positions. He interested himself particularly in the American republic, and that interest was reciprocated: Humboldt represented a living link to the founding fathers, that great age of Washington, Jefferson, and Gallatin; and he was also cast as the "second Columbus," the hero who had discovered America to the scientific world, "the moral discoverer of the southern new world."[21] His writings on Cuba provided fodder for expansionist Southerners, while his outspoken liberal politics led to his appropriation by the antislavery movement, particularly during the 1856 election which pitted the proslavery candidate, James Buchanan, against Humboldt's own protégé, John Charles Frémont.[22] Additionally, his refusal to credit or even allude to the Deity gave his work a "Cheshire cat" quality, with wide pluralist appeal; from the smile on the page different readers could reconstitute the rest of his theology according to their own predilections. Protestant natural theologians were quick to embrace Humboldt as one of their own, who proved by exquisitely Baconian methodology the oneness and glory of God's Creation.[23] Samuel Tyler, for instance, used the *Princeton Review* to trumpet Humboldt's grand work as proof that science is now lifting its view from the utilitarian to the "cosmical aspect of nature," showing how the earth "had been elaborately prepared as an abode for man, and a theatre for human societies," and how all things had worked together under the guidance of unity (383). On the other hand, both Agassiz and Edward Everett felt they had to counter charges that Humboldt's failure to acknowledge God the Creator showed his atheistic tendencies; and the freethinking lecturer, lawyer, and professional atheist Robert Ingersoll used Humboldt as the standard-bearer of enlightened science against religious superstition: Humboldt stood for "the sublimest of truth: 'THE UNIVERSE IS GOVERNED BY LAW!'"[24]

Most important of all, Humboldt's practical, empirically based science seemed to give the United States the key to the continent: "the empire building days were at hand, and so was Humboldt, with his expert knowledge of mines and roads, of commerce and settlements south and west of the new border," of resources and agriculture, that would allow America to throw off European dependence.[25] It seemed that Humboldt's partisan support for American progress, and his practical advice toward its accomplishment, were endorsed even by his scientific theories: he had proposed the existence of a special isothermal zone which traversed the northern hemisphere and along which most of the world's population had moved. American expansionists concluded that the world's great civilizations had all arisen along this zone, in a westward sequence, and the United States was next in line. Humboldt proved that empire was America's Manifest Destiny.[26]

In the climate of 1850s America, Humboldt's empiricism was turned to imperial purpose. He became the valet to empire. Even the daunting realization that most of inland North America was an arid desert found Humboldtian melioration: Joseph Henry of the Smithsonian, among other scientists and meteorological speculators, inverted Humboldt's theory that deforestation resulted in desertification, to arrive at the assurance that *re*forestation would make America's inland desert bloom: rain would, it was said, follow the plow. William Gilpin was certain Humboldt had established scientifically that physical nature, the great organic whole, governed the development of human communities and determined the fate of great civilizations—and determined that the greatest of all was to be the United States.[27] This cosmic foreordination justified the course of expansion and set the scientific seal of doom on, for instance, Native American populations. Victor Cousin's popular idealism had assured American readers that victory was self-justifying and the vanquished had by definition met the fate they deserved. Did not that great preexisting idea, the Manifest Imperial Destiny of the United States, determine the fate of extraneous subject populations, those inconsiderable "parts" to be subjected to the "whole"? Humboldt's empiricist philosophy was absorbed wholesale into the framework of empire, and the legacy that his staunchest and most vocal American proponents established for him was that he made empire scientific. Figuring Humboldt as "the Napoleon of Science" ceases to be merely honorific; he became the figurehead for the conquest of North America by right of science, progress, and knowledge.

In the alliance of science with power, Humboldt indirectly assisted with the conquest of Mexico and the West, although when he learned that his name and work were being used to assert America's claim to Cuba, Humboldt could and did protest with every means at his disposal. But his voice was aging and distant, and the determination of what Humboldtian science would mean to America in the 1840s and 1850s went on in his absence. Thoreau, in distancing himself from the warrior politicians and the promoters of imperial utility, had to reinterpret Humboldtian science for himself. As Humboldt's name became a watchword in the 1850s for western conquest, for the Baconian "reclamation" of the fallen wilderness into the new Eden, it was increasingly difficult for Thoreau to identify with allegedly "Humboldtian" ideals. Having worked out his own quite different version, he had to claim a different kind of heritage and genealogy.

If the idolization of Humboldt and the enthusiastic appropriation of his science found so firm a base in the political, social, and religious ideology of mid-nineteenth-century America, then it remains to be explained why his fame was so quickly eclipsed after his death in 1859. The publication of Darwin's *Origin of Species* a few months later, and the onset of the Civil War—not to mention the failure of the plow to bring rain after all—help account for the shift in attention away from Humboldt. He was being written into the past, most damagingly by Alfred Dove, who used the platform given him in the definitive 1873 biography to bury Humboldt's ghost.[28] Dove finds Humboldt's reputation "adventitious," and determined that his alleged masterwork was out of date before it even saw daylight:

> Science had in the meantime undergone many important changes, the epoch of the intellectual development of mankind, as evinced in the aspect of science in 1834, of which "Cosmos" was the record, had already passed away.[29]

Margarita Bowen suggests that Humboldt's death, so long in coming, brought to some "considerable relief from embarrassment," and that the excessive adulation by his admirers helped undermine his serious reputation. She also shows that Humboldt's attempt to develop a new model of physical geography was widely misunderstood and regarded as "retrogressive" by many geographers; "*Kosmos* was labeled as the last of the great cosmographies rather than as a book pointing the way to the future."[30] Humboldt had intended to prove in *Cosmos* that a popular description of the universe was not only necessary, but possible—ironically, *Cosmos* seemed to prove the opposite, contributing to his own demise. Worse, though he initiated major developments in a multitude of scientific fields, to none of them was his own name attached. His career was spent under the patronage of Prussian monarchs rather than a major university, so he founded no schools, no departments, left no students behind to carry his name into posterity. Agassiz himself recognized this in his memorial address of 1869: pointing to a map of the United States, he lamented that, though "all its important traits are based" on Humboldt's investigations, and his methods are familiar to "every schoolboy," yet "he does not know that Humboldt is his teacher. The fertilizing power of a great mind is truly wonderful; but as we travel farther from the source, it is hidden from us by the very abundance and productiveness it has caused" (*Address* 5–6).

Ironically, Agassiz himself took great care to secure his name institutionally. Through his efforts at Harvard and with the American Association for the Advancement of Science, he settled his reputation as the founder of American natural science. What Agassiz could not acknowledge in his eulogy of Humboldt, since he himself was one of its architects, was the conservative turn scientific and social thought was taking toward a dualistic and deterministic view of a nature divided from, even hostile to, man.[31] In a post-Darwinian world, Humboldt's optimism and his persistent assertions of harmony in nature could readily be dismissed as merely naive. Once he was out of favor, the "Cheshire cat" quality proved his undoing. Positivists who defined science as objective, resting on "facts" and devoid of inference, were quick to dismiss Humboldt (together with followers like Arnold Guyot) as a romantic throwback to an outgrown past. Idealists, or those who inherited the Coleridgean humanism that defined itself against materialistic science, were able to dismiss Humboldt as a sterile, fact-bound empiricist, a German gone French. Thus the two cooperated in suppressing an alternative each found too inimical to comprehend. How ironic it is today that current approaches to science, which stress the role our own knowledge plays as part of the world we seek to understand, have lost sight of Humboldt's work. Today, Humboldtian concepts like plant communities, isotherms, and magnetic storms are routine, the "ecology of ideas" is an exciting new concept—and Alexander von Humboldt's once-glorious name has long since subsided into the dim afterglow of the footnote.

But meanwhile, in the 1840s, Humboldt's name was just beginning its ascent, and his legacy for America had not yet been set. Humboldtian ideas had been percolating through the educated, scientific literature for a generation, but the process of popularization which began with the publication of *Cosmos* brought them, for the first time, directly into the realm of general culture. Humboldt's alternative philosophy was now available to intellectuals already steeped in natural theology, radical idealism, and material reductionism, not to mention the classical tradition and Oriental religion. His appropriation, interpretation, and use by different ideological factions, sketched so briefly here, remind us that Thoreau met these ideas in a contested cultural field. Among all the possibilities, he had to struggle to realize his own definition of the Humboldtian legacy and what it would mean to his own practice. Once he entered the debate, he made it the central focus of his notebooks, journals, and writings through the 1850s; and that is the story to which we now turn.

Fronting Nature at Walden, 1845-1847

Thoreau brought to Walden Pond his dissatisfaction with nature as a mere veil, to be pierced en route to a reality beyond—or, it brought *him* there, as he declared in some of his most eloquent writing: if life *was* mean, he meant "why then to get the whole and genuine meanness of it" (WA 91). Two days after moving into the cabin, he put it to himself like this: "I wish to meet the facts of life—the vital facts, which where [*sic*] the phenomena or actuality the Gods meant to show us,—face to face, And so I came down here" (2:156; 7/6/45). Only a year out of Harvard, he had declared he was fathoming "unceasingly for a bottom that will hold an anchor, that it may not drag" (1:51; 8/13/38); or as he put it fifteen years later, for "wildness, a nature which I cannot put my foot through" (V:293; 6/22/53). *Walden*, among other things, is about searching for a bottom that will "hold an anchor"—and finding it, as he eventually concluded: "Let us not play at kittly-benders. There is a solid bottom every where" (WA 330). One of his most famous passages begins, "Let us spend one day as deliberately as Nature. . . ." It continues:

> Let us settle ourselves, and work and wedge our feet downward through the mud and slush of opinion, and prejudice, and tradition, and delusion, and appearance . . . through poetry and philosophy and religion, till we come to a hard bottom and rocks in place, which we can call *reality*, and say, This is, and no mistake. . . . (WA 97–98)

Against *A Week*'s loose and grab bag structure of layers sliding against themselves unceasingly without rest, Thoreau in *Walden* wrote a dramatically centered and focused text that proclaimed a fixed point, an anchor that could hold the rest of the globe.

Amidst the oscillation between shifting planes of reality in *A Week*, Thoreau developed a new metaphor: measurement. Emerson had postulated natural perfection as a "fixed point" against which we could "measure" our departure (CW 1:39). Thoreau used the concept as a way to stabilize the flux of polar oscillation, and having located firmly a "*point d'appuis*, below freshet and frost and fire," triangulate to points elsewhere. Measurement applies a fixed standard to a shifting reality in hope of revealing stable or "universal" relationships. Thoreau found that this fundamental scientific principle translated well into the moral realm—like musical measure, it made the physical and spiritual worlds commensurate. He had seen the metaphor's utility abundantly demonstrated in Lyell, and as noted earlier, he had already poeticized it in an essay, "Bravery," written in December 1839. The soldier marches to music because the rhythm of his pulse and step corresponds to the movement of the heavenly bodies: "To the sensitive soul, The universe has its own fixed measure, which is its measure also," like the pulse of a healthy body; and the brave man's universal tunefulness compels discord to concord everywhere (*Journal* 1:96–97). When Thoreau reworks this passage into *A Week*, he makes the identification complete: the hero's heart beats "in unison with the pulse of Nature, and he steps to the measure of the universe" (WK 175). In *Walden*, this becomes, famously: "let him step to the music which he hears, however measured or far away" (WA 326). Built of mathematical relationships, music, "the science of melody and harmony," is "the sound of the universal laws promulgated" (WK 175). It forms a triad with science and mathematics, all presenting pure forms of "naked and absolute beauty" (1:197; 11/12/40). Any strain of music, from the beating of a village drum to wind through the telegraph wires to the wood thrush's song, is enough to propel Thoreau into torrents of prose on transcendental harmonies, as it shows him, again, the true "measure" of the universe.

When Thoreau first took up surveying he literalized the metaphor of measure, reaping a famous set of significant numbers in his survey of Walden Pond. In a passage near the end of *A Week*, Thoreau shows enormous confidence in the theory and instruments of measurement:

> The process of discovery is very simple. An unwearied and systematic application of known laws to nature, causes the unknown to reveal themselves. Almost any *mode* of observation will be successful at last, for what is most wanted is method. Only let something be determined and fixed around which observation may rally. How many new relations a foot-rule alone will reveal, and to how many things still this has not been applied! What wonderful discoveries have been, and may still be, made, with a plumb-line, a level, a surveyor's compass, a thermometer, or a barometer!

> Where there is an observatory and a telescope, we expect that any eyes will *see new worlds* at once. (WK 363; emphasis added)

The tools of measurement were not pernicious in themselves, but only as they were used for degenerate ends: to reiterate and reassure ourselves with the familiar, to reduce the unknown to our small measure, rather than to open our eyes to new worlds beyond the reckoning of tradition.

Thoreau, then, brought this debate between real and ideal with him to Walden, in the form of literary projects and verbal abstractions. As he searched for "granitic" truth through life in the woods, his project acquired a new dimension, a material one. The dailiness of his intimate association with lives neither ready-made nor socially encoded necessarily raised some new questions not available to the old Concord anthologist and pencil maker. His confidence in the possibility of "reading" nature and, equally, in the willingness of nature to be read, was being put to the test—as when he confronted "Jean Lapin" on his doorstep and found his sentimental assumptions rebuffed. The old measures were no longer adequate, yet the method seemed sound: Thoreau began to push the details, to pursue his text into finer print, smaller and more precise units, even, as it were, into the unmeasured space between the lines. Three events during the two-plus years at Walden can frame the emerging complexities of his experience: surveying the pond; climbing Mt. Ktaadn; and collecting specimens for Louis Agassiz.

Six years after first purchasing and employing "a levelling instrument and circumferentor combined" in a survey of Fairhaven Cliffs (1:197–98), and nearly a year after moving to Walden, Thoreau applied the instruments to his symbolical pond. As it happened, he was reading, at that same moment, a brand-new book that might have suggested some new methods of reading nature, even though in *Walden* he shrugs it off as one of his "shallow books of travel" (100): John Charles Frémont's *Report of the Exploring Expedition to the Rocky Mountains in the year 1842* (1845). Captain Frémont was applying his own background in mathematics and surveying to the literal discovery of the new world of the American West. The report of this, his second of five exploring expeditions, was printed by the U.S. Government in a massive run of 10,000 copies, and it set the pattern for succeeding topographical survey reports. Two characteristics of this report are of interest here. The first is Frémont's explicit insistence on accurate observation "to show the face and character of the country" (5–6), complete to printing tables of topographical and meteorological observations. Measurement even achieves its own heroic fable; Frémont relates a heart-stopping narrative of the loss of the all-important barometer, just as they approached the Rockies: "[T]he snowy peaks rose majestically before me, and the only means of giving them authentically to science, the object of my anxious solicitude by day and night, was destroyed" (62). But the resourceful American repairs the delicate instrument with his powder horn, some wood, animal skin, thread, and buffalo glue, and the team triumphantly ascends and proves the peak to be the highest of them all, at 13,570 feet. (It was named, inevitably, Mt. Frémont.)

Second, the heroic scientific journey unfolds through a day-by-day narrative combining careful annotations of weather conditions with poetic descriptions of notable plants, encounters with Indians, traders, and buffalo, and the exploits of Kit Carson.[32] "November 1.—Mount Hood is glowing in the sunlight this morning, and the air is pleasant, with a temperature of 38°" (185). "The *artemesia*, absinthe, or prairie sage, as it is variously called, is increasing in size, and glitters like silver, as the southern breeze turns up its leaves to the sun" (14).

Thoreau, who was already casting his own life as a heroic journey to the "west" or wild, found in Frémont's book one possible model for the heroic life of exploration, which cast the close observation of particulars and the tedium of painstaking measurement into a life of high aspiration. What is more, the volume itself modeled a format for conducting, preserving, and communicating observation to the world. Thoreau's own comment suggests he took the hint: he "read" such "shallow" books "till that employment made me ashamed of myself, and I asked where it was then that *I* lived" (WA 100). However directly or indirectly Thoreau took this particular lesson, he did conduct a survey of Walden ("where [*he*] lived") still remarkable for its accuracy, and in reading the measurements read the confirmation of the identity of scientific and moral law: "As I was desirous to recover the long lost bottom of Walden Pond, I surveyed it carefully, before the ice broke up, early in '46, with compass and chain and sounding line" (285). Superstition may have held the pond to be bottomless, but better yet, measure proves it deep: "I am thankful that this pond was made deep and pure for a symbol" (287). Pushed harder still, the numbers yield more:

> When I had mapped the pond by the scale of ten rods to an inch, and put down the soundings, more than a hundred in all, I observed this remarkable coincidence. . . . that the line of greatest length intersected the line of greatest breadth *exactly* at the point of greatest depth. . . . (289)

In a truly scientific spirit, he verifies this observation by surveying Walden's only rival, White Pond. He then triumphantly reads the lesson: "If we knew all the laws of Nature, we should need only one fact, or the description of one actual phenomenon, to infer all the particular results at that point" (290). From there, it is a short step to the moral application:

> What I have observed of the pond is no less true in ethics. It is the law of average. Such a rule of the two diameters not only guides us toward the sun in the system and the heart in man, but draw lines through the length and breadth of the aggregate of a man's particular daily behaviors and waves of life into his coves and inlets, and where they intersect will be the height or depth of his character. (291)

Measurement shows the relative proportions of one thing to another, enabling one to relate things to each other, anything to oneself.[33] It is the very instrument of connection and differentiation. Thoreau's instruments of measurement, applied to the unyielding landscape, have produced a new world of insight—they have found for him the long-lost bottom where his anchor will not drag. After this triumph he can serenely assert, "there is a solid bottom every where" (330).

But a few months later, another encounter would defy his transcendental certainty. To complete his experiment at Walden he traveled north to a country "uninhabited by man," for "we have not seen nature unless we have once seen her thus vast and grim and drear" (2:277; after 9/10/46). What he saw astonished him.

> Coming down the Mt perhaps I first most fully realized that that this was unhanselled and ancient Demonic Nature, natura, or whatever man has named it.
>
> The nature primitive—powerful gigantic aweful and beautiful, Untamed forever. (2:278)

As he adds, "The main astonishment at last is that man has brought so little change," even though he so "overtops nature in his estimation" (2:278) —nature sets human names and measures at nought. After this preview, Thoreau drafts a narrative of the journey, circling around in the chronology back to the mountain, to "front the true source of the evil" (2:294). Here was Thoreau's most "frontier" experience, where the "fact" is not to be nurtured garden-like into poetic truth, but bullies him into a battle for his very scalp. Pushing on alone he climbed "a mile or more—still edging toward the clouds— The mtn was a vast conglomerate or aggregation of loose rocks—as if sometime it had rained rocks—." At first he thinks of chimneys, factories, and old epic poets, but as he gropes for language the social connections drop away:

> It was vast titanic & such as man never inhabits. Some part of the beholder, even some vital part seems to escape through the loose grating of his ribs as he ascends— He is more lone than one . . . Vast Titanic inhuman nature has

> got him at disadvantage caught him alone—& pilfers him She does not smile on him as in the plains— She seems to say sternly why came Ye here before your time . . . Why seek me where I have not called you and then complain that I am not your genial mother. (2:339–40)

Against the smug correspondence and neat, satisfying fit to wholeness Thoreau had accomplished at Walden, this wild mother undoes all standards, bleeding all hope of wholeness out of him, rendering him "less than one." Here man is the measure of nothing. In the published version Thoreau dramatized his emotional reaction: with the loss of wholeness he spins off like a fragment, a loose falling rock, experiencing a disembodiment that wrenched transcendent ascension inside out, delivering the very extinction that it promised so sublimely, but which whiplashed instead into the soul of terror:

> I stand in awe of my body, this matter to which I am bound has become so strange to me. . . . I fear bodies, I tremble to meet them. What is this Titan that has possession of me? Talk of mysteries!—Think of our life in nature,—daily to be shown matter, to come in contact with it,—rocks, trees, wind on our cheeks! the *solid* earth! the *actual* world! the *common sense! Contact! Contact! Who* are we? *where* are we? (MW 71)

Transcendence is built on the dualism that underwrites alienation from the body and so necessitates that nature will always be "elsewhere." It splits open a gulf between body and spirit, elevates the spirit then annihilates the body, so the spirit may cry out perpetually for loss of a home, of bodily touch, of warmth—of "*Contact!*" Thoreau experienced his transcendence not while crossing a bare common but while climbing to a mountaintop—and it turned out to be his worst nightmare.

Finally, a very different set of circumstances found him, in spring of 1847, collecting little pieces of the life in and around the pond, killing them, boxing them up, and sending them off to Louis Agassiz, who was newly installed at Harvard. Agassiz had come to America (through the influence of none other than Alexander von Humboldt) in October of 1846. He was so enthusiastically received that what had been planned as a visit and a lecture tour matured by late 1847 into a permanent position as professor of geology and zoology, created just for him in Harvard's Lawrence Scientific School. Agassiz's first concern in America was to establish a massive collection of zoological specimens, to support his anatomical and embryological research. As he wrote to Spencer Fullerton Baird in April 1847, recruiting his assistance:

> I prefer to have a great number of specimens of *the most common species in all their ages*, than to have few specimens of many rare species. I will mention as an example, that I should collect as many as twenty and more specimens of all your salamander, frogs, toads, and have besides the tadpoles in all their different states, the whole preserved in spirit. So with other reptiles—

—and also fish, and "bats, mice, rats, moles, shrews, weasels, squirrels, etc."[34] Baird obliged and organized a collecting network, with Agassiz's assistant and secretary James Elliot Cabot, who must have contacted Thoreau immediately: in a letter of May 3, 1847, Cabot is already thanking Thoreau for several shipments of specimens, including breams, pout, painted tortoises and snapping turtles, with which Mr. Agassiz was "highly delighted" —although he regrets having to turn down Thoreau's invitation to a "spearing excursion" (CO 177–78). The exchange of specimens and letters continued for some weeks. On May 8, Thoreau offers to take "toll" for his contribution "in the shape of some, it may be, impertinent and unscientific inquiries." He follows with an inventory of Concord River fish, offering to collect new specimens: "there are also minks, muskrats, frogs, lizards, tortoise, snakes, caddice-worms, leeches, muscles, etc., or rather, *here they are*" (179–80). Cabot's reply of May 27 mentions Agassiz's surprise and pleasure at the extent of Thoreau's subsequent collections, including a fox (!) which "is doing well" in Agassiz's back yard. Among the fish are one or possibly even two species new to science. June 1 brings word of one or possibly two *more* new species: in all, a bream, a dace, and two new minnows. Cabot's letter apparently answered Thoreau's of the same date:

> I send you 15 pouts, 17 perch, 13 shiners, 1 larger land tortoise, and 5 muddy tortoises, all from the pond by my house. Also 7 perch, 5 shiners, 8 breams, 4 dace? 2 muddy tortoises, 5 painted do., and 3 land do., all from the river. One black snake, alive, and one dormouse? caught last night in my cellar.

Thoreau goes on to take some of his "toll" in questions: "What are the scientific names of those minnows which have any? Are the four dace I sent to-day identical with one of the former, and what are they called? Is there such a fish as the black sucker described,—distinct from the common?" (CO 177–83). Then the letters, and apparently the collecting, stop, except for one last letter from Thoreau nearly a year later inquiring if Cabot's journal might pay for an article (210); and one letter from Agassiz himself, in July 1849, turning down yet another invitation from Thoreau, this time for two or three lectures before the Bangor Lyceum (244).

What are we to make of a Thoreau who so cheerfully trapped, packed, and shipped so many of his Walden "friends" and neighbors to Harvard's halls of science? One observation is that Thoreau seems to have lost interest quickly: after this flurry of excitement, extending across perhaps five weeks in spring 1847, he no longer mentions such activity, even though Agassiz stated his need for "many more specimens than most naturalists would care for" (CO 181). Thoreau's flirtation with institutionalized science, at least *à la* Agassiz, was apparently at an end. The story becomes hard to follow: the *Journal*, already fragmenting, breaks down completely after December 1846, until April 1850. All that remains are a few leaves, from which the editors conjecture the existence of two notebooks, apparently cannibalized in the process of composition of *A Week* and *Walden*. From the dates alone one can surmise Thoreau was busy working as a writer, not a naturalist: he was polishing and trying to place *A Week*, working up the first draft of *Walden*, drafting, delivering as lectures, and publishing "Ktaadn" in fall 1848, and "Resistance to Civil Government" in May 1849. He was also acting as surrogate father for Emerson's family and affairs, from the time he left Walden in October 1847 until July 1848, while the great man toured abroad.

Yet a couple of notes to himself suggest one direction his thoughts had taken, in the midst of his pondside experiences with unmediated nature: Be vigilant. After drafting the first version of the Walden survey, he asked, "What are these pines & these birds about? What is this pond a-doing? I must know a little more—& be forever ready" (2:242; after 4/18/46). Late in the year this note recurred: he told himself, "No method or discipline can supersede the necessity of being forever on the alert," and reminded himself that for a "*Seer*," "to keep his eye constantly on the true and real is a discipline that will absorb every other" (2:357; after 12/2/46). The urge to "know a little more," the command to "discipline," suggests his realization, through living with wild nature on his doorstep, that there was more to be seen in nature than he had hitherto suspected, if he learned how to look.

In his walks, surveys, travels, and reading, Thoreau was moving away from a grand and abstract transcendentalism toward detailed observation of the specifics of nature, in all its unaccountable diversity. He was learning the extent of error and of ignorance, and beginning to sound out both the potential and the pitfalls in the systems of science that investigated such matters. In traveling to Maine, he found the maps were mistaken. He redrew them, with corrections, in his own notes. Those brave men journeying west to the Rockies had declared their dependence on strict, disciplined observation and precise measurement. A Walden with a measurable bottom signified more to him than a Walden that was bottomless. Familiar Concord fish were altogether unknown to scientific taxonomies. In the grand Wholes, where was there room for information like this? For real nature, in all its unpredictable variety and diversity? In the period from spring 1846 to spring 1847, he was taking his first steps toward a methodized approach to nature: measuring the chivin, surveying ponds and taking their temperature, collecting specimens as part of a scientific network. These steps are significant not because they are necessary but because they are *not*—Thoreau is attaching, as Susan Faye Cannon would say, "unnecessary" science to useful activities, suggesting an outside influence at work.[35] Certainly these were steps toward finding what pines and birds were "about," but he also regarded what he was doing with uncertainty and suspicion, warning himself that method alone was insufficient, that measurements diminished the sublime, that "fact" and accuracy were gained at the expense of "genius," or of "the freshness and vigor and readiness to appreciate the real laws of Nature" (WK 364).

The oscillation that characterized *A Week* permeates Thoreau's work and thought throughout the 1840s. Which was more "real"—to which should he be pledged—the "real laws" or the actual and granite-hard "facts"? Finally they seemed incommensurable, both philosophically and in lived experience: what possible moral good could come out of counting a chivin's fin rays? Ideally all nature would prove out to the last detail, as the success of the pond survey argued: taken to the highest resolution of which the human eye was capable, nature would still yield moral truth. But when driven into a corner, nature could also prove hostile to such calculations, fundamentally unknowable as on Ktaadn, or eternally elsewhere as in the banal familiarities of the Concord woods. Thoreau's Janus vision had produced a yea for today, and a nay for the morrow, and no way to reconcile so massive a contradiction.

After Walden: Old Worlds and New

The crisis came in 1849, when Thoreau at last published the book he had exiled himself from his family to write. Emerson had encouraged him to believe that *A Week* would secure his reputation, even to the dangerous step of staking his personal fortune on its success. For it turned out that no publisher would take the book unless Thoreau assumed the financial risk—which at length he did, bringing it out through James Munroe & Co. in May 1849. The reviews were not unfavorable, but the book stalled, selling only a handful of copies. The carefully crafted mosaic of the best he had to offer, out of twelve years' work, was a commercial failure. It succeeded only in plunging him deep into debt, at just the point when he felt acutely the need to make his life count, to compensate for the deaths both of his brother John, to whom the book was a tribute, and now, two weeks after its publication, his sister Helen.[36]

The book's failure also tore into Thoreau's increasingly difficult friendship with Emerson. Upon his return from England in July 1848, Emerson had discovered that Thoreau had become disturbingly close to his wife, Lydian, and his children. Nearly a year afterward, Thoreau rhapsodized in his *Journal* about "my sister," Lydian: "you are of me & I of you I can not tell where I leave off and you begin.—" (3:17; after 5/26/49). But his friend Emerson, "whom I heartily love," appears so "transfigured" that Thoreau dares not "identify thy ideal with the actual." He continues: "(I was never so near my friend when he was bodily present as when he was absent) and yet I am And yet I am indirectly accused by this friend of coldness and disingennuousness— When I cannot speak for warmth—& sincerity" (3:18-19). By September anguish has turned to bitterness:

> I had a friend, I wrote a book, I asked my friend's criticism, I never got but praise for what was good in it—my friend became estranged from me and then I got blame for all that was bad,—& so I got at last the criticism which I wanted.
>
> While my friend was my friend he flattered me, and I never heard the truth from him, but when he became my enemy he shot it to me on a poisoned arrow
>
> There is as much hatred as love in the world. Hate is a good critic. (3:26; after 9/11/49)[37]

Though the bitterness was mitigated in later years, it never disappeared entirely. Meanwhile the *Journal* records a painful and passionate struggle against feelings of loss and disconnection. Instead of publishing *Walden*, which he had regarded as ready and for which he had earlier received offers, Thoreau was forced to withdraw it from the market. He set to work earning hard cash to pay off the debt, laboring mornings in the pencil factory and expanding his casual surveying into a professional business, complete with a program of books to study, new tools, and a printed handbill advertising the accuracy of his services. By November 1849 he had enough business to purchase and open a separate notebook for survey records.[38] Literalizing his metaphor of measurement gave him one way to keep, as they say, body and soul together. Walking and observation would be turned to account.

For increasingly, as the *Journal* and the letters record, he was walking and observing for a large part of every day. A letter of November 1849 to H. G. O. Blake reports that "within a year my walks have extended themselves, and almost every afternoon . . . I visit some new hill or pond or wood many miles distant" (CO 250-51). And in the *Journal* after May 1849, between the remonstrances and meditations on friendship, appear notes on places visited and measured, and plants found:

> The Copper mines—the old silver mine now deserted—the holt—the great meadows— The Baker Farm—Conantum—Beck-Stows swamp—the Great Fields—Poland . . . The old lime-kiln—the place where the cinnamon stone was found—Hayne's Island. . . .

He notes he has "measured a chestnut 23 feet in circumference at a foot from the ground"—and goes on with his inventory:

> For brooks we have Cold brook—Pantry Brook—Well meadow brook—Nut meadow Brook Wrights brook—Nagog-brook . . . For hills—Nagog famous for huckleberries where I have seen hundreds of bushels at once—Nashoba —of Indian memory . . . For ponds Walden—Flints or Sandys White. . . . (3:23–24; after 9/11/49)

He enters 1850 alternating such notes with comments on friendship, and it becomes clear the two are intertwined, even interchangeable, around the concept of mutual "invitation" or "affinity": "If you are inviting in this sense the rocks and trees will come to you." If his "neighbor" does not, then there can be no such "affinity" between them: "I do not draw him strongly enough—& hence I have no right to go to him. / My love for another is my affinity for him . . ." (3:45; after 1/5/50). Successive pages of fragmentary notes mix "May-Flowers," hemlocks, "epigaea repens" and alder catkins with still more notes on the loss of friends. The conclusion becomes inescapable: Thoreau is, at least metaphorically, "falling in love" with nature, displacing his passion from his unresponsive and remote human friends to those who do respond to his "affinity"—rocks and trees. In an undated fragment from this period he seems to recognize what is happening: "When our companions fails us we transfer our love instantaneously to a worthy object. As the sunlight which falls on the walls and fences, when those are removed falls instantaneously on the mountain & spires in the horizon" (3:58; 1848–50). The removal of *human* artifacts or barriers frees the sunlight, the beams of his vision, to fall on "mountains and spires," now beloved in their stead, and granted to be "worthy objects" of his love. After John's death he had searched nature for his "brother," in a trope that emerged only briefly. Now it becomes difficult to separate the tropes from their literalization. One torn fragment reads: "Actually I have no friend I am very distant from all actual persons—and yet my experience of friendship is so real and engrossing that I sometimes find myself speaking aloud to the friend I" (3:58). With one "object" of his love removed, Thoreau directed his fierce energy to another. This paragon of isolated individualism could not conceive of himself in isolation, but only in sympathy with and response to another, as part of a community of others.

One can also track these same months along a different axis, that of his reading. Under this stress, he turned, once again, to the Hindus. In September 1849 he petitioned in person for library privileges at Harvard, and while the final (favorable) decision was pending, checked out two books on Hindu philosophy.[39] One motivation was apparently to recruit himself to a wise, Hindu stoicism, as expressed in the November 1849 letter to Blake:

> "Free in this world, as the birds in the air, disengaged from every kind of chains, those who have practiced the *yoga* gather in Brahma the certain fruit of their works."
>
> Depend upon it that rude and careless as I am, I would fain practise the *yoga* faithfully. . . .
>
> To some extent, and at rare intervals, even I am a yogin. (CO 251)

Shortly thereafter he summarized the "inquiry in Hindu philosophy" as "how to commit suicide in an effectual and worthy manner"—by "pure wisdom or contemplation—" (3:48-49; after 1/5/50). Had he pursued this direction much farther or longer, one wonders how such an annihilated self would have found the means to interrupt "pure wisdom" for the sullied act of creation.

Hindu writing was also embraced as part of a wider search through cosmic literature generally. Against his acute fear of loss, separation, and fragmentation, Thoreau was reading intensively in theories of the All. Perhaps the bifurcation that divides *A Week* now seemed doubly inadequate. It is as if, in the vacuum partly produced and partly abetted by its ignominious and agonizing collapse, Thoreau is reading around for a way to reconnect his broken universe into a whole. The final half dozen entries in his literary notebook (which he then discontinued) show that around 1849 to 1850 he worked through John Bernhard Stallo's *General Principles of the Philosophy of Nature* (1848), an article, "The Philosophy of the Ancient Hindoos," by none other than James Elliot Cabot (September 1848), Coleridge's *Theory of Life* (1848), and—last but for a brief mention of Allston's *Lectures on Art* (1850)—Otté's translation of Alexander von Humboldt's *Cosmos* (1849).

The three works of science in this grouping, published within a few months of each other, all attempt to present a theoretically unified view of the cosmos, but the third differs radically from the other two. Stallo, who emigrated from Germany at sixteen years of age in 1839 and by the mid-1840s was teaching mathematics, natural philosophy, and chemistry at St. John's College, New York, offered his American readers a "delineation" of German "philosophy of nature" or *Naturphilosophie*. Echoing Emerson's *Nature* (and using the same sources), Stallo declares that "Nature is the absolute symbol of the Ideal,—the indispensable figure by which the Spiritual demonstrates itself" (48). He is concerned not with physical nature but with the metaphysics of a world founded on the sundering of polar forces that "eternally love each other" and are eternally striving toward reunion (21). Indeed, the lines Thoreau copies concern the love of the "magnetized subject" for the "breast of the telluria parent" (LN 357). In 1852 Thoreau dismissed "Stallo the German" by linking him with "theological dogmas" (IV:295; 8/12/52).

His extracts from Coleridge are considerably longer, and are concerned largely with Coleridge's "principle of Individuation" and its ultimate manifestation in man; the polarities which govern life are invoked to assure that "liberty" be balanced with an equally intense "reverence for law," "independence" with "submission to the Supreme Will!" (LN 359-62). While these extracts fit in well with Thoreau's preoccupation with resignation and philosophical reconciliation, neither they nor the *Theory of Life* as a whole points toward the direction Thoreau was shortly to take.[40] That is, Coleridge's *Theory of Life* gives Thoreau nothing new, but (as argued in Chapter 2) reiterates and reinforces transcendental idealism by offering in its service a highly technical argument based on *Naturphilosophie* and on seventeenth-century metaphysics—an armchair argument which considers nature in the abstract and arranges matter as a hierarchy of organization, beginning with metals and ending in man.

Thoreau's extracts from Humboldt's *Cosmos* are much more perfunctory, consisting of secondhand quotations from Aristotle and the Maha-Bharata, both from Volume II.[41] But even this slight mention warrants two observations: first, Thoreau incontestably did read Humboldt's most important work, in which Humboldt established both a theoretical basis and a practical methodology for linking science and imagination; and Thoreau read it, almost immediately after it first became available in English, at a "pivotal point" in his career.[42] Second, he was reminded, as the extracts demonstrate pointedly, that certain modern scientists—such as Humboldt and Lyell—explicitly placed their work in a tradition which honored and embraced classical and Oriental thought. While his reading in Coleridge could only have confirmed the pattern within which he had been struggling for over a decade, Humboldt contributed a program for connection, not estrangement, which came with the highest endorsements: Greek and Hindu classics, even Goethe himself. In Humboldt Thoreau found the catalyst for which he was so manifestly ready.

Thoreau continued his reading of Humboldt. He was already familiar with Humboldt's name: he had used it in the opening of his 1842 essay "A Walk to Wachusett," recalling how "we sat down" with Homer, "roamed" with Virgil, "or with Humboldt measured [!] the more modern Andes and Teneriffe" (NHE 31).[43] His reading of Lyell had introduced him at second hand not only to Humboldt's work and reputation (which Lyell cites freely), but to Humboldtian methodology. So had his reading of Frémont, although the captain's report did not make explicit his debts to Humboldt. Indeed, in the intellectual culture of 1840s New England, Thoreau could hardly have avoided encountering Humboldt's name and acquiring some familiarity with his reputation, but it was not until 1849 that anything actually authored by Humboldt was readily available to him in English. Once he encountered *Cosmos*, he read everything that came out, as soon as it came out: both translations of *Ansichten*, the three-volume *Personal Narrative* (1852-53), Klencke's *Lives of the Brothers Humboldt* (1853). By April 1850, he was quoting from Otté's translation of *Ansichten der Natur*, published barely three months before under the title *Views of Nature*.[44] In May he copied out from *Views* another extract pertinent to his topic at hand:

> Nothing memorable was ever accomplished in a prosaic mood of mind. The heroes & discoverers have found more than was previously believed only when they were expecting & dreaming of something more than their contemporaries dreamed of—when they were in a frame of mind prepared in some measure for the truth[.]
>
> Referred to the world's standard—the hero, the discoverer—is insane. . . . Humboldt says, speaking of Columbus, approaching the New World—[that he supposed from the evidence of his senses] "that he was approaching the garden of Eden." (3:67; 5/12/50; CC 95)

From this point forward, Humboldt becomes part of Thoreau's own "standard," a figure for his ideal, the heroic dreamer, explorer, and rediscoverer of a new world.

I believe that Humboldt was able to give Thoreau both the theoretical framework and the practical methodology that he needed to unite the individual and particular with the Cosmic Whole. Reading these new translations and editions of books by Humboldt himself must have crystallized ideas already present in Thoreau's mind, but unformed. That is, Humboldt arrived in Thoreau's hands at the very moment Thoreau both needed him and was prepared to apprehend him—as he wrote, "in a frame of mind prepared in some measure for the truth." Or as he reiterated in 1860, after his revelatory encounter with Darwin's *Origin of Species*: "A man receives only what he is ready to receive, whether physically or intellectually or morally. . . . We hear and apprehend only what we already half know" (XIII:77; 1/5/60). Humboldt's books gave Thoreau a path to follow, a path he was already half on, at a time when the old approaches had failed him and he needed to strike out in a new direction.

Much of Thoreau's physical energies in 1850 were taken up with remodeling the new family house on Main Street (the "Yellow House"), assisting with the pencil (and now also plumbago) factory, and attending to his growing surveying business. He surveyed sixty house lots in Haverill in May, and the new courthouse lot in Concord in June; by the end of the year, his new notebook recorded over twenty surveys.[45] He also traveled widely, following up his first trip to Cape Cod, with William Ellery Channing in October 1849, with a second trip (by himself) in June 1850. July found him on the beach again, this time searching unsuccessfully for the remains of Margaret Fuller Ossoli and her family in the tragic shipwreck off Fire Island, in New York. In August the family at last moved into the remodeled house, and Thoreau took over the roomy attic, which from then on was his base of operations. In September he and Channing were off again, this time to Quebec, on an excursion that spurred Thoreau to open a new notebook on Canada. Another landmark date fell on December 18, 1850, when Thoreau was elected a corresponding member of the Boston Society of Natural History, an honorary position bestowed in gratitude for a rare American goshawk Thoreau had been given in December 1849 and which he had presented to Samuel Cabot (James Elliot Cabot's brother), the society's curator of birds (3:41; CO 252). This gave Thoreau access to an important library, and from then on Thoreau visited the society's rooms regularly on trips to Boston, withdrawing books and consulting with its curators.[46]

Thoreau seems to have settled down after the Canada trip, for the *Journal* records an interesting development: in early November the pattern of fragmentary and inconsistently dated entries gives over, in the space of a few weeks, to regular, dated entries that soon become daily, recording that day's observations in both the natural and contemplative lines. The *Journal* changes character, becoming not just a record of things seen, but more important, a tool for seeing. Thoreau begins to use it as a training ground for his new orientation to the natural world. Entries typically open with a formulaic expression: "I saw in Canada . . ."; "I am attracted by a fence . . ."; "I pluck" the wild apples; "This afternoon I heard a single cricket"; "I notice that. . . ." In each case the initiating instant of perception is followed by a paragraph detailing the object or phenomenon that has come forward to his senses, and exploring possible associations.

> I saw an old bone in the woods covered with lichens which looked like the bone of an old settler—which yet some little animal had recently gnawed & I plainly saw the marks of its teeth—so indefatigable is nature to strip the flesh from bones—and return it to dust again. No little rambling beast can go by some dry and ancient bone but he must turn aside and try his teeth upon it. An old bone is knocked about till it becomes dust— Nature has no mercy on it. It was quite too ancient to suggest disagreeable associations—it was like a piece of dry pine root. (3:138; 11/11/50)

Thoreau the rambler is turning aside and trying his own teeth upon such objects as catch his attention. Nor will this specimen ever grace a display cabinet in town: instead of seeing it as a dry, hard, and detachable object, Thoreau is looking at a process, taking place in a context which includes his own actions. He also begins to ask questions of his surroundings. On November 16 he not only records that the shrub oaks retain their leaves through the winter, he wonders, "why do they?" (3:142). Three days later he investigates, breaking off a leaf to observe that the stalk has not quite died down to the point of separation (3:145; 11/19/50). Soon an analogy occurs to him: "You might say of a very old & withered man or woman that they hang on like a shrub-oak leaf almost to a second spring. There was still a little life in the heel of the leaf-stalk" (3:175; 1/8/51). By such a steady march of minute observation is he beginning to establish a new relationship with nature, turning the command to vigilance and discipline toward specific yet highly transcendental ends. What drives him is desire. He has found a way to cultivate his love:

> My Journal should be the record of my love. I would write in it only of the things I love. My affection for any aspect of the world. . . . I have no more distinctness or pointedness in my yearnings than an expanding bud. . . . I feel ripe for something yet do nothing—cant discover what that thing is. I feel fertile merely. It is seed time with me— I have lain fallow long enough. (3:143–44; 11/16/50)

He enjoys, it seems to him, "an unusual share of happiness" (3:144).

To see an object becomes, for Thoreau, both an act and an expression of love. He will soon explore the ways in which this leads to, or becomes, knowledge, both private and public. The goal was to learn, and communicate, how seeing or sensual experience, "love" or "sympathy," and knowledge could form a single complex process. Temperament might set the terms for this desire, but circumstances shaped the form it took, directing him how to "love" pines, and to love pines rather than people. Given this direction, and its expression in multiple sustained acts of "contact" with the natural objects of his love, natural history offered institutional means of knowing, expressing, and communicating, ways to further shape his "yearnings" into a program. He was not wholly alone after all, as the books he checked out of the BSNH library and the conversations he had there with Cabot, Thaddeus W. Harris, and others assured him. What was more, his special expertise increasingly conferred on him local respect, as his fellow townspeople (and their children) brought to him their questions and their findings.[47]

Paradoxically, the enabling move is not toward but away from knowledge—to forget what one thinks one knows, to defamiliarize the familiar. Shedding preconceptions or previous "understanding" reopens the capacity to see things anew, yet it still leaves Thoreau with a mystery: "I saw Fair Haven Pond," with its island, meadow, water, hawks, and

> I did not see how it could be improved— Yet I do not see what these things can be. I begin to see such an object when I cease to *understand* it—and see that I did not realize or appreciate it before—but I get no further than this. How adapted these forms and colors to my eye—a meadow & an island; what are these things? Yet the hawks & the ducks keep so aloof! and nature is so reserved! I am made to love the pond & the meadow as the wind is made to ripple the water. (3:148; 11/21/50)

Once emptied of "understanding," of previous knowledge, what comes next? Static contemplation may have been the Hindu ideal, but Thoreau wanted not a pure and mystical state of mind, a static "enaction" of his world across bipolar antagonisms, but a creative interaction, "poetry." His question became how to act on the love he felt, that sense of congruity and uncanny intimacy, without becoming masterful or dominant and hence destroying the vital "reciprocity." Nature "invited" him, but still remained "aloof."

Across the spring of 1851 Thoreau developed an answer. First he worked with the idea of knowledge as "a novel & grand surprise on a sudden revelation of the insufficiency of all that we had called knowledge before. An indefinite sence [*sic*] of the grandeur & glory of the Universe. It is the lighting up of the mist by the sun" (3:198; 2/27/51). Or as he asked some months later, "With all your science can you tell me how it is—& whence it is, that light comes into the soul?" (3:306; 7/16/51). The revelation of insufficiency initiated a state of readiness or "expectation." The goal of this heightened state of alertness was to expand on the revelation of newness through an intensity of experience, applying *all* the senses, not just seeing: "he that hath ears let him hear—see—hear—smell—taste—&c while these senses are fresh & pure" (3:323; 7/21/51). Thoreau was not being drawn away from poetry, from romantic or transcendental nature, toward a dry, resistant, and threatening form of scientism that was drowning his epiphanies with facts. He wished his epiphanies to happen *through* facts, through sharp and actual experience with real things. "And we are enabled to apprehend at all what is sublime and noble only by the perpetual instilling and drenching of the reality which surrounds us," he wrote in *Walden* (97). In the spring of 1851 the evidence of that "instilling and drenching" begins to build the *Journal* into a new form, as he uses writing as a tool, not of record, but of perception. The more he learns how to see, the fuller the entries become.

And having set himself this challenge, it is not the poets but the naturalists who show him how much there is to be seen that he had never thought to see before. On June 7, 1851, over a year after his first annotations from Humboldt, Thoreau notes that he has been reading Charles Darwin. Soon he is copying out lengthy extracts from Darwin's "'Voyage of a Naturalist round the World,'" and the great British naturalist's acuity of vision seems to stimulate Thoreau to new feats of awareness: "I wonder that I even get 5 miles on my way—the walk is so crowded with events—& phenomena. How many questions there are which I have not put to the inhabitants!" (3:245; 6/7/51). Now that he has realized his ignorance it seems he has everything to learn. A telling passage occurs at the end of Darwin's book, as he is evaluating the benefits of such a voyage. The first is beholding the scenery—but not just that:

> there is a growing pleasure in comparing the character of the scenery in different countries, which to a certain degree is distinct from merely admiring its beauty. It depends chiefly on an acquaintance with the individual parts of each view: I am strongly induced to believe that, as in music, the person who understands every note will, if he also possesses a proper taste, more thoroughly enjoy the whole, so he who examines each part of a fine view, may also thoroughly comprehend the full and combined effect. Hence, a traveller should be a botanist, for in all views plants form the chief embellishment. (*Voyage* 500)

Darwin is describing the classic Humboldtian spiral, from awe before nature's beauty, to inquiry, to awe enhanced with knowledge; from whole, to detail, to a finer whole. He who wishes not only to love, study, and behold the beauty of nature but to recreate it should follow the same pathway—and he, too, should be a botanist, as Thoreau would write years later:

> How much of beauty—of color, as well as form—on which our eyes daily rest goes unperceived by us! No one but a botanist is likely to distinguish nicely the different shades of green with which the open surface of the earth is clothed,—not even a landscape-painter if he does not know the species of sedges and grasses which paint it. (XIV:3; 8/1/60)

The implied reference is to the paintings of Frederic Church, whose sweeping sublimity and botanical faithfulness were legendary. On July 4, 1855, Thoreau visited the Athenaeum gallery in Boston on the way to Provincetown, where he saw one of Church's new Humboldtian paintings, *The Andes of Ecuador*. Richardson notes that although he did not record his impressions, "Thoreau's future work would have interesting similarities with Church's."[48] Indeed: they were all drinking from the same stream.

Hence Thoreau's desire is not to become one who "knows" what it is he is looking at, and who comes by this certainty through a detachment of observation that allows him to master, subordinate, and categorize the field. Rather, Thoreau places himself as a link in the connecting chain of community, that member of it whose nature it is to read the community into language, to be "the scribe of all nature," its poet—he wants to be "the corn & the grass & the atmosphere writing" (4:28; 9/2/51). His active and continual investment of his self, bodily and intellectually, in that community will bring all the myriad detached pieces of it into a whole, a *cosmos*—not in the mind, but connected through mind as participant, the agent of "love," connection, sympathy. To the staged, but nevertheless genuine, howl of fear he attributed to himself on the top of Ktaadn, Thoreau is constructing his answer. He will know the world not through the abstraction of dry and barren systems, but through involvement with it to the uttermost limit of his capacity to see; he has designed an epistemology of contact.

Through the summer of 1851 he was developing these heady insights in, and by means of, his *Journal.* He experiments with a present tense, stream-of-awareness narrative technique that seems to carry the reader on his shoulders as he roams the landscape:

> I hear some whipporwills on hills . . . I now descend round the corner of the grain field—through the pitch-pine wood in to a lower field, more inclosed by woods—& find my self in a colder damp & misty atmosphere, with much dew on the grass— I seem to be nearer to the origin of things. . . . Now I go by the spring and when I have risen to the same level as before find myself in the warm stratum again. (3:251; 6/11/51)

To contain and shape the sprawling narrative he is beginning to generate, he devises a seasonal progression. This also gives him a tool of measurement, a restrictive lens which enables him to focus his observational energy. "No one to my knowledge has observed the minute differences in the seasons—" he writes; "Hardly two nights are alike. . . . A Book of the seasons—each page of which should be written in its own season & out of doors or in its own locality wherever it may be—" (3:253; 6/11/51). On June 29 he marvels, "How different is day from day!" (3:279); on July 6 he analyzes what he has just witnessed: "Now June is past. June is the month for grass & flowers . . ." (3:287). Ten days later he has a new season: "Methinks this is the first of dog-days. The air in the distance has a peculiar blue mistiness or furnace-like look—though, as I have said it is not sultry yet— It is not the season for distant views—" (3:310; 7/16/51). As the seasons multiply, the entries grow longer: nearly seven text pages that day; nine text pages on July 21. Another new era opens on July 22: "The season of morning fogs has arrived I think it is connected with dog days . . ." (3:326). The next day he is out early, advising himself to alertness—and beginning to sound a little daunted:

> You must walk so gently as to hear the finest sounds—the faculties being in repose— Your mind must not perspire True, out of doors my thought is commonly drowned as it were & shrunken pressed down by stupendous piles of light etherial influences—or the pressure of the atmosphere is still 15 lbs to a square inch— (3:329; 7/23/51)

Then, after so many yeas, comes the inevitable nay, the final entry for the long day:

> But this habit of close observation— In Humboldt—Darwin & others. Is it to be kept up long—this science— Do not tread on the heels of your experience Be impressed without making a minute of it. Poetry puts an interval between the impression & the expression—waits till the seed germinates naturally. (3:331)

Humboldt and Darwin have shown him the world-revealing potential of the mere "habit of close observation," but what he was learning through the pulse of his own thoughts and the measure of his own stride was how to process such a potentially overwhelming overload of information, how not only to gather it but to form it into meaningful wholes.

"Worlding," the various strategies he developed to sort and present the wealth of "facts" he was collecting so promiscuously, will be discussed in the following chapters, for this became one of his prime concerns. Meanwhile, it is clear that he went through cycles of enthusiasm and exhaustion, as he worked on his walks and in his chamber to accomplish his huge ambition. The intensity of his labor fatigued him: "I have the habit of attention to such excess that my senses get no rest, but suffer from a constant strain. Be not preoccupied with looking. Go not to the object; let it come to you" (IV:351; 9/13/52). "I have almost a slight, dry headache as the result of all this observing. How to observe is how to behave. O for a little Lethe!" (V:45; 3/23/53). As he came to think of his activity as science in action, it discouraged him to encounter that science promoted in America:

> Ah what a poor dry compilation is the Annual of Scientific Discovery. I trust that observations are made during the year which are not chronicled there. That some mortal may have caught a glimpse of Nature in some corner of the earth during the year—1851. One sentence of Perennial poetry would make me forget—would atone for volumes of mere science. The astronomer is as blind to the significant phenomena—or the significance of phenomena as the wood-sawyer who wears glasses to defend his eyes from sawdust— The question is not what you look at—but how you look & whether you see. (3:354-55; 8/5/51)

Surely science was a method, "*how* you look," not merely an object or compilation of things seen; and science that resulted in dust-catching compilations of "knowledge" only erected a barrier to real knowledge, which had as its goal not answers that ended questions but answers that produced *more* questions. Early on he had written that "A true answer will not aim to establish anything, but rather to set all well afloat" (1:139; 6/23/40). The "Annual of Scientific Discovery" was full of answers that established "facts" and so shut down inquiry, asserting that their "significance" was completely accounted for. Thoreau wanted not answers but the process of asking, and he grew angriest with the institution of science when it sought to close down rather than open up discussion, the flow of activity. A. Hunter Dupree asserts a similar distinction for Asa Gray, the botanist, in comparison with his rival Agassiz: "Where Gray looked at nature he saw questions. Agassiz saw only answers." Gillian Beer observes that "Ideas pass more rapidly into the state of assumptions when they are *unread*. Reading is an essentially question-raising procedure."[49] Thoreau is learning, as he reads science and practices his readings of nature, how to raise questions about science rather than accept its cultural assumptions. In effect, he is using his reading to keep himself just a little off balance with the universe, and the results show in the play of his writing. He suggested his own vision of true scientific affiliation in a bit of wordplay: in answer to the Society for the Diffusion of Useful Knowledge, Thoreau proposed the "society for the diffusion of useful Ignorance" (3:184; 2/9/51) as finally more conducive to knowledge.

It is from this perspective that one must consider the famous passage, often quoted to cap the argument that Thoreau knew his turn to science would be fatal, and written toward the end of this extraordinary summer:

> I fear that the character of my knowledge is from year to year becoming more distinct & scientific— That in exchange for views as wide as heaven's cope I am being narrowed down to the field of the microscope— I see details not wholes nor the shadow of the whole. I count some parts, & say "I know." The cricket's chirp now fills the air in dry fields near pine woods. (3:380; 8/19/51)

As he adds the next day (upon realizing how readily he could distinguish the various notes of the crickets), "I am afraid to be so knowing" (3:381). But he is differentiating his surroundings in new ways, trying to learn how he can make the process of connecting details into wholes work, without succumbing to the weight of accumulating detail. Humboldt too had warned that the worker in the field risked losing sight of the higher ends of his labor. However, this hazard was not intrinsic to the enterprise, but arose only from human limitations in an infinite universe, and it could be avoided by keeping the cosmos in view: the grand whole to which one's own labors were contributing. Nor was knowledge to be feared: the increase in knowledge only increased the beauty and wonder of the cosmos, as the horizon expanded ever wider around us. In a more cheerful mood Thoreau echoed Humboldt:

> I fear that the dream of the toads will not sound so musical now that I know whence it proceeds. But I will not fear to *know*. They will awaken new and more glorious music for me as I advance, still farther in the horizon, not to be traced to toads and frogs in slimy pools. (IV:31; 5/7/52)

Typically, his passages of discouragement and exhaustion were soon followed by passages recording the renewal of joy and wonder. In Thoreau's excitement they are often hasty and broken, seldom as finely rounded (and as crisply quotable) as his dolorous expressions of despair:

> Is that galium? . . . Compare that at Lee's. I should like to know the birds of the woods better, what birds inhabit our woods? I hear their various notes ringing through them. What musicians compose our woodland quire? They must be forever strange and interesting to me. How prominent a place the vireos hold! It is probably the yellow-throated vireo I hear now. . . . (VI:337; 6/9/54)

Some new incident or observation would be enough to shake off the fear that dust was gathering on a world gone stale: "How sweet is the perception of a new natural fact!—suggesting what worlds remain to be unveiled. That phenomenon of the Andromeda seen against the sun cheers me exceedingly. . . . At sight of any redness I am excited like a cow" (4:471; 4/19/52). Such findings could seem to Thoreau admonitory, against the presumption that he could ever use up the world. Some events, he wrote, plainly were allegorical, as "when I thought I knew the flowers so well," a beautiful purple azalea he had never heard of was shown him by a hunter:

> Ever and anon something will occur which my philosophy has not dreamed of. The limits of the actual are set some thoughts further off. That which had seemed a rigid wall of vast thickness unexpectedly proves a thin and undulating drapery. The boundaries of the actual are no more fixed and rigid than the elasticity of our imaginations. (V:203-4; 5/31/53)

For although Thoreau liked to advise that one "learn science & then forget it" (4:483; 4/22/52), the emphasis in the early years of this decade was not on the latter but on the former. As he began working out his new way of seeing in 1851, he purchased such books as Asa Gray's new guides to plant classification, and Agassiz and Gould's landmark text *Principles of Zöology* (1848). He began to read widely in science: in 1851 alone he read, in addition to Darwin's *Journal of Researches* (1846), *The Earth and Man* (1849) by Humboldt's disciple Arnold Guyot, Cuvier's *Animal Kingdom*, Robert Hunt's *Poetry of Science* (1850), and the controversial books on evolution by Robert Chambers and Hugh Miller, engaging heatedly, at least in his *Journal*, in the wider debate they had ignited. His reading gave him new words to see by, and new tools to see with, tools he borrowed from other workers in the field even as the very act of buying these books, or traveling to the building which housed the Boston Society of Natural History to check them out, consolidated his sense of belonging to a community of people engaged in parallel pursuits.

Several years later Thoreau paused to reflect on the course of his studies. Though twenty years before he had learned a little botany, he had lacked any system, and so forgot it all. "But from year to year we look at Nature with new eyes. About half a dozen years ago I found myself again attending to plants with more method, looking out the name of each one and remembering it." He began to bring them home, and "gaze with interest" at swamps, wondering if he could become so familiar with their plants as to know each one. Though the botanies made it seem "like a maze to me, of a thousand strange species," he soon knew them all: "I wanted to know my neighbors, if possible,—to get a little nearer to them" (IX:156–57; 12/4/56). And having begun to "look at Nature with new eyes," he experienced a transformation. For all the books and all the science, the globe was still perennially young and undiscovered. We each of us have "new eyes": "each object appears wholly undescribed to our experience—each field of thought wholly unexplored— The whole world is an America—a *New World*. . . . The end of the world is not yet" (4:421; 4/2/52). Humboldtian science had given Thoreau new eyes, and in giving him new eyes enabled him to see "new worlds at last."

4

Cosmos

Knowing as Worlding

Descriptions of nature, I would here repeat, may be sharply defined
and scientifically correct, without being deprived thereby
of the vivifying breath of imagination. The poetic elements must be derived
from a recognition of the links which unite the sensuous with the intellectual;
from a feeling of the universal extension, the reciprocal limitation,
and the unity of the forces which constitute the life of Nature. . . .
He who, familiar with the great works of antiquity, and in secure possession
of the riches of his native tongue, knows how to render with simplicity
and characteristic truth that which he has received by his own contemplation,
will not fail in the impression which he desires to convey;
and the risk of failure will be less, as in depicting external nature,
and not his own frame of mind, he leaves unfettered
the freedom of feeling in others.
—Alexander von Humboldt, *Cosmos*

Critics of American transcendentalism have long discussed the reasons for Thoreau's departure from Emerson. Ostensibly Emerson's "disciple," Thoreau yet exhibited differences which do not grow from Emerson's theoretical framework, and which are generally attributed to temperament. F. O. Matthiessen established the important distinction between the two men when he defined Thoreau as having "a more dogged respect for the thing" and a greater interest "in the varied play of all his senses," not just the visual. Both features prevented him from disappearing into "the usual transcendental vapor," and he succeeded as an artist when, and because, he *balanced* means and end, fact and meaning; yet his balance became shakier in later years, and his vision "narrowed."[1]

This notion of balance carries forward from Matthiessen's day to our own, and much of Thoreau criticism has been concerned with whether, or how well, he maintained it. Inheritors of what Ben Knights calls the post-Coleridgean "suspicion of material nature," which subordinates science and craft to true culture and art, will tend to distrust Thoreau's turn to material data, and find in his statements of doubt evidence that he knew he had taken the lower of two paths. From this point of view there are three likely positions: triumph, defeat, or truce. Triumph, Thoreau's achievement against heroic odds of a precarious balance between material and ideal worlds; defeat, as Thoreau lost the war between them; truce, as Thoreau achieved a compromise on a lower level which allowed him at least a measure of contentment. All three presume Thoreau's inability or refusal to affiliate himself with "legitimate" science, and rely on the narrative that casts poetic and scientific ways of knowing as opponents in a pitched battle.[2]

How might this picture change if the two-sided imagery is complicated? Even Joseph Wood Krutch, who understands the potential of Thoreau's science, laments that he was excluded by the pathetically materialistic science of his own day, caught as a literal nondescript between transcendental idealism and a barren positivism dominated by Louis Agassiz, between arid compilations and sentimental vaporings. This guarantees that Thoreau's stance will be a lonely one indeed, and assures that any science he did conduct was necessarily "bad," amateurish and outside the community of mainstream scientists—as if, as Robert Sattelmeyer comments, "his unfitness in this field were a necessary precondition to taking him seriously as an artist."[3] While I might not claim Thoreau altogether for science, I do doubt he was quite as isolated as this picture assumes. The two poles inscribed by these assumptions have been defined against each other: poetry, imagination, and intuition against science, hard facts, and objective reality. This exclusionary division widened during Thoreau's lifetime, as romanticism's dualistic framework split down the middle, obscuring the alternative which proposed that literature, imagination, and science were integrally related and that the division (though not the *difference*) between man and nature, subject and object, was an illusion. The split wrote Alexander von Humboldt out of intellectual history altogether, and backed Thoreau into the anachronistic dualism invented for him by orthodox, post-Coleridgean literary criticism.

For, finally, "temperament" cannot account for *what* Thoreau did, either in his private walks and researches or in his public acts of communication with and participation in the scientific community, however indirect it often was. That is, why did he cast his "temperament" into the particular forms which characterize his life, activity, and *Journal* after 1850? It seems true enough that he always did have a stubborn respect for facts: as far back in his literary career as the Harvard essay on Locke, he insisted that the philosopher in search of truth must take a "fresh and unprejudiced view of things" in order to build "for himself, in fact, a new world" (EE 103). This turn to the empirical always sat uneasily with the Emersonian mandate to dissolve the empirical, to absorb and remake nature into spirit. For Thoreau, reality could not consist wholly and utterly in the rational mind. "How can you walk on ground when you see through it?" he wondered (4:389; 3/13/52). While he found in the transcendental version of rational holism an exciting and productive approach to the natural world, still he was conflicted. The question becomes, then, why did Thoreau not simply carry to his grave the pattern of conflict that he exhibited throughout the 1840s? What happened during the years 1849 to 1851 to so radically alter his entire pattern of activity and his self-conception of his career? Why did his new pattern take the particular form of the peripatetic poet-naturalist?

For everyone seems pretty well agreed that something happened, and that the turning was manifested in the new form the *Journal* had taken by late 1850. Not that he shed his old self entirely, but he added something new to it. Recently a couple of descriptive terms have been coined to indicate the distinctive nature of Thoreau's new enterprise: Robert Richardson calls it a kind of "practical transcendentalism"; the editors of *Journal 3* reverse the polarity to call it a "higher empiricism."[4] Both terms resemble, strikingly to my mind, the phrase by which Margarita Bowen designates Humboldt's program, "thinking empiricism," and the phrase Humboldt himself used to designate his interactive approach: "rational empiricism." Gillian Beer, grappling with a similar dilemma in Darwin studies, calls Darwin's approach "romantic materialism." Even William James finds he must construct a new term for his holistic pluralism, calling it "radical empiricism."[5] All of these terms grope for some way to designate a seemingly *anomalous* combination of empiricism with something "higher"—indeed, the two coined for Thoreau are meant to emphasize his isolation in the culture of his time, his unique status as a nondescript.

Yet clearly there is a convergence at work here. A coherent grouping that links—for starters—Thoreau, Humboldt, Darwin, and William James is not negligible. Could Thoreau have been participating in a contemporary but *non*-Emersonian tradition? This would link Thoreau explicitly and purposefully with others, as part of a larger social community, and suggest his activity was more than merely idiosyncratic—that he saw himself participating in a wider community and that in fact part of his new energy and extraordinary sense of commitment originated with his renewed sense of being, not alone and isolated, but allied across time and space with others who shared his vision.

My own contrived label, "empirical holist," is designed to indicate the nature and extent of the system of alliances to which Thoreau linked himself, and to suggest why that activity seems "scientific"—that is, not just because it is "empirical," but because it is *social*, deliberately patterned along certain lines shared with the similarly patterned activities of others, whether or not they, or he, possessed a college degree in science, joined the American Association for the Advancement of Science (AAAS), or regularly attended scientific meetings. Viewing Thoreau in this way permits us to see him as a social being, connected directly and indirectly to the intellectual movements of his time, rather than as a hermit confined by his idiosyncracies to a sterile universe of lonely contemplation. I contend that what distinguished Thoreau was precisely the way in which knowing became for him a social activity, grounded in his (and our) relationship to material objects: he developed an epistemology of *contact* rather than transcendence. That this entailed risks as well as pleasures will become increasingly clear in the following chapters.

Thoreau as Humboldtian

At the turn of the 1850s, Thoreau moved through the crisis produced by his external circumstances and the internal contradictions of rational holism, to a new program for his career, one which would enable him to fuse poetry and science, natural particulars and spiritual aspirations. Thoreau's own version of the Humboldtian project was distinctive, true to his own sense of purpose and the cosmos, and shaped by his very different circumstances. For it was, finally, the Humboldtian program that Thoreau enacted: explore; collect; measure; connect.

First, explore. Thoreau had long insisted that nature was never to be studied in the hothouse, the laboratory, or the museum:

> Where is the proper Herbarium—the true cabinet of shells—and Museum of skeletons—but in the meadow where the flower bloomed—by the sea side where the tide cast up the fish—and on the hills and in the valleys where the beast laid down its life—and the skeleton of the traveller reposes on the grass. (2:78; 1842–44)

Where Humboldt and Darwin had traveled the globe, Thoreau's exploring was done not globally but locally: he "travelled a good deal in Concord" (WA 4). In mid-1851, in the midst of copying several pages of extracts from Darwin's *Voyage of the Beagle*, Thoreau speculated: "There would be this advantage in travelling in your own country even in your own neighborhood, that you would be so thoroughly prepared to understand what you saw— You would make fewer traveller's mistakes" (3:259; 6/12/51). Two months later he wondered why a traveler must first get away from home, and asked himself, "Why not begin his travels at home—! Would he have to go far or look very closely to discover novelties." Such an approach would have a tremendous advantage: "Now if he should begin with all the knowledge of a native—& add thereto the knowledge of a traveller— Both natives & foreigners would be obliged to read his book. . . . It takes a man of genius to travel in his own country—in his native village . . ." (3:356–57; 8/6/51). Thoreau's dance between knowledge and ignorance was an attempt to achieve this double stance: to know the ground enough to initiate the desire for more knowledge; to defamiliarize the familiar and enable the "impartial eye" of the traveler, who "may see what the oldest inhabitant has not observed" (3:384; 8/20/51). As he notes, he finds he is most interested "in the things which I already know a little about—a mere & utter novelty is a mere monstrosity to me" (3:357). The notion became a settled point in his philosophy. In 1859 he advised H. G. O. Blake to "live at home like a traveler. It should not be in vain that these things are shown us from day to day. Is not each withered leaf that I see in my walks something which I have traveled to find?—traveled, who can tell how far?" (CO 538). Thus his walks became the central activity of his day. In "Walking," the late essay on which he worked all through the 1850s, he warned his readers that "the walking of which I speak has nothing in it akin to taking exercise, as it is called . . . but is itself the enterprise and adventure of the day" (NHE 97).

Although his aim was to practice the Humboldtian program in his own countryside, Thoreau did travel as widely as his resources permitted. He never took these excursions casually, but prepared himself in advance by reading extensively and gathering and studying maps. He set out in plain and rugged clothes, carrying his specially constructed knapsack with partitions for his botany, plant press, "guidebook, spyglass, and measuring tape."[6] These travels gave him contact with other regions, confronting him with new customs, dialects, and local histories, strange flora and new fauna, all different enough from Concord to set going the process of comparison Humboldt and Darwin had modeled. They helped him to set "the limits of the actual . . . some thoughts farther off" (V:203–4), as when, rather than speculate in the study, he sought to experience wild nature in Maine, and the vast, bleak ocean of Cape Cod. He also traveled through his reading. John Aldrich Christie has documented a "minimum" of 172 separate travel accounts, many in multiple volumes, which Thoreau read fully and carefully.[7] Typically Thoreau refused to settle for condensed versions, too, insisting on reading the full-length original: he spurned, for example, the popular abridgement of Humboldt's *Personal Narrative*, titled *Travels and Researches* (1833; Emerson owned a copy), holding out instead for the appearance of the new, three-volume translation in 1852.

Second, wherever Thoreau traveled, whether on the ground or in books, he collected. From his travel and science reading he collected many notebooks full of a hodgepodge of information and insights: the "Fact Book," the Canadian notebook, the twelve Indian notebooks. Richardson writes that the common denominator in the "travel" books was actually the natural history they contained: Thoreau loved the "detail" and filled his extract books and *Journal* with similar local detail. The details he gathered provided materials for comparison, allowing him to mix constantly the local and the global, keeping both rooted in specific data.[8] He alludes to this methodology in "Huckleberries," when he quotes "at length the testimony of the most observing travellers . . . for it is only after listening patiently to such reiterated and concurring testimony, of various dates—and respecting widely distant localities—that we come to realize the truth" (NHE 231). By such means he was able comfortably to compare the waters of Walden and the Ganges, Humboldt's Orinoco River to a rill in an April meadow.

Thoreau also collected from his own original exploration, amassing, classifying, and organizing into patterns the same kinds of material objects as Humboldt and Frémont. His specialty was plants: Harding notes, "He began collecting, drying, labeling, and classifying botanical specimens until in a period of ten years he was able to locate more than eight hundred of the twelve hundred known species of Middlesex County." At his death, his mother and sister donated to the Boston Society of Natural History "his collections of more than one thousand pressed plants, his collection of New England birds' eggs and nests, and his collection of Indian antiquities, consisting of the arrowheads and other implements he had picked up in Concord."[9] It is revealing that this collecting and classifying was done along the patterns established by the field sciences: Thoreau sought to classify or organize these objects according to the systems then being established by the community of scientists. As he reflected in 1856, what he had lacked in his earliest studies was "system," or method; learning the system enabled him not to gather isolated and forgettable facts, but to participate in forming a shared body of knowledge. Thus it became an important question, in classifying plants, whether to give preference to the Linnaean "artificial" system or to John Lindley's "natural" system, and Thoreau informed himself on this technical debate like any professional (he favored the latter).

The fundamental tool in collecting was, of course, language. The first step toward designating a plant as a collectible object is to give it a "scientific" name, that is, a place and significance not only in a scientific system but in the social system composed by scientists. Thoreau learned "the language of naturalists," as he said, "to be able to communicate with them" (V:42; 3/23/53). Channing reported him saying that knowledge grows by communication; "I can now learn what *others* know about the same thing." But Thoreau also understood the value of naming as a tool for seeing: Sharon Cameron writes that for Thoreau to be bereft of a name was to be unable to identify the experience, and his frustration with the absence or inadequacy of common names helped lead him to his concern with classification. Ellery Channing recalled how Thoreau had known a particular rush by sight for twenty years, but was prevented from describing its peculiarities because he did not know its name: "With the knowledge of the name comes a distinctive knowledge of the thing. That shore is now describable, and poetic even." To this end Thoreau added to his library handbooks and field manuals for identifying birds, mammals, insects, reptiles, flowers, trees, lichens, and mosses.[10]

The worlds that were opening to him, and his recognition of a potential still to be realized, are suggested by a *Journal* passage from August 1851: "How copious & precise the botanical language to describe the leaves, as well as the other parts of a plant. Botany is worth studying if only for the precision of its terms—to learn the value of words & of system. . . . Suppose as much ingenuity . . . in making a language to express the sentiments" (3:382; 8/20/51). He understood that induction into the system itself composed and created new forms of knowledge—though as ever, he warns himself against too much of a good thing: when his eye is caught by fungi in a lamp, he recoils, startled to find himself stooping so "low," and worried that he "should next be found classifying carbuncles and ulcers" (V:51; 3/25/53). Yet there was an advantage in the "hard & precise terms" used by, for instance, a lichenist: "No one masters them so as to use them in writing on the subject without being far better informed than the rabble about it." No man can write intelligibly on lichens without having something to say, while "every one thinks himself competent to write on the relation of the soul to the body . . ." (4:368–69; 3/1/52). Of course, this insight too is attended by its consequent warning, which he adds the next day: scientists could also use their language to conceal how slight their "true knowledge" is; any truly significant discovery can be reported in the language of the newspapers (4:370). Years later, Thoreau's own scientific paper would be reprinted in Greeley's *New York Tribune.* Science, Humboldt maintained, should be the property of all, not guarded by the few.

Third, Thoreau's comprehensive understanding of the theory and practice of measurement has already been suggested. One of Thoreau's ways of seeing a thing was to measure it—to this purpose his walking stick was notched in inches—but to regard this behavior as merely compulsive is to overlook its purpose, which was to reveal patterns that enabled insight into all sorts of unsuspected connections. As Richardson states, Thoreau began to assemble data "not because he was becoming unimaginative, but because he had enough imagination to see what kinds of fresh generalizations were possible through statistics"—as his later studies were to show.[11] And as Humboldt taught, the only way to investigate pattern and periodicity was to establish "certain fixed points" as a ground for comparison (*Views* 363).

Humboldt had propounded accuracy in measurement, a point on which Thoreau prided himself professionally: in the handbill advertising his services as a surveyor (in which he repeated the word "accurate" three times), he claimed his work was "warranted accurate within almost any degree of exactness," a claim which later surveys have in fact confirmed.[12] As with scientific naming, though, this insight comes with its attendant danger: one might lose sight of the higher ends in the routine of means, and so descend into isolation and fragmentation, or into the rigidity and blindness of knowledge congealed into false certainty. Knowledge could be "the lighting up of the mist by the sun" (3:198), but wrongly used it could destroy that light. Thoreau was apt to name the former poetry, the latter science:

> The scientific startling & successful as it is, is always some thing less than the vague poetic—it is that of it which subsides—it is the sun shorn of its beams a mere disk—the sun indeed—but—no longer phosphor—light bringer or giver. . . . Science applies a finite rule to the infinite.— & is what you can weigh & measure and bring away. Its sun no longer dazzles us and fills the universe with light. (3:44; 1/5/50)

Science weighs and measures *in order* to bring away: Thoreau here grasps that the process of science consists not just in a state of mind, but precisely in the activity of "bringing away" objects and data from the site of collection to a center of connection.[13] This process of collecting to connect was the lesson he had learned from participating in Agassiz's network, gathering specimens from Walden, packing them into cartons, and shipping them to the Cambridge center. Though he rejected a subordinate position in the network Agassiz was instituting through Harvard, Thoreau understood that gathering and mobilizing was the activity of the scientist, and it became his activity, too, as he roamed the woods and fields with higher ends in view. He can bring away actual objects—plants, nests, arrowheads—or symbolic objects: names, measurements, observations, field notes. Through his *Journal* he symbolically brings away what he can from the field to his study. Or, more accurately, he brought the field to his study by means of inscriptions in his field notebooks (none of which, apparently, survive), in which he kept his "out-of-door notes" taken on folded papers. From these notes he wrote up at home the full-scale *Journal* entries. Channing, Thoreau's favorite walking companion, was amazed and fascinated with Thoreau's methods. The field notebook was Thoreau's "invariable companion," and in it he wrote, using pencil, "but a few moments at a time, during the walk." Into the notebook Thoreau jotted

> all measurements with the foot-rule which he always carried, or the surveyor's tape that he often had with him. Also all observations with his spyglass (another invariable companion for years), all conditions of plants, spring, summer, and fall, the depth of snows, the strangeness of the skies, —all went down in this note-book. To his memory he never trusted for a fact, but to the page and the pencil, and the abstract in the pocket, not the Journal. I have seen bits of this note-book, but never recognized any word in it. . . .[14]

Humboldt would have approved, *Ansichten* having originated from notes made in the field: "Detached fragments were written down on the spot and at the moment, and were afterwards moulded into a whole" (*Aspects* v). Later in his book Humboldt reminds his readers that writings made "on the spot," in the immediate presence of the phenomenon or soon after, can "lay claim to more life and freshness" than can recollections (210).

Given this method of composing wholes from field-collected fragments, organization becomes a critical problem, one that Humboldt himself, as I have mentioned, did not fully solve. Thoreau's mosaic method of quilting together thoughts and observations had not proved completely successful in *A Week*, yet he was still faced with the same problem: How to sort through and connect the data and meditations into a coherent whole? It occurred to him that the very collection had, in its raw state, a kind of completeness:

> I do not know but thoughts written down thus in a journal might be printed in the same form with greater advantage—than if the related ones were brought together into separate essays. They are now allied to life. . . . It is more simple—less artful— I feel that in the other case I should have no proper frame for my sketches. Mere facts & names & dates communicate more than we suspect— Whether the flower looks better in the nosegay —than in the meadow where it grew—& we had to wet our feet to get it! Is the scholastic air any advantage? (4:296; 1/27/52)

In a sense, extracting passages from his *Journal* pages was like bringing away flowers from their native fields. In both cases—one literary, the other scientific—the generating context was lost and replaced by one artificially constructed, a necessarily fictive structure. One solution was to attempt to preserve the context intact, to bring nothing away: to think of the *Journal* itself as one great whole, itself a fully realized work, as Cameron has argued.

But Thoreau did attempt other ways of composing "wholes of parts," for he recognized that "Thoughts accidentally thrown together become a frame—in which more may be developed—& exhibited." "Disconnected" thoughts juxtaposed by chance "suggest a whole new field in which it was possible to labor and & to think" (4:277–78; 1/22/52), he concludes. Although he may not have read Whewell, he is working instinctively with the idea of consilience.

Thoreau was experimenting with various ways of "subjecting isolated observations to the process of thought, and by combining and comparing them" to discover the nobler relations within the cosmos (*Cosmos* I:60–61). His ultimate aim, like Humboldt's, was to communicate his findings to others. Thoreau too cast some of his data into visual forms: his meticulous surveys, the map of Walden Pond, his numerous maps redrawn and corrected from other sources.[15] He drew up hundreds of tables and charts which still await analysis. And though untrained as a visual artist, he peppered his *Journal* pages with small sketches and diagrams which show an uncertain but careful hand, and an alert eye. He was more successful grouping his "facts" into the connective wholes of the lecture: while not so widely known as Emerson, he appeared frequently before the public from 1838 on. And though he might have let his reputation rest on the unmediated mass of the *Journal*, simply accumulating page after page without shaping new essays, it is worth recalling that this is just what he didn't do. In his self-appointed role as a mediator, he worked hard to compose various subsets of what he was amassing into socially available forms, not just to see new worlds himself, but to present those worlds, and the means for seeing them, to others. To "know" in splendid isolation was as bad as deluding oneself that an object could be isolated and unconnected: Thoreau conceived of himself as part of the community of knowing, which in one of its forms was also the community of scientists.

Thoreau, it could be argued, may have picked up Humboldtian methodology superficially without subscribing to any of its underlying assumptions, simply applying it to his own very different agenda. As the foregoing has indicated, there were certainly differences in their placement in time, situation, locale—in, they both might have said, their "histories." Yet what connects them are their shared assumptions, which link both of them to romanticism in general, as surely as they differentiate both of them from the rational holism represented here by Coleridge and Emerson. Both Thoreau and Humboldt avowed that nature formed a whole, infinite in extent, not moved by divine law imposed or "imbreathed" from without into a dead or passive nature, but rather animated and generated from within by "a chain of connection" (*Cosmos* I:23), linking all parts and rendering them mutually interdependent. The whole was knowable, then, not in spite of but through its individual and constituent members; Thoreau can conceive of knowing a swamp not by direct revelation of its essence, but by naming and knowing each of its elements, through "drenching" all his senses in it, in an act that he imagines as "friendship." Far from disregarding the anomalous, he cherishes it as a source of potential illumination. "Any anomaly in vegetation makes nature seem more real & present in her working," he observed (4:6; 8/21/51). In an Emersonian mood Thoreau may "fear" such intimate knowing, but in a Humboldtian mood he feels no threat, but exults in an infinitely expanding horizon.

Thoreau's affirmations can be seen to follow the outlines of empirical holism, affiliating him to Humboldt and to his widening circle of explorers, naturalists, artists, and scientists—even, however tenuously, to the cooperative network of international science. To review the six features: First, matter—"nature"—is not a veil that keeps us from a higher reality, but is itself that reality, and can be known only through experience, through the senses, through "contact." Second, such contact, by moving us emotionally, makes us aware of harmony and order in the universe, to which both Humboldt and Thoreau give the name "morality."[16] But Thoreau is hesitant to claim such moral truth for science, which instead of offering some "occult connection" with the truth (*Cosmos* I:50), offers limited and descriptive laws. "Science affirms too much," Thoreau protests; it shows *why* lightning strikes a tree, "but it does not show us the moral *why* any better than our instincts did" (IV:157; 6/27/52). Furthermore, the immorality of science could be dangerous: "The inhumanity of science concerns me, as when I am tempted to kill a rare snake that I may ascertain its species. I feel that this is not the means of acquiring true knowledge" (VI:311; 5/28/54). It is as risky to assume science and moral truth are identical, as to seal them off from one another, allowing science to justify moral barbarity. The solution is to link the physical and moral universes, to see how each contributes to the other, in a process of relational knowing that Thoreau calls "reciprocity." Or, as he put it somewhat gnomically in 1851: "Obey the law which reveals and not the law revealed" (3:201; 3/30/51).

For, third, our understanding of scientific "law" is subject to change. The field of nature is ever new, defying the laws we so belatedly construct. "The universe is wider than our view of it," he declares in *Walden* (320); therefore we must experiment with different views: "The sucker is so recent—so unexpected—so unrememberable so unanticipatable a creation. . . . The world never looks more recent or promising—religion philosophy poetry—than when viewed from this point," Thoreau exclaims (4:450; 4/15/52). Yet the flux of nature leaves readable traces of itself, presenting us with its own history—in the strata of the rocks, the soil, the rings of a tree—if we learn how to read them. Fourth, while the movement of that flux seems, in Thoreau, generally progressive, it comes to be characterized less by a strenuous goal-directed upward striving than a principle of "succession," by which multiply generated interactions combine to a series of stages. It is unclear, in Thoreau, whether men are always "better" than foxes, or whether men and foxes are simply different nations. Nor whether civilized Europeans necessarily are "better" than the "salvages" or Indians. Neither is Thoreau compelled by the mystique of polar forces, the great engine of division that drives Emerson's vision. The problems that engage Thoreau most deeply are, instead, those of distribution, pattern, and wholeness.

Fifth, it is remarkable how consistently in Thoreau nature acts as an expressive artist. The works of man pass out of his hands "into the hands of nature, to be perfected" (WK 62); "Walden" from the first is a presence that "speaks" to him, in a language that "only the practised ear can catch" (1:47; 6/3/38). The voice of the wood thrush declares "the immortal wealth and vigor that is in the forest"; shall we not look into "the face of Nature? Let me know what picture she is painting, what poetry she is writing, what ode composing, now" (IV:190–91; 7/5/52). The human artist who wishes to speak the truth of nature will, as Humboldt recommends, not dress it with artificial ornament but represent "the simplicity of individualizing truth" (*Cosmos* II:81). Simplify, simplify, pleads Thoreau; "A true account of the actual is the rarest poetry" (WK 325). All elements of nature speak, even though we have trouble hearing them: to the silence of the eternal stars, he prefers the "fertile and eloquent silence" of the forest, for "I must hear the whispering of a myriad voices" (IV:471; 1/21/53). The great tragedy is man's refusal to *listen*: when the Bellerica dam destroys shad returning to spawn, he asks, "Who hears the fishes when they cry? It will not be forgotten by some memory that we were contemporaries" (WK 37). The poem of creation, we are told in *Walden*, is uninterrupted—"but few are the ears that hear it" (85).

Sixth and finally, the moral law Thoreau seeks is a communal accomplishment. As we have seen, and will see more extensively in the next chapter, knowledge is sought for its "significance" in the wider community of knowers, and in Thoreau this significance always bears on the questions "How to Live. What to Do."[17] Knowing must be, then, not an isolated but a relational act. One should state "facts" not as if they were isolated but as they relate to ourselves: "A fact stated barely is dry. It must be the vehicle of some humanity in order to interest us. . . . It must be warm, moist, incarnated,—have been breathed on at least. A man has not seen a thing who has not felt it" (XIII:160; 2/23/60). Facts that are "warm, moist, incarnated" are facts conceived as part of the larger community:

> Nature must be viewed humanly to be viewed at all; that is, her scenes must be associated with humane affections, such as are associated with one's native place, for instance. She is most significant to a lover. A lover of nature is preëminently a lover of man. If I have no friend, what is Nature to me? She ceases to be morally significant. (IV:163; 6/30/52)

From the human side of this relation, meaning or significance is created not by the lone consciousness, in contemplative isolation, but by and within a commonwealth of meaning. Nor can it be assumed that man is the hierarch. The trick of perspective makes every point seem to itself a center (3:298; 7/10/51), which is why, to overcome this illusion, the observer must "love," feel with or "sym/pathize" with his object, must realize that that "object" is itself the "subject" of its own knowing. "Could a greater miracle take place than for us to look through each other's eyes for an instant?" (WA 10). Man must realize that the gaze does not belong to him alone: nature *looks back*. Thoreau can even feel this about a raindrop: "We two draw nearer and know one another" (1:120–21; 3/30/40). He thinks of a muskrat, then thinks of the muskrat thinking of *him*: "While I am looking at him I am thinking what he is thinking of me. He is a different sort of man, that is all" (3:151; 11/25/50). Flowers should be named for men only if the flowers "may be supposed thus to reciprocate" their love (4:26–27; 8/31/51); the trouble with the villagers is they don't know a spruce from a fir—"Neither do the spruce trees know the villager" (VI:22; 12/22/53). While it is characteristic of any system to arrange a hierarchy, that is the danger of systems, which needs to be recognized and counteracted:

> In the true natural order the order or system is not insisted on. Each is first, and each last. That which presents itself to us this moment occupies the whole of the present and rests on the very topmost point of the sphere, under the zenith. The species and individuals of all the natural kingdoms ask our attention and admiration in a round robin. We make straight lines, putting a captain at their head and a lieutenant at their tails. . . . It is indispensable for us to square her circles, and we offer our rewards to him who will do it. (XIV:119–20; 10/13/60)

In this way, countering the tyranny of "system or arrangement," does community become the basis for Thoreau's theory of knowledge, his epistemology of contact: true knowledge is generated and maintained by the community of knowers, a "round robin" in which the center rotates, which includes all as subjects and all as objects. In this way Thoreau breaks down the dualism embedded in the foundation of "rational" holism, which assumes that knowing can take place within the isolated, rational mind. "Shall I not have intelligence with the earth? Am I not partly leaves and vegetable mold myself?" (WA 138). The result is a sustained argument for and demonstration of *relational* knowing, out of which Thoreau proposes a kind of science that Humboldt would have recognized and honored (as he honored Goethe's) but which in America was a novelty.

Thoreau developed these insights over the 1850s until his death, into a version of social ecology strongly reminiscent of Humboldt's ecology of mind, where man was not simply "in" nature but where man and nature were at every level dependent on and expressive of each other, and the "facts" of nature were energetic co-productions of the human mind operating with and within the field of natural objects. Such a proto-ecological vision is moral to the core, in its insistence on co-production, yet it also arose out of the mainstream scientific insights of Thoreau's day, which claimed that the unseen and unknown could be imagined by thoughtful observation of the familiar and everyday. This was Lyell's most basic argument, and Thoreau was recurring to it late in his life: "It is the discovery of science that stupendous changes in the earth's surface, such as are referred to the Deluge, for instance, are the result of causes still in operation, which have been at work for an incalculable period." Such steady progress according to existing laws characterizes the living realm, too; science knows that plants are not spontaneously generated, but come from seeds, or causes that are still in operation, however slow and unobserved. While it is a truism that "'little strokes fall great oaks,' . . . such, too, is the rise of the oak; little strokes of a different kind and often repeated raise great oaks, but scarcely a traveller hears these or turns aside to converse with Nature, who is dealing them the while" (XIV:311–12; 1/14/61; DI 36–37). Science, if properly conceived by those who *do* listen, who "turn aside to converse with Nature," can reveal the moral truth that each of us, in each of our actions, is continually building the social and natural world we inhabit. In this republic of particulars, change cannot be dictated from above, but only generated from below, by the responsive actions of every individual member. Building from the roots: Thoreau was a radical even to his science.

What then of Thoreau's fortunes with his fellow scientists—the community of knowledge as it was taking shape in the antebellum United States? Thoreau felt himself connected, but found himself pushed to the margins. For instance, after his exchanges with Agassiz in 1847, Thoreau and Agassiz had gone very separate ways. Agassiz had arrived in America already a world-renowned zoologist and paleontologist, whose theory of Ice Age glaciation had put him in the first rank of international science. He was lionized in Boston, quickly achieving status as one of the great and powerful men of the country. His goals at Harvard were to build and house the best zoological collection in America (he succeeded—by the 1860s it was one of the best in the world), and to train young scientists to the most modern standards of the profession. Agassiz effectively remade Harvard's old-fashioned zoology department, bringing in the latest system of classification and the newest anatomy and embryology—in the process overshadowing the American botanist Asa Gray, who had been appointed professor of natural history in 1842, and who had been quietly instituting his own rather different reforms.

Surely Agassiz's revolution was sheer and obvious progress? A look at the textbook Agassiz published in 1848, with Augustus A. Gould, is suggestive. While Thoreau's touchstone would be the living relationship of animals and plants to each other, to himself, and to mankind, *Principles of Zoölogy* presents the most arid anatomical facts in terms of a rigid scheme of classification. Thoreau bought the book and read it carefully, as part of his scientific reeducation, but from his experiential and phenomenological point of view it must have seemed deadly. Agassiz believed his task was to "render prominent the more general features of animal life, and delineate the arrangement of the species according to their most natural relations and their rank in the scale of being . . ." (5). The "true character and dignity" of natural history is to indicate "to us, in Creation, the execution of a plan fully matured in the beginning, and invariably pursued; the work of a God infinitely wise, regulating Nature according to immutable laws, which He has himself imposed on her" (10). As Agassiz concludes, to study the succession of animals in time and their distribution in space "is therefore to become acquainted with the ideas of God himself" (206)—ideas made manifest in physical creation around us, which to be read properly must be read just as one reads the catalog of specimens collected in their rank and file in the cabinets of a great museum. Thoreau had had, in 1843, his encounter with nature stuffed and pickled "in spirit." His response was predictable: "I hate museums, there is nothing so weighs upon the spirits. They are catacombs of Nature. They are preserved death" (1:465; 9/24/43). As Thoreau commented acidly fifteen years later, "A dead specimen of an animal, if it is only well preserved in alcohol, is just as good for science as a living one preserved in its native element" (XI:360; 11/30/58).

Agassiz was fighting to define the mainstream science of his day as idealist and Coleridgean. As he developed Harvard's program through the 1850s, he was moving to consolidate and professionalize American science in general. This meant, first, relegating natural history field studies to the position of a second-rate science: Agassiz attempted to shut out, for example, such active field researchers as S. F. Baird and Asa Gray from the inner circle of the AAAS, which he helped form in 1848.[18] Second, the move to professionalize science through specialized university training and the formation of national societies meant turning attention and resources away from local organizations like the Boston Society of Natural History, to which Thoreau remained loyal until his death. Sally Gregory Kohlstedt details the effect on the local amateur community of the growing support of science by government and the training of scientists by colleges. In the 1840s, research and study groups at Harvard were superseding the cooperative enterprises and research efforts of the BSNH, although their specimen collection was still "among the best in the nation," and Agassiz himself joined as a member. But by mid-century, "Attendance dropped and finances were strained." Emphasis on systematic training of specialists in the colleges and on graduate training abroad "minimized the possibilities of the casual, mutual-help training of older members"—the kind of training, that is, of which Thoreau himself had taken advantage. The BSNH survived, but only by changing its goals from research to public education, opening the doors of its own Museum of Science in 1867.[19]

Thoreau practiced the local, cooperative, field-based model of natural history even as Harvard and the AAAS were succeeding in denigrating such a model to "amateur" status, effectively taking "science" away from its naturalist practitioners. Baird, then the secretary of the AAAS, had met Thoreau on a visit to Emerson in 1852, and he was impressed enough to propose Thoreau for membership, sending him a letter and a questionnaire in March 1853 (CO 310).[20] Thoreau's first reaction to this recruiting effort was to fulminate in his *Journal*:

> Now, though I could state to a select few that department of human inquiry which engages me, and should be rejoiced at an opportunity to do so, I felt that it would be to make myself the laughing-stock of the scientific community to describe or attempt to describe to them that branch of science which specially interests me, inasmuch as they do not believe in a science which deals with the higher law. (V:4–5; 3/5/53)

But note the tenor of his rejection: not that what he was doing was not "science," but that the keepers of professional science in America in 1853 would not grant that it was science. In this light, it is revealing to look again at Thoreau's reply to Baird, which he waited until December to write, and in which he politely begged off with the excuse that he could not attend the meetings (held in Washington, D.C.). On the questionnaire Thoreau entered his occupation as "Literary and Scientific, combined with Land-surveying"; he expanded on this under "Remarks":

> I may add that I am an observer of nature generally, and the character of my observations, so far as they are scientific, may be inferred from the fact that

> I am especially attracted by such books of science as White's Selborne and Humboldt's "Aspects of Nature." (CO 309–10)

How carefully chosen are his two names: in the person of Gilbert White, the Anglican parson-naturalist of Selborne, he invokes the long and once-honored tradition of amateur natural history in which he knew he also participated.[21] With Humboldt's name, he counters the idealist/positivist science being institutionalized in America by Agassiz (and promoted by their mutual friend Emerson), and aligns himself with the tradition of empirical holism which was being continued in America by naturalists like Baird and Gray, and which had been established by Alexander von Humboldt.

Relational Knowing: Thoreau's Epistemology of Contact

In the years after 1850, what Thoreau strove to create was, in a sense, a new form of science, a *scientia* that would be relational rather than objective. This "relational knowing" extended and applied the possibilities opened up by the disintegration of subject/object dualism, which encouraged the subject to "know" by seeing correspondence in the world's objects, as if they were the mirror of the self, or by "reading" the book of nature as if it were a text ready-made for decoding. By contrast, knowing as an active process in Thoreau's sense becomes no less than what H. Daniel Peck calls "worlding," the making of world "by the interaction—the 'dance'—of the creative self and the world."[22]

This way of seeing the world anew enabled Thoreau to refine and develop his early idea of the "poet" as the hero who joins science to philosophy, who "uses and generalizes the results of both" (2:53; 1842–44). William Ellery Channing captured the hybrid flavor of this enterprise when, in the title of his biography of Thoreau, he dubbed his friend "The Poet-Naturalist." As Channing testified, Thoreau's seeing required a daily investment of energy and study. It was a sustained interaction among senses, mind, and object, an activity of the mind engaged in a reciprocal process: the mind enters nature, nature is taken into the mind; self and nature react on and finally make, and remake, each other. Once nature is seen as an active agent, a full contributing partner in this enterprise, "she" becomes meaningful in all her actions and expressions. Thoreau's quest for meaning carried him, with patient fascination, forth into the wild, day after day for a decade, gathering, identifying, and naming all the elements by which he composed his world. What energized him was his conviction that nature does not lie passive and ready-made, to be "read" by his educated eye; rather, it is continually creative, improvising in front of his very eyes everywhere and at every moment, in an unfathomable wealth of experience. In the midst of this wealth he found (or made) in Walden a stable center around which he could fuse and organize such bewildering plenitude—a "fixed point" that served to anchor and make significant facts that were otherwise confusing, isolated, and sterile. He made it his fertile ground, his "garden," in which he could help nurture his facts toward their flowering into truths.

Hence, over the years, "knowledge" became a particularly vexed and troubling term in Thoreau's lexicon. Thoreau was at a crux, in which the old distinctions signified by the traditional terms were breaking down. When he is facing the past, "knowledge" tends to mean an Aristotelian philosophy that distributes facts into categories, jarring with a Platonic "truth," a higher reality associated with poetry. His unease is registered in comments that deny the ultimate value of knowledge: "The highest that we can attain to is not Knowledge, but Sympathy with Intelligence" ("Walking," NHE 128). But as already discussed, "sympathy" desires as well as fears "knowledge." So to keep the old terms useful, Thoreau renewed them through redeployment, aiming to bring "science" (the intellect, natural history, and material facts) into harmony with "philosophy" (the spirit and higher truth). Their harmony is created through the agency of the poet/naturalist (he uses both terms but favors the former), who *joins* intellect with spirit, nature with mind, in an act of relational knowing that, having tempered dualism into difference, links the two through sympathy in an act which creates "truth." This conscious mediation sends the mind into nature and brings nature into the mind, bringing them together in a heroic, but unstable, union.

The breakdown of this union yields "knowledge" *without* relation, a perverted knowledge with two degenerate forms: "the man of science" who turns nature into objects to be collected, pickled, and abstracted; the "scholar" or aesthete who prefers dry and dusty volumes of poetry to actual life, or Greek columns to the trunks of living and neighboring trees, or who evaporates into the "sublimo-slipshod" raptures Thoreau so pointedly deplores in Channing (4:188; 11/15/51). Of the two perversions, the former is far more dangerous, for by objectifying the world to its own designs it numbs mankind to those "volatile" but crucial experiences that cannot be commodified. Mankind, by falling into this error, reduces objects to their market value, rendering the world withered, cold, dry, dead. A complicit science acts as the handmaiden to this evil, but that science need *not* be complicit—need not objectify nature in order to arrive at either truth or useful knowledge—is the guiding idea of his own scientific research and thought.

The figure of "the Naturalist" who joins poetry and philosophy, spirit and fact, or philosophy and science, is not solely Thoreau's invention. In an 1851 essay titled "The Naturalist," the Boston critic, essayist and poet Henry T. Tuckerman sets forth as his ideal one who "weds nature to humanity," who recognizes "the connection between the ideal and the material world, and the office of imagination as well as that of reason in the interpretation of her mysteries" (66). Such a one will draw into "the sweetest union poetry and philosophy," for it is obvious what "mutual service" literature, "the art of communicating truth," and science, "the investigation of nature," are capable of giving to each other (69). Tuckerman's model for this unifying ideal is, naturally, Alexander von Humboldt. Though there is no record of Thoreau's having read this essay, sometime after 1851 Thoreau copied into his Fact Book a similar analysis by the Boston botanist Edward Tuckerman.[23] Students of nature may be "observers," who start from nature and accumulate facts; "philosophical observers" start at the opposite end with human "reason";

> or finally, in some large and beautiful minds, we can discern neither of these ways by itself, but only what seems their real and original union, wherein the divine reason appears, . . . and facts are observed not only, but eternal laws are prescribed to science—the naturalists. (FB I:58–59).[24]

Years earlier, before both Henry and Edward Tuckerman published their separate accounts of the naturalist as the mediating figure between science and poetry, natural fact and divine law, Thoreau had already envisioned such a figure, calling him the "poet" who "uses and generalizes the results" of both the collector of facts (who walks and acts) and the philosopher (who sits and thinks) (2:53; 1842–44). In 1839 he had imagined the poet as "another nature—Nature's brother" (1:69; 3/3/39), children of the same father in a reciprocal relationship sustained by mutual love. Twelve years later the metaphor is not sisterly but matrimonial: the poet can paint "the most barren landscape and humblest life in glorious colors," while

> The intellect of most men is barren. They neither fertilize nor are fertilized. It is the mariage [*sic*] of the soul with nature that makes the intellect fruitful—that gives birth to imagination. When we were dead & dry as the high-way some sense which has been healthily fed will put us in relation with nature in sympathy with her—some grains of fertilizing pollen floating in the air fall on us—& suddenly the sky is all one rain bow—is full of music & fragrance & flavor— (4:3–4; 8/21/51)

The "prosaic" man of intellect, he continues, is a "*barren*" flower, the poet a flower that is "fertile & perfect." The naturalist too can pursue his study with love, but he is liable to barren dryness; the typical botanist's manual is "dry as a hortus siccus.—Flowers are pressed into the botanist's service" (4:306; 1/30/52). Such work is sterile because it presumes a nature that is passive and quiescent, that can be "pressed into service"—and a student of nature willing to do the pressing. Thoreau's ideal is far different, even erotic: "There must be the copulating & generating force of love behind every effort destined to be successful. The cold resolve gives birth to—begets nothing. . . . The poet's relation to his theme is the relation of lovers. It is no more to be courted. Obey—report" (4:307; 1/30/52). The poet must yield to love, ravish and *be* ravished, if the marriage is to be consummated.[25]

Thus, although his investment in nature marked him as a "naturalist" to his friends, in Thoreau's own lexicon the heroic and beautiful spirit who can marry the sundering halves of man and nature, philosophy and science, is "the poet": "Every man will be a poet if he can—otherwise a philosopher or man of science. This proves the superiority of the poet" (4:436; 4/11/52). However, Emerson's mystification of "the poet" haunts Thoreau even as he struggles to work out a way to enact the ideal in practice: not only is he aging physically, necessitating a supposed "loss," but he is practicing a "knowledge"-based, or empirically grounded, program that automatically looks suspicious to his transcendental training (and his transcendental friends). So he regularly complains that his poetic faculty is declining, even as his own actions and prose bely him. After a 9 A.M. walk with Channing, he laments:

> Too late now for the morning influence & inspiration.— The birds sing not so earnestly & joyously—there is a blurring ripple on the surface of the lake.— How few valuable observations can we make in youth— What if there were united the susceptibility of youth with the discrimination of age. Once I was part and parcel of Nature—now I am observant of her. (4:416; 4/2/52)

Even as, on one level, he acknowledges that observation requires knowledge, "discrimination," and time, he must condemn the value of observation precisely because it requires patience and discipline. After this bout of romantic self-flagellation, Thoreau continues directly with the very kind of observation he feels compelled to regret even as he relishes it: "What ails the pewee's tail? It is loosely hung, pulsating with life. What mean these wag-tail birds?" (III:378). In one paragraph, he is an Emersonian "poet" in decline, while in the next he is a poet-naturalist routinely—but joyfully—practicing his craft. Yet in the terms available to him, the latter position seems unstable, difficult to sustain and defend, even, at times, to himself. If the youthful "poet" in him has died, what can be left but the "naturalist"—like that "learned & accurate naturalist" who told him sadly that no one wrote "biographies" of the flowers, only catalogued them like his library books? (4:330; 2/6/52). Against these two devolved terms, "poetry" that is callow and "science" that is withered, Thoreau continually works on his synthesized ideal, one which honors the slow accretions of experience and sustained practice. "All wisdom," he reminds himself, "is the reward of a discipline conscious or unconscious" (4:47; 9/5/51). Or, metaphorically: "Youth supplies us with colors age with canvass" (4:292; 1/26/52).

The precise point at which Thoreau's poet can actually succeed in "marrying" man and nature, or the disciplines of poetry and science, is the individual fact. Some facts seem already to inhabit, simultaneously, both worlds. That is, they are not just material vehicles ferrying spiritual truth (like his "amphibious" boat in *A Week*), but simultaneously both object-facts and spirit-facts, luminous with mutually enhancing meaning. The fish that inhabit Walden Pond also "inhabit" Thoreau's mind, yet are no less materially fish for all that. The radical metaphor, according to Charles Feidelson, Jr., was "a mode of perception that united past and present, idea and material fact, in the objectively given." These "radically metaphoric" objects need no translation on the poet's part, but stand equally luminous in both worlds, needing only to be recognized to be seen: "Corn grows in the night."[26] "We look to windward for fair weather." Properly seen, could not every fact face both ways? How necessary were the traditional distinctions?

> I have a common place book for facts and another for poetry—but I find it difficult always to preserve the vague distinction which I had in my mind—for the most interesting & beautiful facts are so much the more poetry and that is their success. They are *translated* from earth to heaven— I see that if my facts were sufficiently vital & significant—perhaps transmuted more into the substance of the human mind—I should need but one book of poetry to contain them all. (4:356; 2/18/52)

An isolated "fact" could stand just as dry and barren on the spiritual side as on the natural side, like the dusty poems in Harvard's library, or the arid superstition that Walden was bottomless. The poet's act would be to take the otherwise-isolated fact from *either* side and by "transmuting" it "into the mind," bring it into connection with the complementary realm. The poet thus stands at the "center," as mediator, bringing "facts" from their disconnected positions on the periphery back into their web of relationships. The poet's view is like that from a mountaintop—in a distinctively Humboldtian visual metaphor:

> It was not to see a few particular objects . . . that I ascended the mountain, but to see an infinite variety far and near in their relation to each other, thus reduced to a single picture. The facts of science, in comparison with poetry, are wont to be as vulgar as looking from the mountain with a telescope. (IV:392; 10/20/52)[27]

The task of the poet, then, is to draw facts together into a single composition, sustaining them in a "warm," "living" environment. "Fact" and "truth" turn out to be not stable binary terms but stages in a process of comprehension fostered by the receptive, sympathetic, or fertile mind. Thus facts are *made*, not found like objective "things"; they are made through this process of composition, in which the mind itself is the "fact," the "seed," that, pollinated by the world, flowers with it into truth. Those facts merely "found" are like those in the dry compilation of the Annual of Scientific "Discovery," which are barren; or like those in "Gilpin on the Picturesque," who "talked as if there was some food for the soul in mere physical light and shadow, as if, without the suggestion of a moral, they could give a man pleasure or pain!" (VI:103; 2/6/54). But the same facts of "light and shadow," composed by the poet in an afternoon landscape, create the truth of the vision at the end of "Walking," where late evening sunlight "gilding the withered grass" awakens us to the reality that we cannot walk to the Holy Land, for we are already there (NHE 135–36).

This process of sustained attention requires unremitting discipline and energy. The poet who allows himself to be lazy or distracted by, say, politics or surveying learns that facts once perceived as "truth" can devolve back into their isolated and atomized forms. But much more dangerous is the deliberate deployment of the tools of relational knowing to *reduce* rather than enhance connectivity: that is, to "account for," or explain away, phenomena "scientifically." Thoreau cares for what the beauty of clouds "suggests and is the symbol of . . . and if, by any trick of science, you rob it of its symbolicalness, you do me no service and explain nothing. . . . What sort of science is that which enriches the understanding, but robs the imagination?" Yet before we assume this sentiment is anti-scientific, note Thoreau's turnabout: "Just as inadequate to a pure mechanic would be a poet's account of a steam-engine." After all, Thoreau was an engineer as well as a poet, and if he condemned the "trick of science" in robbing the imagination, he also resented the "trick of poetry" in robbing the *understanding*. Neither view should be sacrificed to the other: "Ah give me pure mind—pure thought. Let me not be in haste to detect the *universal law*, let me see more clearly a particular instance. . . . By perseverance you get two views of the same rare truth" (4:223; 12/25/51). This passage surprises because we are conditioned to imagine "pure mind" subsists in the "universal," but here it lies in the particular. Proper science (and proper poetry) will not divorce but connect the "two views," the both/and of our binary perceptions—though it takes, as he says, "perseverance."

Thoreau leaves no doubt that the poet must throw his body into the process: "We cannot write well or truly but what we write with gusto. The body, the senses, must conspire with the mind." The intellect needs the heart, the liver, every "member": "Often I feel that my head stands out too dry—when it should be immersed. A writer a man writing is the scribe of all nature—he is the corn & the grass & the atmosphere writing" (4:28; 9/2/51). Energy must be directed because the *process* of affinity will end in a *product*, writing: the point of fusion happens *in* the mind, but *through* a text. Or, the act of seeing is completed by the act of saying: "First of all a man must see, before he can say."

> A fact truly & absolutely stated is taken out of the region of commonsense and acquires a mythologic or universal significance. . . . See not with the eye of science—which is barren—nor of youthful poetry which is impotent. But taste the world. & digest it. . . .
>
> . . . At first blush a man is not capable of reporting truth—he must be drenched & saturated with it first. What was *enthusiasm* in the young man must become *temperament* in the mature man. (4:158; 11/1/51)

In this epistemology of contact, significant "seeing" demands total immersion to the point of saturation. To really see a swamp, become a swamp: soak up its juices, right "up to one's chin" (WK 300). Here lay the solution for the romantic poet who had the tragedy to outlive youth: make truth not your revelation but your profession, part of you, your "temperament." Work at it, discipline yourself to both joy and wisdom, and you will grow like corn in the night. One of Thoreau's most important metaphors emerges at this time: ripeness. After setting the first fruit of *Walden* he settled in to study the mythology of ripening trees, seeds, and fruits. As a poet, he was refining his specialization in the radical metaphor, taking his linguistic technology for knowing into the material realm, fusing language and thing, thoughts and facts, into a mythology of the material.

Not everyone understood Thoreau's new project. Channing could not follow his friend's field-note-taking methodology. He tried it, then abandoned it, proclaiming "a little petulantly" (according to Thoreau), "'*I* am universal; I have nothing to do with the particular and definite.'"[28] Thoreau turned to his *Journal*:

> I, too, would fain set down something beside facts. Facts should only be as the frame to my pictures— They should be material to the mythology which I am writing. . . . I would so state facts that they shall be significant shall be myths or mythologic. Facts which the mind perceived—thoughts which the body thought with these I deal— (4:170; 11/9/51)

The poetic fusion, in which object and mind interpenetrate and the resulting fact is nurtured into truth, happens when the mind goes out to nature *and* when nature, so "invited," shows its "affinity" by reciprocating. Instead of passively mirroring or "corresponding" to man, nature actively "answers" to his call. That is, instead of corresponding to each other, they *articulate* each other: each, as Thoreau said, "publishes the other's truth" (1:69).

A defining moment occurs in April 1850 in an extraordinary passage, covering several pages, in which Thoreau described himself coursing over the fields, pursuing a puzzling birdlike tone: "tink tink tink too regular for a cowbell." He finally located it in a boy's water wheel in a meadow rill. The image haunted him: a fortnight later, at night from his window in the village, he could still hear the boy's bell, the sound wafting over the water, the boy speaking the stream to his ears. By day he returned, absorbed in what the *stream* spoke of the *boy*:

> some country boy whose house was not easy to be seen—some arkwright or Rennie was making his first essay in mechanics. . . . It was the work of a fabulous, farmer boy such as I never saw To come upon such unquestionable traces of a boy when I doubted if any were lingering still in this vicinity, as when you discover the trail of an otter.

Is the boy natural or human? Is the waterwheel a work of nature, or art? The division dissolves. "It seemed that nature sympathized with his experiments." Thoreau's writing is tentative, fitful, luminous, tenacious. As Thoreau wrote, he was articulating the meaning of the event to himself, publishing its truth even as he recognized it published *his* truth. Then one day the sound was gone, into history. "You could not hear it—you could not remember it. and yet the fit ear could hear it ever—the ear of the boy who made it. . . ." (3:49–52; after 4/1/50).[29]

The shape of his new career emerged through the remainder of that spring of 1850 and through the following year, as Thoreau accepted the call to articulate nature directly, without the mediation of the child's bell: to *be* the bell, sounding the stream of nature, or, as he wrote the year after, to be "the corn & the grass & the atmosphere writing" (4:28; 9/2/51). Or as he wrote in "Walking": "He would be a poet who could impress the winds and streams into his service, to speak for him" (NHE 120).

In this new mythology, nature becomes the "raw material of tropes and symbols":

> If I am overflowing with life, am rich in experience for which I lack expression, then nature will be my language full of poetry,—all nature will

> *fable*, and every natural phenomenon be a myth. The man of science, who is not seeking for expression but for a fact to be expressed merely, studies nature as a dead language. I pray for such inward experience as will make nature significant. (V:135; 5/10/53)

Science and poetry are both forms of language; in this sense they are twins, vulnerable to each other. "Every poet has trembled on the edge of science," Thoreau wrote (IV:239; 7/18/52). What distinguishes them is the use to which language is put. Scientists who construct the fact as objectively given imagine it can express a fact "merely," extracting "dead" facts (or a "dead language") from the living system, and reducing them into words instrumental toward their own rational designs. Thoreau's "poet-naturalist" participates, rather, in a living linguistic system, which includes but is not limited to his consciousness, and therefore refuses manipulation by his will. Yet it might willingly *answer* his will:

> Is it not as language that all natural objects affect the poet? He sees a flower or other object, and it is beautiful or affecting to him because it is a symbol of his thought, and what he indistinctly feels or perceives is matured in some other organization. The objects I behold correspond to my mood. (V:359; 8/7/53)

Here, constructing nature as "symbol" echoes Emerson's—or the scientist's—instrumental use of nature to express his own organization, but Thoreau's qualifier clarifies his hope that his own thought might be realized, expressed, or "matured" in the organization of *another*: he, and they, will answer to each other. The risks attendant on that hope are many: nature may not answer, or answer in unexpected ways; correspondence to one's mood may open possibility, or limit vision to the command of the self. Thoreau works within all these possibilities.

Thoreau's greatest success in making nature "fable," in seeing every natural phenomenon as a "myth," was of course *Walden*, in which he took his historical experience at a small glacial lake in eastern Massachusetts and turned it into a timeless American fable—a "myth" which, in the organization of a Massachusetts landscape, matures an indistinct feeling or perception taken to correspond to a characteristically "American" or "modern" mood. Thoreau makes his reader believe that nature at Walden Pond does fable, that he has not appropriated its "facts" to his symbolic convenience, but has truly participated in a living language, has let nature articulate herself, through him, the poet of Walden. "My thought is a part of the meaning of the world, and hence I use a part of the world as a symbol to express my thought" (IV:410;

11/4/52). Here is his ideal—not so much taking nature "up" into spirit, as drawing nature and spirit "into the sweetest union," through the agency of the self who is, and inhabits, both.

In the months after *A Week*'s publication, Thoreau had written of "seeds beginning to expand in me, which propitious circumstances may bring to the light & to perfection" (3:43; 1/5/50). Two years later, with his reeducation as a naturalist well underway, he took out the manuscript of *Walden* he had once thought complete, and began again to work on it. To the early polemical chapters on economy and right living he began to infold his new observations, reflecting the turn his interests had taken since 1850. Between January 1852 and April 1854, the material he added—mostly to the later chapters—doubled its length, and revolutionized its character: in Sattelmeyer's words, "the most important major change between the early and later versions of *Walden* lies in the more learned and scientific cast of the later additions and revisions. . . . In the finished version of *Walden* Thoreau is a scientist," in the sense that he "believed that the results of his investigations into nature expressed actual and not merely 'poetic' truth."[30] In this new book the loose, sliding pieces of *A Week* would meet and fuse, and the once vague and miscellaneous character of his natural history musings would be grounded in reality and transfigured into a new vision. The work that was growing under his hands made him think of ripening and maturation. On August 7, 1854, he wrote:

> Do you not feel the fruit of your spring and summer beginning to ripen, to harden its seed within you? Do not your thoughts begin to acquire consistency as well as flavor and ripeness? How can we expect a harvest of thought who have not had a seed-time of character? Already some of my small thoughts—fruit of my spring life—are ripe. . . . (VI:426; 8/7/54)

Thoreau had found his fixed point and the measure of his own stride: from the vague and undirected sense of "fertility" he had reported in late 1850 (3:143–44), he had found both a purpose and a medium. He had gone from reading other men's cosmologies to writing, in *Walden*, a cosmos of his own seeing. Two days later his *Journal* entry reads, in full:

> *Aug. 9. Wednesday.*—To Boston.
> "Walden" published. Elder-berries. Waxwork yellowing.

There was, literally, to be life after *Walden*: it was the first fruit, not the last fruit, of the hard-won consilience of poetry and natural science enabled by the Humboldtian program of empirical holism. After all, *Walden* had had its genesis nearly ten years before. Even as Thoreau was finishing it, his studies were opening up new problems and questions, and the completion and relative success of *Walden* itself would propel his thinking in a new direction. He had learned to see new worlds, yet *Walden* would not exhaust the worlds he was seeing, nor the tantalizing difficulties unfolded to him in the very process of "worlding." The writing and publication of *Walden* would complete and close one line of endeavor, but would also enable—by demonstrating what it could *not* do—the work of the years that followed. Thoreau's world was hardly yet at an end.

Writing the Cosmos: Walden

Thoreau and Walden created each other. Every able-minded reader knows how Thoreau invested his self in the pond, until both he and Walden become "earth's eye," revolving the globe around its deep vision. Thoreau's artistry creates a fiction of wholeness, roundness, an eternal cycle of withdrawal and reunion. The reified Walden seems so inevitable that it gives a bit of a jar to read Thoreau in "The Pond in Winter": "What if all ponds were shallow? Would it not react on the minds of men? I am thankful that this pond was made deep and pure for a symbol" (287). It was a close call, as the fervent tone of the source passage makes even clearer: "I thank God that he made this pond deep & pure—for a symbol" (4:291; 1/26/52). Thank God, indeed—there's a curious interplay here between fate and chance, the overt certainty that all nature *will* symbol (reinforcing, not contesting, our desire) and the sneaking suspicion of a certain arbitrariness behind it all. His Walden might just as well have been made shallow and muddy.

The very suggestion could put all "significance" at nought, but instead it throws responsibility onto the percipient self. Here is the thrill and peril of the notion that nature "answers" to us: then it is up to us to put the right question. "The universe constantly and obediently answers to our conceptions; whether we travel fast or slow, the track is laid for us" (97). The optimism echoes Emerson's in the opening of *Nature*: "whatever curiosity the order of things has awakened in our minds, the order of things can satisfy" (CW 1:7). Nature always answers us—and that's the problem. Wherever we lay the track, sure enough, that's where the track will take us.

So, for example: Is Walden bottomless? No: sounding the depths gives Thoreau a bottom, answering his need for a nature he "cannot put his foot through." And measuring its shoreline gives moral law, with the enticing conclusion that, given one fact, we could "infer all the particular results at that point." Except this would require knowing "all the laws of Nature," and we know only a few. As Thoreau continues, the hope that we will ever encompass them all fades:

> Our notions of law and harmony are commonly confined to those instances which we detect; but the harmony which results from a far greater number of seemingly conflicting, but really concurring, laws, which we have not detected, is still more wonderful. The particular laws are as our points of view, as, to the traveller, a mountain outline varies with every step, and it has an infinite number of profiles, though absolutely but one form. Even when cleft or bored through it is not comprehended in its entireness. (290–91)

Emerson's convergent single law disintegrates into an infinitely multiplying number, sliding into new outlines with every new viewpoint, in a universe "wider than our views of it" (320). The viewpoint becomes the only available unifying center, and from this center, planted at Walden, Thoreau constructed a deliberately beautiful organic whole. Yet he also built in the undoing of that whole. "The Pond in Winter" opens with a puzzle:

> After a still winter night I awoke with the impression that some question had been put to me, which I had been endeavoring in vain to answer in my sleep, as what—how—when—where? But there was dawning Nature, in whom all creatures live, looking in at my broad windows with serene and satisfied face, and no question on *her* lips. I awoke to an answered question, to Nature and daylight. . . . Nature puts no question and answers none which we mortals ask. (282)

His dreaming question is answered: there is no question, and no answer. Yet being a mortal himself, he goes on to derive both, making the incommensurate opposites of man and nature commensurable by sounding and surveying the pond. Regarding the outline of the pond, he maps his answer: the rounded coves, the incised lines straight as arrows, appear on the page before the reader's eyes, stabilizing the infinitely shifting profile of the universe. Nature has answered to his conception. The image is entirely self-contained, a universe to itself, except for (as H. Daniel Peck points out) one detail: the erratic diagonal slash of the railroad, rudely slicing off one of the coves. Thoreau's fiction of the magic circle is held whole against intrusions across space, like the railroad and the ice cutters; and across time, like Thoreau's shocking aside that the "woodchoppers" have since "laid waste" Walden's shores (192). These give what Peck calls the "cross-currents" in this pastoral world apart.[31] The achievement of *Walden*, both in the living and in the writing, was willfully made: against its serene and effortless tone and wide-margined days runs the cross-grained writer who hammered out every sentence, defied the traditions of his village, and forced the earth to "say beans instead of grass."

The only recourse in such a universe is to establish and insist on a "viewpoint," for the only way to hold the multifariousness together is to be the poet-self who so heroically mediates between nature and spirit, drawing "into the sweetest union" the wild woods and the civil township, the unspoiled garden and the ravaging engine of progress, history and eternity, narrative and myth. What that self demonstrates is not a universal and suspended truth, nature's ultimate answer, for there is none. There is only a way of seeing, a mode of vision and following from it a mode of life: no theory, only method. "How to Live. What to Do."

The oscillation that had characterized *A Week* has not disappeared, but in *Walden* it is compressed into a tight dialectic. Images of division dominate, for example, the climactic passage at the conclusion of "Where I Lived and What I Lived For": from shallows to whirlpools, "hard bottom and rocks in place" to "Time is but the stream I go a-fishing in," head and hands, burrowing deep and rising like vapor. The fact is a "cimeter" whose "sweet edge" divides you; the intellect is a "cleaver" that "rifts its way into the secret of things" (97–98). What had been disruptive oscillation is here tempered into a tight interplay of contraries, which instead of establishing then undercutting either side centers on the very question of dividing and divining. Divisions into shifting opposites permeate the book, so much so as to become a principle of organization. Chapters proceed in pairs: "Reading" and "Sounds," "Solitude" and "Visitors," "Higher Laws" and "Brute Neighbors." Thoreau's Janus vision controls the book, as assertion is followed by inversion, and paradox insists that both are true. Wherever he casts his line, he will catch "two fishes as it were with one hook" (175).

What helps fuse such contrary elements into unity is that "fixed point" from which everything can be measured, the fixed center of the pond. Thus Thoreau used the Pond to anchor all the myriad particulars of his world. All the contending metaphors and multiple directions finally run through it, grounded in the image of pond, woods, and cabin, focused through the stability, and utter mundaneness, of its actual existence: this real place is on the map (one reason Thoreau maps it for us is to make this quite clear), and it is no frontier lake or exotic Utopia but here, home. The centered image of the pond becomes the standard—physical and moral—which measures all else, and to which all else is referred. Amidst the confusion, in which "Nature and human life are as various as our several constitutions," there cannot be only one way. There are, he reminds us, "as many ways as there can be drawn radii from one centre" (10–11). That there is only one *center* is the foundation and enabling premise of the book—one reason it is titled, simply, *Walden*.

There are many other possible centers, but here they are only implied. Indeed, the point of his exaggerated specificity is to *force us out*: Thoreau's

Walden is exactly the one place we cannot be. The writer will not be everyone, everywhere, but only Thoreau at Walden (and in the nick of time too).[32] The only saving grace to his self-centered audacity is the sly implication that each reader cannot help but be similarly placed. Thoreau puts it to us to ask where it is, then, that *we* live. To him, the shore of Walden Pond merely looked like "a good place for business . . . a good port and a good foundation," though not ready-made: you still must "build on piles of your own driving" (19–21). His gesture of self-sufficiency and exclusion is so total that, rightfully or not, it has come to stand for his entire life and career. However, by setting up his own place of "business," his center for calculating and comparing cosmic accounts, Thoreau by extension *does* include all his transactions and, through the network of inscriptions, any other transactions that he might draw to his center.

Though *Walden* was first conceived as a sequel to *A Week*, one inspiration for its revision was surely Humboldt's *Cosmos*. Thoreau was in effect rewriting *Cosmos* and doing it better, because he was directing it to higher ends, and from an individual, "first-person" consciousness. First-person narration had worried the scientist Humboldt, because it seemed to him insufficiently representative and therefore uninteresting (*Personal Narrative* [1852] I:xiii–xiv, xix–xx). The artist Thoreau immediately cut off such objections by asking who *else* we might suppose is authoring the book (3). He refuses to fade into transparency, deferring authority to nature or tradition. From his opening paragraph, with its repeated phrases of withdrawal and self-sufficiency, Thoreau makes clear that this book will *not* be scientific. It is, quite simply, insufficiently social.

Science packages experience so that it may be shared, communicated, socially sanctioned. The point in *Walden* is that each person's experience is unique, and at a fundamental level incommunicable: Thoreau can preach at John Field, but cannot reach him. Thoreau's "experiment" in itself does neither Field nor us any good, for it is of no value as just one more brick in the social edifice of knowledge. It is of value as we repeat his experiment in our own lives—acting, ironically, like good skeptical scientists, validating experimental results in our own laboratories. So *Walden* is social after all—but built using an alternative network, one authorized by individual, empirical, relational experience rather than by experience as socially determined, the consensus of the many mystified as the One and disseminated as Tradition. For the latter is precisely the problem. It overdetermines what any of us believe we can make of our individual experience.

Thus Thoreau takes himself ostentatiously off the world network altogether, to build his own parallel world: he has, he says, "my own sun and moon and stars, and a little world all to myself" (130). Nor is his world hidden in some wilderness: to be socially useful, it must be visible, public, a spectacle. As Stanley Cavell reminds us, Thoreau positions himself just far enough away "to be seen clearly."[33] Since an exemplary life is useful only as it parallels our own lives, Thoreau gives elaborate accounts of all the mundane business details entailed by building and maintaining a life at Walden. Nor does he fail to link his life to ours, whether covertly through the book we hold in our hand, or explicitly through the railroad, by which he is "related to society" and which whirls society past his door, each to the delight and amusement of the other. Indeed, the railroad men who take him for a fellow employee are correct, for "I too would fain be a track-repairer somewhere in the orbit of the earth" (115). What needs repair is less the *machine* (which he rather admires), than the tracks on which the machine travels, tracks built, he reminds us, on the bodies of those Irish "sleepers." These tracks—the local material link in the growing world network—conduct passengers not to heaven but to Boston, thereby determining the network on which the public will travel: "and thus one well conducted institution regulates a whole country" (117–18). To counter-act in such a world, we must all become practicing scientists.

"Whether we travel fast or slow, the track is laid for us" (97). Who, then, laid that track? The building of the railroad instances the social injustice of man against man, but the locomotive itself looms less as an arbitrary invention of man than as a new force of man-nature, relating human and natural possibilities like the boy's waterwheel, a hybrid artifact that in Thoreau's imagination bespoke both man and nature and reprised the beginning of the industrial revolution: the "devilish Iron Horse" (192). The industrial revolution had in a sense just begun at Walden. Thoreau moved to its shores only a year and a month after the first train arrived in Concord, in June 1844.[34] The slash of the railroad across the landscape introduced historical change into the timeless realm of nature, cracked the timeless bowl of Walden: so runs the symbolic pairing, the human machine violating the garden of nature. The irony, as Leo Marx shows, was that it was the machine that created the pastoral garden, and transported people to it; Marx found in the machine the "root metaphor" of the nineteenth century. What drew the nineteenth-century imagination was the machine's transformative power, the way technology hybridized the tremendous forces of nature with the precise designs of human purpose. In Emerson's project, *how* would nature be transformed into man, fate into power? By the agency of the machine; the evidence was as close as the Fitchburg Line. Technology and metaphor merge: recall that the engine of fate was powered by polar forces, be they matter and energy, fate and freedom—or water and steam.[35] What connects those polar opposites? Flow and circulation, under many names: wind, water, spirit, language. Circulation between the poles powers the universe into a dynamic system, and the metaphor holds as true for Emerson in the realm of spirit as it does for the engineer in the realm of coal and steam: cosmic forces were coming together to propel the universe forward. One couldn't argue with the railroad: "We have constructed a fate, an *Atropos*, that never turns aside. (Let that be the name of your engine)" (118). There was no turning back: the process was irreversible. History wasn't just happening; men and nature were producing it.

The railroad drew Walden into history, out of the timeless and into the nineteenth (and then the twentieth) century, and its power could not be revoked. Lyell had erected the eternal return of cycling time to stave off the threat of irreversible change, the same threat which met Thoreau on the Fitchburg Railroad. Thoreau understood both the power of the engine, and its fragility. The power as it extended along the tracks was invincible: it never turns aside. But it never turns *aside*; step ten feet from the tracks, and it cannot follow. All Thoreau had to do to defeat the railroad was cross the tracks "like a cart-path in the woods" (122).

Yet one can't so easily step off the track of time, and Walden was caught in its web. The surge of the seasons across the polar forces of summer and winter produced the explosive energies of "Spring," and with them irreversible change; the penultimate chapter closes with the moment after which the Pond was for Thoreau no longer a present but a historical fact. Imaginatively Thoreau could protest that the pond was pure, eternal, and self-renewing, and in fable he could create a cycle of the seasons that repeated endlessly without change; yet the pages of *Walden* itself not only recorded change, they engineered it. Writing is also a form of technology, another way for man and nature to bespeak one another. Thoreau's writing put Walden on the map—literally; he packaged it up into a neat whole, just like the fish he sent to Agassiz, set the pond on the tracks and sent it whirling away to the world like the pastoral valley blowing past (121–22).

What introduces change into the eternal discursive present of *Walden*'s narration is the onset of otherness—natural or human. At the end of "Brute Neighbors," the hitherto timeless present of endless and unspecified summer suddenly tumbles through a swift fall to the closure of winter. The cascade is initiated by the howl of the loon, who, tiring of Thoreau's neighborly games, calls up a wind from the east that ripples the smooth water and drives Thoreau away, off the "tumultuous surface" of the pond (234–36). Nature has generated history. The mirroring moment occurs in the thawing tumult of spring, when "the sun shines bright and warm . . . re-creating the world" into its aboriginal innocence (314). The generative "principle of all the operations of Nature" is catalyzed by none other than the railroad, for it is in the cut bank behind the pond that Thoreau witnesses the "hybrid product" of sand and water, sun and ice, as the heat of the sunshine melts the frozen earth into all the flowing forms of life; human destruction has set in motion the principle of creation and renewal (304–9). Spring is "like the creation of Cosmos out of Chaos" (313), reassuring Thoreau with the "tonic of wildness," of absolute otherness, such that he can rejoice in vultures feeding on carrion, the rotting of dead horses, the "rain of flesh and blood," and the "inviolable health of Nature" that can afford that myriads be sacrificed and suffer. *Atropos*, indeed: "The impression made on a wise man is that of universal innocence" (317–18).

The tight dialectic of cosmos and chaos, doing and undoing, nearly finishes the book, except that before ending, Thoreau must reassert our role in such a universe. His final insistence on "universal innocence" is necessary, for it makes the perennial dawn of possibility he invokes in the "Conclusion" available for anyone, anywhere, at any time. In the tale of the "Artist of Kouroo," if the work is "perfect," limits of time and space vanish. Yet his final image is a more quizzical one, drawing on the New England fable of a "strong and beautiful bug" burrowing its way out of an old table. Thoreau had started his adventure by burrowing down and in to granitic truth. Now the direction of the fable has reversed: up and out to the warmth and light. This asserts Thoreau's greatest theme, making the story the appropriate close for his central work: each one of us, however bound and encrusted by time and tradition and circumstances, retains the ability to burrow up and out to our own light. Few of us would aspire to be the Artist of Kouroo, but the image of the bug equalizes us all. If one advances confidently in "the direction of his dreams" he will "pass an invisible boundary" as new laws establish themselves around and within him; he need only be "*extra-vagant*" enough (323–24). If we made our world, we can always *re*make it, both individually and collectively. In another form the same image will dominate the writings after *Walden*, in which the bug becomes a seed, and renewal is not an isolated achievement but a communal act, the collective accomplishment of all the seeds in the ground. As Richardson says, "The movement is from economy to ecology."[36] Hence Thoreau's late essays would dwell not on singular but on collective wholes: ever-renewing communities composed of a multitude of entities, drawn into coherence by the perceiving and participant mind.

Ironically, in order to make the truth of *Walden* possible, Thoreau had to put the facts of Walden behind him. So long as he stayed in the field, collecting and observing, there could be no end to the material, no point at which he could close the circle. The author of the *Cosmos* must draw the line somewhere, withdraw from the field to the study to write, sacrifice process to product. Humboldt the scientist couldn't do this. As he wrote he added new findings as they came in, and his project grew beyond the limit of even Humboldt's life. Three planned volumes ballooned to five, which even then were incomplete; he died while working on the fifth. To finish *Walden*,

Thoreau had to do deliberate violence to his subject: cut it off, draw an arbitrary line and sacrifice whatever lay beyond it. Thoreau could justify this in literary terms: after all, poetry "puts an interval between." That crucial interval allows the poet to shape the now-distanced object, to render it into fiction, to articulate the translation from "fact" to "truth." Yet that act was suspect, for it inevitably translated the other into the self.

Making Walden socially available had meant cutting corners to create a rounded, shapely, fictional entity, even one energized by deconstructive crosscurrents. This is the paradigmatic aesthetic act, rendering nature into an "organic" whole. But Thoreau was alert to the consequences: first, the artist's tendency to read the shapely whole back into nature, smooth out anomaly, demand conformity. This act denied his fundamental belief in nature's vital lack of completion; so in the *Journal* he would make amends, following nature beyond the rounded whole he had made of her in *Walden*.[37] Second, his fiction had reified a once-lived reality, commodified it and quite literally shipped it to market. To make nature into scripture, he had to make his scripture into a commodity; and no matter how closely he studied "how to avoid the necessity of selling them," the baskets of delicate texture he wove *had* been woven, and he *did* take them to market (WA 19).[38] The act of taking up Walden into a symbolic system meant making it into a social fact, a new word in the language. By thus making it mobile he took it from its own orbit and put it into his own, sending the pond, like the pastoral valley he saw going by on the Fitchburg Railroad, whirling off and away to Boston. As with his beloved apples, he watched while the pulp went to market and the fragrance to heaven. The experience was a lesson in the ecology of mind: his "Walden" entered the economy of nature as surely as the railroad, both endangering and preserving the landscape he had sanctified, and in any event changing it forever. Walden Pond was not immutable after all; "the woodcutters, and the railroad, and I myself have profaned Walden," he confessed (197).

Thoreau, then, did not leave the field a second time and write another *Walden*. In April 1852 he may have envisioned a "book of the seasons," in which he would chart the periodicity of Lyell's great year, hoping such a chart would help him "know why just this circle of creatures completes the world . . ." (4:468). But as he walked the fields collecting data for such a chart, the world never completed, the cosmos kept expanding: repetitions he had anticipated did not repeat, or if they did they were wildly, unpredictably different. After years of this he mused, "It takes us many years to find out that Nature repeats herself annually. But how perfectly regular and calculable all her phenomena must appear to a mind that has observed her for a thousand years!" (XIII:279; 5/5/60). But as Cameron comments, such a scale is not human, and the sense it made would not be human sense. And so what the *Journal* documents is not the fixity of meanings in nature, but the fact that meanings cannot be fixed. Therefore no analogy, no myth, can help but falsify the truth of a nature that "resists being symbolic."[39] The *Journal* became open-ended, accumulating new layers year upon year. The periodic phenomena never came around to the closure which would have allowed Thoreau to conclude the *Journal* and write the book. This magic circle kept expanding—endlessly.

Thoreau was capable of feeling frustrated and defeated on this count, but the bulk of the *Journal* and the late essays also show him happy to relinquish the search for symbolic totality. If he cannot delimit the complete picture, he can collect the elements which, he hopes, he can *compose* into the picture "that, at any given moment, he is unable to see."[40] Humboldt, confronting the same dilemma, worked with a similar solution: he wrote of combining "groups" of facts into a whole not found in nature, but composed by the interaction of the observing mind with nature. Such wholes were like "pictures" of nature, which did not pretend to embrace all of it but simply "Aspects" of it, or "Views"—*Ansichten*. Humboldt's word is, as Robert Van Dusen observes, an interesting choice, for it implies "the concept of intuitive contemplation," or intuitive visualization. While the English translations, "aspects" or "views," suggest the presence of an external reality which is represented to an onlooker, the original word suggests that the process of intuitive perception helps to *create* the "picture." This puts far more emphasis on the process of perception and how it affects the process of composition.[41] Accordingly, Humboldt prefaced the collection of *Ansichten der Natur* with a brief description of his methodology: "Several detached fragments, written on the spot, have since been wrought into a whole." His aim was to survey nature, to show proofs of the "co-operation of forces" and renew enjoyment in its "immediate aspect." To this end, "Each Essay was designed to be complete in itself; and one and the same tendency pervades the whole" (*Views* ix). Such "aspects" bring all the tools of observation, analysis, and connection to some single one of the infinite facets offered by the face of nature. Thoreau's late form may have been the essay not because he lacked skill or energy to reprise his masterpiece, but because he was pursuing a fundamentally different form: a Concord version of *Ansichten*, a linked and interwoven sequence, flexible and open-ended, and amenable to revision, reshuffling, and addition—rather like Whitman's solution to a similar problem, in *Leaves of Grass*.

In *Walden*, Thoreau had critiqued the mobilizing network of science/ society by offering a parallel work that stood off, but in view of, that network; yet in its masterfully totalizing cosmic unity, it had paralleled science *too* effectively. It too reified and commodified its object, sending it off to market like the most complicit science of Frémont or Agassiz. What

Thoreau attempted next was a different way of making knowledge social, a new form that might counteract the masterful worlding of *Walden*, and which corresponded with the alternative science he would propose in "The Succession of Forest Trees." This exploratory form permitted Thoreau to conceive of a whole that would be loosely composed of provisionally assembled pieces as they suited his varying mood, and which presented nature as the ever-changing but interlinked unity he knew it to be. And so the late essays were precipitated from the *Journal*, as he collected masses of material around some suitably partial, contingent, but richly connective experiential "fact": the turning of autumn leaves, seeing wild apples, picking huckleberries.

5

A Plurality of Worlds

> We might try our lives by a thousand simple tests; as, for instance,
> that the same sun which ripens my beans illumines at once
> a system of earths like ours. If I had remembered this
> it would have prevented some mistakes.
> —Henry David Thoreau, *Walden*

One of the controversies that enlivened scientific discourse of the 1840s and 1850s concerned the possible plurality of worlds. Was earth unique, or was there, as Thoreau fancied in *Walden*, "a system of earths like ours"? The controversy turned on integrating the findings of the new science of geology with the older science of astronomy, or deep time with vast space. Both sciences suggested that a great deal of the universe had nothing whatsoever to do with human beings. Did this mean that the universe was an infinitude of waste and chaos, within which the earth alone gave sanctuary to life and intelligence—God's one special creation, his Garden? Yes, argued the redoubtable Whewell, in *The Plurality of Worlds* (1853).[1] But how, argued others, could this poor, flawed planet truly be God's single creation? There had to be other, finer worlds in the heavens. To doubt their existence denigrated God, implying His power did not extend across all time and space. Or, proposed still others, had God's law simply, from the first moment of Creation, set in motion the progressive evolution of matter into life, here on earth and everywhere in the heavens? This was the position taken by Robert Chambers in his popular book *The Vestiges of Creation* (1844), the book Whewell was targeting: Chambers had proposed only "a blind process governed solely by mechanical necessity."[2] Thus Thoreau's allusion to Chambers' theory points toward the deepening divides which were already tearing natural theology apart, well before 1859 and Darwin's *Origin of Species.*

Thoreau was already moving beyond the terms set by traditional natural theology, in his sanguine acceptance (at least much of the time) of a cosmos in which man was only a small part. The real point of the plurality of worlds debate was, of course, not life on the stars but life on earth. Were all those extraneous other beings beyond the circle of humanity truly waste and chaos? Or did they evidence the power of life to form worlds other than ours? Thoreau found the latter prospect exhilarating:

> The stars are the apexes of what wonderful triangles! What distant and different beings in the various mansions of the universe are contemplating the same one at the same moment! Nature and human life are as various as our several constitutions. Who shall say what prospect life offers to another? Could a greater miracle take place than for us to look through each other's eyes for an instant? (WA 10)

As he concludes: "there are as many ways as there can be drawn radii from one center" (11). Though he centered vision on the constitutional magic of the self, the self that initiated the world didn't always circumscribe it. In the long meditation ending with his declaration that "The whole world is an America—a *New World,*" Thoreau explores such a vision: "The poet says the proper study of mankind is man— I say study to forget all that—take wider views of the universe— That is the egotism of the race." One has only to get up at midnight to find the whole civilized world, that "gigantic institution" of mankind, slumbering. "Man is but the place where I stand & the prospect (thence) hence is infinite. it [*sic*] is not a chamber of mirrors which reflect me—when I reflect myself—I find that there is other than me" (4:418–20; 4/2/52).

One consequence of his penchant for turning the mirror inside out, imagining the other *as* other yet with eyes like himself, was his political conviction that no man has the right to pursue his pleasure on the back of another. In science, this principle has the effect of breaking down the subject/object dualism that lay at the heart of scientific acts of knowing. Instead, to Thoreau, each object could be the subject of its *own* knowing: "If I had remembered this it would have prevented some mistakes" (WA 10). If we could remember that the eyes of every object are also gazing back at the subject, we could live in not one but *many* worlds—"aye, in all the worlds of the ages" (10). By contrast, to render nature either symbolically or (f)actually instrumental—to ride on nature's back like Emerson's Savior on the meekly receiving ass—one must sever the possibility of nature's answering back by insisting, like Coleridge, that it is not a living and self-generating system but a collection of dead objects, available for animation by our own needs and desires. But Thoreau's reciprocal and interactive process of knowledge breaks down the subject/object duality altogether, and strikes down the notion of "objective" science at its root:

> There is no such thing as pure *objective* observation. Your observation, to be interesting, *i.e.* to be significant, must be *subjective*. The sum of what the writer of whatever class has to report is simply some human experience, whether he be poet or philosopher or man of science. The man of most science is the man most alive, whose life is the greatest event. (VI:236–37; 5/6/54; emphasis in original)

Across the shared ground of language, science is fundamentally the same as poetry and philosophy: none can do more, or less, than report human "experience," and all three will live, or not, according to the fullness of life brought to bear on them. The goal, then, is to *live*, to experience:

> It makes no odds into what seeming deserts the poet is born. Though all his neighbors pronounce it a Sahara, it will be a paradise to him; for the desert which we see is the result of the barrenness of our experience. No mere willful activity whatever, whether in writing verses or collecting statistics, will produce true poetry or science. (VI:237)

All a man can do, Thoreau concludes, is "tell the story of his love," and if he is fortunate he will remain forever in love, keeping "coldness"—*lack* of relation—from reaching the heart.

Not one's external circumstances, but the condition of one's mind and heart, will turn fact into truth, the Sahara into Paradise. But how does one cultivate such a faculty, if "mere willful activity" is inadequate? It takes a pose of sublime, transcendental confidence to condemn "method" and "willful activity." Thoreau, like any successful artist or scientist, must rely on both, and yet he must agonize that such reliance signifies his loss of the health and wholeness from which natural inspiration was to spring as a spontaneous function. It could not be achieved willfully, any more than Calvinist grace, by dint of intellectual effort or knowledge.[3] The need to labor at it signified malfunction, loss, disease, one's fall from harmony with the whole, out of alignment with the flow. And so were set the poles of work and spontaneity which drew Thoreau into creative effort, yielding the twin strategies of knowledge or "intentionality," and ignorance or "seeing with the side of the eye," which worked together in what I have called Thoreau's "epistemology of contact."

Intentions of the Eye

In Humboldt, seeing is both partial and total: partial, in the place-bound, limited view given to the researcher who is in actual contact with nature; total in the way such partial views combine and interconnect to create the grand view, the mountaintop perspective that integrates the multitude of details into a coherent whole. The eagle-eyed view from the alpine or celestial heights transmutes into the totalizing God vision of the scientist, scaling the heights while the rest of us go on with our lives down in the warm valleys:

> Mathematical truths stand aloof from the warm life of man—the mere cold and unfleshed skeletons of truth.
>
> Perhaps the whole body of what is now called moral or ethical truth may have once existed as abstract science, and have been only gradually won over to humanity.—Have gradually subsided from the intellect into the heart.

In Thoreau's 1840 meditation, abstract law from on high descends like Christ to assume human form, even as God's absolute truth must put on the garments of the affections. As he continues:

> The eye that can appreciate the naked and absolute beauty of a scientific truth, is far rarer than that which discerns moral beauty. Men demand that the truth be clothed in the warm colors of life—and wear a flesh and blood dress. They do not love the absolute truth, but the partial, because it fits and measures them and their commodities best—but let them remember that notwithstanding these delinquencies in practice— Science still exists as the sealer of weights and measures. (1:196–97; 11/12/40)

And yet Thoreau ends with a warning. Cold and abstract law from on high "seals" the measure of our lives—science the stern lawgiver. Does science consist of the abstracted, "unfleshed," "skeletal" commandments that delivered from the mountaintop seal our human fate? Or the collective sum of all the bustling confusion that extends from the top of the mountain, to its foot, to the horizon and beyond, into infinity? The latter might seem fuller, richer, warmer, more descriptively responsive to reality "on the ground," but the former offers a tremendous advantage: power. "Holistic explanation bears a sort of mirror image relationship to reductionism," notes the biologist Richard Lewontin.[4] Reductionism does not diminish the field of vision, but expands it beyond the horizon, beyond to all possible horizons. In its chamber of mirrors, it sees itself everywhere, never anywhere new.

Thoreau's problem was to retain the power yet see something other than his own reflection, to look into the infinite from the place where he stands and yet see a connected and meaningful whole. Science, like poetry, is made possible by "putting an interval between"—the saving interval which allows abstraction and composition. This interval separates life from the writing: the fish in the brook from the specimen in the book or the laboratory jar; the experience in the field from the collocation of results in the report; the man on the street from the cadaver on the slab. The interval makes the creative act possible, but also makes it partial, fitting truth to our commodities. However, it has become part of the powerful fiction of science that such willful activity is not partial and selective, but total; that truth exists not out in the fragmented and imperfect world but in the vacuum of the laboratory, which is so carefully constructed and sealed precisely to enable the facts to speak the truth.[5] Hence science is not fictive, accomplished by a human community, but transcendent, absolute, and "objective."

But Thoreau's working experience convinced him that no fact could exist in a vacuum; or, that all facts become visible to us only when we know what it is we seek, in which case we must, as Darwin noted, have some theory or concept that will make a fact "significant." The theory might be, for example, that all facts can be catalogued and stored away: Thoreau notices that books are usually written "willfully," or "as parts of a system, to supply a want real or imagined. Books of natural history aim commonly to be hasty schedules, or inventories of God's property, by some clerk"—and so conduct to ignorance not knowledge (WK 97–98). But that there must *be* a theory or system was unavoidable. For instance, moved by spring's return to theorize that the year is a circle, Thoreau concludes that learning the iterations of the natural system might guide him into calling forth something novel:

> Why should just these sights & sounds accompany our life? Why should I hear the chattering of blackbirds—why smell the skunk each year? I would fain explore the mysterious relation between myself & these things. I would at least know what these things unavoidably are— —make a chart of our life—know how its shores trend—that butterflies reappear & when—know why just this circle of creatures completes the world. Can I not by expectation affect the revolutions of nature—make a day to bring forth something new? (4:468; 4/18/52)

Under his theory that the universe formed a "complete" and closed circle, Thoreau was determined to observe all possible causes and coincidences, but he quickly found himself hard put. Over the next few weeks he tried to keep up with the totality of spring's onset, but by mid-May he was losing track. On the nineteenth he complained that since the fourteenth there had been too much to observe; the next day, he despaired that there were just too many birds arriving and plants leafing: "I must observe it again next year." Or perhaps he could get a jump start on the spring, by catching it twice in the *same* year: "It is worth the while," he decided three days later, "to go a little south to anticipate nature at home" (IV:65–69; 5/19–5/23/52). A few weeks later he envisioned a complementary strategy, a strategic withdrawal:

> Nature is reported not by him who goes forth consciously as an observer, but in the fullness of life. To such a one she rushes to make her report. To the full heart she is all but a figure of speech. This is my year of observation. . . . You are a little bewildered by the variety of objects. There must be a certain meagreness of details and nakedness for wide views. (IV:174; 7/2/52)

It became evident to him that the plenitude of nature must be edited, even for the most (Whitmanically) inclusive observer. Indeed, the eye edits automatically, Thoreau has noticed, just by choosing what to see or not see. One of his earliest narratives records how he *chose* to see an arrowhead, and lo, there it was at his feet (1:8–9; 10/29/37). This limitation, then, can be cultivated into a strength, if one educates oneself in the art of "anticipation," or "expectation." One can call forth something new, as when by sheer force of expectation he became the first to locate a rare or unknown plant. Yet even so, not all one's callings are answered: "We soon get through with Nature. She excites an expectation which she cannot satisfy. The merest child which has rambled into a copsewood dreams of a wilderness so wild and strange and inexhaustible as Nature can never show him." He expects *more*—more than every year the same old dead suckers floating on the river. Coleridgean dejection seizes him: "In me is the sucker that I see. No wholly extraneous object can compel me to recognize it. I am guilty of suckers" (VI:293–94; 5/23/54). If nature lives "in his life alone," no fact exists independently of his consciousness, and he is the guilty perpetrator of all the evil he sees.

But fortunately such exhaustion and its consequent guilt were rare; Thoreau's solipsistic moods tended not to be productive ones, for he needed the energizing assurance that nature was rushing to him to make her report. More typical of the late years was his steady refinement of "anticipation" into a principle he named "intentionality of the eye":

> It requires a different intention of the eye in the same locality to see different plants, as, for example, *Juncaceæ* and *Gramineæ* even; *i.e.*, I find that when I am looking for the former, I do not see the latter in their midst. How much more, then, it requires different intentions of the eye and of the mind to attend to different departments of knowledge! How differently the poet and the naturalist look at objects! (XI:153; 9/9/58)

The counter-intuitiveness of this notion delighted him: "A man sees only what concerns him. A botanist absorbed in the pursuit of grasses does not distinguish the grandest pasture oaks. He as it were tramples down oaks unwittingly in his walk" (XI:153). Earlier he had written that it was "impossible for the same person to see things from the poet's point of view and that of the man of science," though even there he permitted the poet to have science as his "second love" (4:356; 2/18/52). But, importantly, with "intentionality," both the poet's and the naturalist's vision could be available to the *same* person, if that person could learn to refocus his intention not just from rushes and grasses to oaks, but from one discipline to the other: from "poet/naturalist" to "poet-naturalist." After all, there are "manifold visions in the direction of every object" (WK 48).

Thoreau suggested the possibility of such manifold vision two months later: the distance that counted was finally not between poet and naturalist, but between them both together and the "mass of men" who are uninterested in, say, lichens, and could hardly imagine them as "sympathizing companions":

> It is remarkable how little any but a lichenist will observe on the bark of trees. The mass of men have but the vaguest and most indefinite notion of mosses, as a sort of shreds and fringes, and the world in which the lichenist dwells is much further from theirs than one side of this earth from the other. . . . Each phase of nature, while not invisible, is yet not too distinct and obtrusive. It is there to be found when we look for it, but not demanding our attention. It is like a silent but sympathizing companion. . . . (XI:296; 11/8/58)

The poet's love has matured into a form of "attention," alert but undemanding, and steeped in the kind of knowledge that makes such acute responsiveness possible. This form of attention became a key tool of scientific research once Thoreau had mastered both the method and a scientific theory, as we shall see shortly in discussing *The Dispersion of Seeds*. Such insights put Thoreau in the vanguard of scientific theorists of his day. John Herschel and William Whewell (whom Thoreau apparently did not read) had been at work since the 1830s formulating the "scientific method," or that powerful heuristic whereby feedback loops between theory and fact (or call and response?) advance our state of knowledge about the physical universe. Both Herschel and Whewell recognized how fundamentally contingent this process is: we can never be certain, though we can be provisionally sure, since our theories have led us to make only appropriate discoveries. Thoreau, using his poet's lens, put it differently: "The hunter may be said to invent his game, as Neptune did the horse, and Ceres corn" (XIII:140; 2/12/60).

The problem with such a powerful heuristic is that it can eliminate any possibility which you fail to "anticipate." Even just looking too hard will cost: "the more you look the less you observe," Thoreau observes; "Go not to the object; let it come to you." What he needs, he decides, "is not to look at all, but a true sauntering of the eye" (IV:351; 9/13/52), a sidelong vision that will invite the unbidden. In the midst of the fatigue that brought on his call for "a little Lethe!" Thoreau admonished himself that "Man cannot afford to be a naturalist, to look at Nature directly, but only with the side of his eye. He must look through and beyond her. . . . I feel that I am dissipated by so many observations. I should be the magnet in the midst of all this dust and filings" (V:45; 3/23/53). Such an insight is the necessary counterbalance to forms of attention. Thoreau even finds that the occasional distraction is actually productive. After a session of surveying, he was again reminded

> of the advantage to the poet, and philosopher, and naturalist, and whomsoever, of pursuing from time to time some other business than his chosen one, —seeing with the side of the eye. The poet will so get visions which no deliberate abandonment can secure. The philosopher is so forced to recognize principles which long study might not detect. And the naturalist even will stumble upon some new and unexpected flower or animal. (VIII:314; 4/28/56)

Experience taught him that even the walk that seemed "profitless and a failure" was "on the point of being a success, for then you are in that subdued and knocking mood to which Nature never fails to open" (XIII:111; 1/27/60). As he also said, to be lost is the beginning of being found, for "not till we are lost do we begin to realize where we are, and the infinite extent of our relations" (V:64; 3/29/53; cf. WA 171).

These twinned and mutually generating strategies, intentionality and side-of-the-eye, parallel and support the dialectic that runs through Thoreau's work between knowledge and ignorance: Learn science, he has advised; then forget it. Live at home like a traveler. Knowledge guides the eye, but limits it to what it already knows; ignorance informed by knowledge defamiliarizes the known and prepares the mind for the novel, the unanticipatable. Thoreau is establishing, in effect, a way to let chaos sneak into the system, the order and arrangement created by the mind; or, a way to build disorder and incompleteness into the fabric of orderly structures. In some of his last and pleteness into the fabric of orderly structures. In some of his last and richest writing on this matter, he makes clear that the goal of this process is not some flawless mirror of nature, a perfect representation in words, but an altogether new entity, a *new* fact in nature:

> It is only when we forget all our learning that we begin to know. . . . To conceive of it with a total apprehension I must for the thousandth time approach it as something totally strange. If you would make acquaintance with the ferns you must forget your botany. . . . You must be aware that *no thing* is what you have taken it to be. . . . You have got to be in a different state from common. Your greatest success will be simply to perceive that such things are, and you will have no communication to make to the Royal Society. (XII:371; 10/4/59)

To succeed in perceiving that "such things are" is to use science to go beyond it, make some form of communication of no interest to the Royal Society—for one to be redeemed by ferns, he adds, takes a different method than the Aristotelian. He enlarged on this idea, so reminiscent of the late poems of Wallace Stevens, one year later: the truest description (or representation) of an object is the sight of the thing itself, which no scientific description, or representation, can replace. Perhaps, then, the "unconsidered expressions of our delight" are nearer to absolute truth, in their "unconscious affirmations" of existence—Stevens' "pure being." Scientific description sees "mechanically," giving us nothing new, but only an object "mechanically daguerreotyped on our eyes, but a true description growing out [of] the perception and appreciation of it is itself a new fact, never to be daguerreotyped, indicating the highest quality of the plant,—its relation to man . . ." (XIV:117–18; 10/13/60). The closer we are to the object itself, the more completely we can shed the measured or scientific account: we are not distracted from the thing itself to the system or arrangement, our hierarchy of "captains" and "lieutenants"—though this is, as a society, what we reward. Instead, Thoreau conceived someone (himself, surely) "Who describes the most familiar object with a zest and vividness of imagery as if he saw it for the first time, the novelty consisting not in the strangeness of the object, but in the new and clearer perception of it" (XIV:120). His meditations were taking him not just into the ecology of nature, but into an apperception of the ecology of mind. "My thought is a part of the meaning of the world . . ." (IV:410). A thought becomes itself a "new fact," entering the economy of the system and altering it—or as he wrote in 1842, "I dont [*sic*] know how much I assist in the economy of nature when I declare a fact— Is it not an important part in the history of the flow that I tell my friend where I found it—" (1:383; 3/20/42). Or as he wondered in 1840, "Do not thoughts and men's lives enrich the earth and change the aspect of things as much as a new growth of wood?" (1:147; 7/3/40).

Through his initial engagement with all the particulars of his environment Thoreau learned how to *see* "new worlds," and he learned that in seeing—in the eye's intention, or the eye's side vision—he did not just see but *created* his world, even as he was created by it. Doing this required the sustained energy and attention that he called, simply, being "awake": "Morning is when I am awake and there is a dawn in me," as he wrote in *Walden*; "To be awake is to be alive." We must learn to "reawaken" ourselves "by an infinite expectation of the dawn," and so we can elevate our lives "by a conscious endeavor," carving and painting "the very atmosphere and medium through which we look, which morally we can do" (WA 90). So *Walden* became his experiment in the making of life, in exploring just how far "The universe constantly and obediently *answers* to our conceptions . . ." (97; emphasis added). When, toward the end of the book, Thoreau asks, "Why do precisely these objects which we behold make a world?"—Why is man neighbored by "just these species of animals . . .?" (225), we recognize this as the familiar Adamic theme: the heroically innocent American sweeping the world clean for a new beginning, baptizing it anew in the purified waters of the cleansed consciousness.[6] But for Thoreau, for all its world-making power the consciousness can never be cleansed; what makes his writing so interesting is his discovery, which he is too scrupulous to ignore, that preexisting ideas always control the mind, that this tendency finally is not wrong but inevitable, and should be not denied but educated. Nor can the world ever be swept clean, for we are "neighbored" by other consciousnesses than our own. The pull of these twin resistances—the structures of our own perception, knowledge, and ideology, which organize our vision, and the fantastic complexity of an external world that may answer to us but always keeps something in reserve—generated in Thoreau a dialectical understanding which convinced him that none of our "worlds" are wholly new; all are imbricated with traces of their origins. Yet in the poet-naturalist's consilience or "leaping together" of the opposites of mind and nature, something new *is* made, what for Thoreau is the greatest wonder of all: a wholly new fact in nature.

If we call that "new fact" a work of art, none of this seems surprising; Coleridge and T. S. Eliot have taught us to think of art as something "new" that arises organically out of the "old" to reshape, by some small or great degree, the entirety of human culture. Of course the canonical Thoreau can be read comfortably in this fashion: he is a distinctly "individual" talent within the American "tradition." But instead of reabsorbing Thoreau's works into literary tradition, where the "ideal order" they modify hovers above the material facts of social history, science, and technology, I would like to ask how Thoreau's "new world" ramified into *science*. The poet-naturalist's visions, in Thoreau, are precisely the kind of fact which assists in the "economy of nature," since the poet's task is to yoke material and spiritual, and thereby show the infinite interrelatedness of both realms. So, for instance, at Walden Pond the ecological system of Middlesex County was taken up by Thoreau's own linguistic system, merged with the resources of American literature as *Walden*, and added to language as a new concept, "Walden"; thus contributing a new fact to the economy of Concord, the United States, and "nature" both local and global. So long as it was just a body of water in eastern Massachusetts, no one could encounter it unless he or she were also in eastern Massachusetts; but Thoreau went out, inscribed the pond onto paper, and returned with an inscription with which he hoped to win over allies who would believe in his new facts/truths. Until Thoreau translated "the pond" (or any of his observed and collected facts) onto paper, he could not possibly mobilize it (or them) on his behalf, to persuade others. That is, what gave his writing its persuasive power was, as Matthiessen first noted, its specificity. He grounded his metaphors in facts of his experience, which he rendered into facts of our experience, thereby mustering us (if we accede) to his side.

This interpretation attempts to read Thoreau's endeavors through Bruno Latour's account of the process of science. According to Latour, the distinctive pattern of Western science has been the sustained and collective construction of networks or pathways of agreed-upon "facts," through the process of going "out," collecting and inscribing things to render them "mobile," and bringing them back to central locations where they, and calculations about them, can be assembled, presented, and compared.[7] What made Humboldtian science so attractive and powerful in this particular half century in America was its ability to provide the methodology necessary for success in exactly this endeavor, in its exploratory and network-building phase—literally providing a way to "*see new* things."[8] Recall, for instance, that the expeditions whose reports Thoreau read—such as John Charles Frémont's—were directed toward selecting railroad routes, to expanding the networks of material transmission; and some of this insight worked its way into Thoreau's evocations of the Fitchburg Line. As the network expanded, the effort was increasingly collective—as was, for instance, Agassiz's collecting network, in which Thoreau so memorably participated. Thoreau assisted in mobilizing things—literally packing turtles into boxes—for another's use, transmitting them to Harvard, thereby helping to establish Harvard as New England's own "center of calculation."

The power of the methodology was quickly obvious to Thoreau, but having theories of his own, he wished to be his *own* center, not to collect for another's. He cast himself, like the heroic Humboldt, as both his own explorer and the man at the center of his own calculations; like Humboldt, gathering the inscriptions of others (through his notebooks), generating his own (in his personal collections and his *Journal*), and with them crafting the cosmos. Thoreau applied the scientific methodology he was learning to "higher" ends: not the nationalistic appropriation of territory, or the improved commodification of nature into resources, or the transport of products to market centers, but the establishment of the vital importance of those more volatile elements of experience which could not be carried on the tracks of commodification. This meant rousing people to "awaken" and transform their condition from commodified and alienated isolates to communally responsive individuals. In effect, he rejected his position on the network in order to construct a counternetwork. But science is above all social. The individual scientist who is isolated and without allies ceases to be an effective scientist; if he persists, he will be labeled a crank—or, perhaps, a poet. By refusing to be a cooperative and subordinate worker on another's scientific network, Thoreau sought to develop, using similar methodology, a scientific network of his own—an anarchist science to match his anarchist politics. This was his weakness, in that it cut him off from institutional science and limited what he could do, as well as what he would be perceived as doing.[9] It also was his strength, for in adopting so powerful a model to his own purposes he achieved an eloquence and persuasive power that enabled his work to survive, and ultimately to influence the cultural network in ways that outflanked American institutional science.

Neither poetry nor science alone seemed adequate to Thoreau. Ungrounded poetry was slipshod and callow, and the "mass of men," uneducated in science, were liable to the dangerous ignorance of those who had never learned how to see. Even what they *did* know, however vital and rich in relational understanding, could not be shared because it was limited to their isolated experience. On the other hand, science, though calculated to be public, was thereby shorn of that very vitality:

> Science does not embody all that men know—only what is for men of science. The Woodman tells me how he caught trout in a box trap—how he made his troughs for maple sap of pine logs—& the spouts of sumack or white ash which have a large pith. [He can relate his facts to human life][10]
>
> The knowledge of an unlearned man is living & luxuriant like a forest—but covered with mosses & lichens and for the most part inaccessible & going to waste—the knowledge of the man of science is like timber collected in yards for public works which stub supports a green sprout here & there—but even this is liable to dry rot. (3:174; 1/7/51)

Thoreau tried to join the "Woodman" and the "man of science" into something new: literary science, perhaps; not literature-and-science but science seen as literature, in its fictive constructions of the world, and literature seen as science, in its operational effectiveness in the world. Thoreau's consilience of an Emersonian insistence on higher or spiritual ends with a Humboldtian, worldly empiricism resulted in not just a new "fact" or a new literary work but an experimental new genre, conceptually avant-garde even in our own time.

Perhaps there could be some science, Thoreau implies, which is *not* suffering from "dry rot." The object of such science would be not the "husk," the specimen pickled and preserved and shipped to Boston alongside apples, huckleberries, and timber. The "effluence" such science cannot perceive and so leaves behind is still out there in the field, in the strong but evanescent relationships between the scientific observer and the subjects of his investigations, and in the interactions among the subjects themselves. The lesson of Thoreau's science would be not division and sterility, which confirmed the mournful descent Thoreau feared into the "evil days." It was instead the lesson of renewal and continuance. For all that was gone, his science taught Thoreau how much yet remained in the beauty, abundance, and complexity of the animal and plant life around Concord. The hills the woodcutters laid bare were instantly reclothed with huckleberries. Nature had long ago anticipated and provided for such an emergency, healing the scar and compensating for the loss with fruits the forest could not produce; so "Nature rewards with unexpected fruits the hand that lays her waste" (NHE 227–28). What was more, the forest "laid waste" around Walden in the early 1850s was growing back. Creation, Thoreau was convinced—even before reading Darwin's *Origin of Species* (1859)—was going on at every moment and all around him. If the evil days *were* coming, perhaps a new kind of science could even fend them off. In the late 1850s, Thoreau turned his investigations to a systematic study of regeneration. The fact that thus flowered into the most compelling truth of all was, at last, the seed; and it was to the dispersion and growth of seeds that Thoreau turned, in the studies of his final years.

Worlds without End: *The Dispersion of Seeds*

One of the enabling assumptions of Coleridgean and of Emersonian romanticism was that matter in itself lacks the power to "organize." That power, which makes dead matter alive or "organic," comes from above nature—"supernatural"—as spirit which circulates through matter, imbreathes or "animates" it, making it body forth the thoughts of the Creator. The creator as poet or hero or scientist bodies forth his own thoughts in alignment with the creative spirit that subtends and circulates through all: command through obedience. To the mind that shapes the universe in solitary splendor belongs the triumphs of the spirit, and the burden of despair, for the creative transformation of the universe into mind begins and ends with the originating self. Thus "intentionality of the eye" organizes the chaos of the universe into meaning, in a momentously powerful technology of control; the power of that control is generated by the force of contending opposites which initiate and sustain its grand progressive dialectic. The polar opposites describe between them the field as a whole, in its totality; the universe is closed and secured against change, even as it is haunted by the irrational chaos it has constructed itself against.

The system seems impermeable, except for one lurking suspicion voiced in Kant but quickly suppressed: what if matter organizes *itself*? What if living nature does not require the power of preexisting mind—whether divine or human—to lift it from chaos, but creates its own quite independent order, forming and building it from within according to its own purposes? If it was not created *for* us, or for anyone, but just is? For one thing, the singleness of a nature ultimately unified by the one and only one would be interrupted, deflected into a vision of the universe more like Humboldt's "rich luxuriance of living nature and the mingled web of free and restricted natural forces" (*Cosmos* I:79). For another, man would be deflected from the organizing center to somewhere on the side—indeed, there would be no clear center, only a shifting one responsive to one perspective or another. Humboldt organized his *Cosmos* to emphasize this point: he deliberately began his narrative with the stars, worlds beyond earth, to displace not just man but all things terrestrial from our illusory center. "Here, therefore, we do not proceed from the subjective point of view of human interests. The terrestrial must be treated only as a part, subject to the whole. The view of nature ought to be grand and free, uninfluenced by motives of proximity, social sympathy, or relative utility" (*Cosmos* I:83).

As Thoreau stood on the sandbank of the Deep Cut, admiring the thawing forms that flowed down its sides, he could look in both directions. The released energy of the earth flowed into the forms of life:

> Innumerable little streams overlap and interlace one with another, exhibiting a sort of hybrid product, which obeys half way the law of currents, and half way that of vegetation. As it flows it takes the forms of sappy leaves or vines, making heaps of pulpy sprays a foot or more in depth, and resembling, as you look down on them, the laciniated lobed and imbricated thalluses of some lichens; or you are reminded of coral, of leopards' paws or birds' feet, of brains or lungs or bowels, and excrements of all kinds. (WA 305)

The animation is a direct creation of the sun, as he can see; the shaded bank will remain "inert," while on the sunny bank the luxuriant foliage forms are "springing into existence." It is as if he were in "the laboratory of the Artist who made the world and me," who was "with excess of energy strewing his fresh designs about."[11] Here near the "vitals of the globe" he can see "in the very sands an anticipation of the vegetable leaf. No wonder that the earth expresses itself outwardly in leaves, it so labors with the idea inwardly. The atoms have already learned this law, and are pregnant by it" (306). The law is the law of the leaf, in an insight recalling his Goethean ghost leaves from twelve years before: "Even ice begins with delicate crystal leaves" (307); "The Maker of this earth but patented a leaf" (308). The flowing sand forms into fingers and blood vessels, arteries that "glance like lightning" from stage to stage before being swallowed up into the deltoid fan at their base. Thoreau imagines the human body expanding and flowing into these rivers of life, boundaries dissolving, matter yielding to the flow of energy, poetry, life:

> The earth is not a mere fragment of dead history, stratum upon stratum like the leaves of a book, to be studied by geologists and antiquaries chiefly, but living poetry like the leaves of a tree, which precede flowers and fruit,—not a fossil earth, but a living earth; compared with whose great central life all animal and vegetable life is merely parasitic.

We—the earth, ourselves, our institutions—are "plastic like clay in the hands of the potter" (309).[12]

In the midst of this ecstatic vision—a Thoreauvian answer to Emerson's transparent eyeball—Thoreau sounds one small note of sharp-eyed difference: "It is wonderful how rapidly yet perfectly the sand organizes itself as it flows, using the best material its mass affords to form the sharp edges of its channel" (307). The flowing sand is organized by the single law of the leaf, but somehow it also crafts itself as it descends, selecting material with which to turn the boundaries of its forms as it translates itself from form to form, and finally to the rippled stasis at Thoreau's feet.

In the workshop of nature, central law was organizing raw matter into the forms of life—yet perhaps even raw matter had some power to "organize itself" as it was carried to its Maker's end. In the sandbank passage, earth's poetry was articulated in Thoreau's poetry, as his mind took the sandbank up into spirit and rendered its truth. This was a distinct and highly particularized phenomenon, readily linked with one of the oldest patterns in Thoreau's thought, and he made the most of it. Meanwhile in his *Journal* other, newer patterns were emerging—and matter itself became an acute problem, as the *Journal* amassed itself around his daily practice of studying, walking, and writing. Apart from the value of the *Journal* as an accumulating timescape of experience in the land, coherent by virtue of its totality, smaller patterns were emerging and clustering around phenomena less conducive to literary convention's single convergent law: continuance, rather than narrative closure; boundary breakdowns, as wild nature rooted itself in the street gutter and domestic productions ran wild; new boundary formations, as leaves didn't flow but ripened and fell like fruit. Such observations cued Thoreau into acts of attention which, instead of spiraling him up and beyond the material world, lured him ever more deeply into it.[13]

It had all started with that call to vigilance: if you wait for something to attract your attention, "you are not interested at all about it, and probably will never see it" (DI 127). Nature may call, but you will never hear it until you attend, until you *look* for a thing instead of waiting for it. It amazed Thoreau how many things had been so little attended to. And once he was looking with intent, he found himself a practitioner of science: observation led to question, question to theory, theory to new observation. It was a self-conscious process, as when Thoreau noticed how the plumes of the pitch pines were gnawed off every fall. Mere noticing was not enough: "I resolved last fall to look into the matter." Accordingly, one night, he thinks it over, and reasons that squirrels must gnaw off the twigs "to come at the cones, and also to make them more portable. I had no sooner thought this out than I as good as knew it." He returns to the forest, looks for twigs, observes the signs they have indeed been cut off, carried away, and collected together by squirrels for ease of transport, consumption, and storage. "Thus, my theory was confirmed by observation" (28–29).

It's a simple case, but that's what makes it typical of Thoreau: though so obvious, no one had ever thought to look before. It "called" but no one heard. Thoreau's emerging questions are exactly of this type: the obvious, mundane, and trivial, so easily and always dismissed as "*little* things" (NHE 211). Such as: why do just *these* plants and animals live exactly *here*? Such a question—the distribution of local inhabitants, their relations to past inhabitants—had started Charles Darwin "patiently accumulating and reflecting on all sorts of facts which could possibly have any bearing on it": result, the origin of *Origin of Species* (1). While Darwin had ranged widely, Thoreau used the centering device of Concord, relating everything to "our village" and using labels of human ownership: Lee's Cliff, Beck Stow's Swamp. The more global your question, the more local must be your address. Thoreau's question became, not what unifying law governed the distribution of life, but how do plants and animals distribute themselves across the landscape? It was a Humboldtian, a Kantian, move: we cannot address the law or system until we understand what the components are and how they interact.

Darwin's attention had been directed by unexpected changes in space, while Thoreau's was directed by unanticipated changes in time: he observed patterns of growth shifting. The landscape exhibited a history of change that, unlike geological change, he could observe within his lifetime. A favorite blackberry field imperceptibly metamorphosed into a pitch pine wood—which perhaps he would "survey and lot off for a wood auction, and see the choppers at their work" (DI 33). An open grassy field east of the Deep Cut turned into "Thrush Alley," a favorite walk in deep pinewood where he hears the wood thrush sing in the shade (34). In his experience, human action (including his) was a regular part of the process of change. In *Walden* he had mourned that since he left its shores "the woodchoppers have still farther laid them waste," ending his rambles through the woods and silencing his Muse: "How can you expect the birds to sing when their groves are cut down?" (192). But two paragraphs later the destruction lays the conditions for hope: "where a forest was cut down last winter another is springing up by its shore as lustily as ever" (193). In the millennial narrative, earth is coming to an end, the land is "laid waste," Apocalypse darkens the sky. But, as Thoreau marvels over and over in the 1850s, the world didn't end, couldn't be coming to an end. Nature refused the Christian narrative, incorporated destruction into continual regeneration. A white birch springs up in the gutter on main street, and Thoreau sees how quickly the forest "would prevail here again if the village were deserted" (45). His attention turns to what comes *after*, even after ripeness and fruition: dispersal, circulation, renewal. Not death, but "succession." The *agent* of succession becomes the key, a material metaphor, matter which fables before his eyes: the seed. "As time elapses and the resources from which our forests have been supplied fail, we too shall of necessity be more and more convinced of the significance of the seed" (DI 23–24).

As Thoreau admitted in a letter of 1856, "I am drawing a rather long bow."[14] It would take years of dedicated labor before he could have much to show. But by 1860, he was forming the mass of writing around a series of clusters, and beginning to draft his answer to what would come after *Walden*. At his death he left 631 pages of manuscript on "Wild Fruits," thousands of pages of notes, and a 354-page first draft for a book with the working title *The Dispersion of Seeds*. Unpublished for generations, this manuscript now forms the bulk of the "new" Thoreau volume, *Faith in a Seed*—a significant addition to the Thoreau canon, and one which makes clearer and more accessible the nature of Thoreau's natural history studies.[15]

In these studies, the dynamic principle that drives nature remains the same, but the singular center is dispersed and distributed across multiple lines of connection. "Law" has given way to pattern. In *Walden* Thoreau parodied the philosopher as lawgiver: the pompous "Hermit" advises the "Poet" where to dig for fish bait, "for I have found the increase of fair bait to be very nearly as the squares of the distances" (224). Compare the hard-earned observation Thoreau makes in *Dispersion*: "I have many times measured the direct distance on a snowy field from the outmost pine seed to the nearest pine to windward, and found it equal to the breadth of the widest pasture" (27). The measure of a pine seed's blown distance over snow gives the measure of maximum pasture width: pasture size is not random but the result of the interaction, in real space, between a tree, its seed, the snow field, and the wind—all made commensurate by Thoreau's act of measure, which relates or makes relative the apparently arbitrary. Measurement has in this case related two networks—tree distance, or surface topography, and the distance a pine seed can blow when it has the snow on which to "sled" (27). Helped along like this, a pine seed can perhaps travel many miles from its source, and pines be spread from Concord to the far end of the continent. Why is this important? This small fact points to the multitude of networks that interconnect everything in nature (including man). The basic question is "how the seed is transported from where it grows to where it is planted," and the principle "agents" are wind, water, and animals (24). There is no longer just the one network, but as many networks as the combined ingenuity of these four elements can devise—not all roads lead to Boston. Their combinations will account for the nature and distribution of life on the face of the earth.

The plant universe divides in two—not a polar dualism this time, but a distributional distinction, along the uniting factor of transport. Some seeds, such as pine, birch, and maple, have built-in wings and fly and plant themselves. Other seeds are heavy and lack wings, and so borrow the wings or feet of animals and are planted by them. Cherries, for instance, lack vegetable wings, so "Nature has impelled the thrush tribe to take them into their bills and fly away with them; and they are winged in another sense, and more effectually than the seed of pines, for these are carried even against the wind" (68). As this suggests, these two classes correlate with two general patterns of dispersal and growth: windblown seeds form distinct, confined patches or stands, as the seeds were blown altogether; planted seeds form irregularly bounded woods of varying size and purity (163). Blown seeds are distributed everywhere, but spring up only if conditions are favorable (28); heavier, planted seeds encounter certain constraints (they go only where the animals go), but also certain advantages (animals often put them exactly where they want to be).

A million seeds blow everywhere so that one may grow. Such fecundity proves nature's economy is not economical but spendthrift, extravagant: "apparently only one in a million gets to be a shrub or tree. Nevertheless, that suffices; and Nature's purpose is completely answered" (61). This is a sign of health: "I love to see that Nature is so rife with life that myriads can be afforded to be sacrificed and suffered to prey on one another," Thoreau had written in *Walden* (318). "You must try a thousand themes before you find the right one—as nature makes a thousand acorns to get one oak" (4:41; 9/4/51). He is agog at the calculations of catastrophe if such vitality were unleashed. If every white willow seed became a tree, "in a few years the entire mass of the planet would be converted into willow woods, which is not Nature's design" (DI 61). In nature's design, mass profusion entails mass destruction—the more of the one, the more of the other, as Darwin detailed. From the sensible utilitarian view that seeds exist only for reproduction, such proliferation can only seem wasteful. But nature knows better: "She knows that seeds have many other uses than to reproduce their kind. If every acorn of this year's crop is destroyed, or the pines bear no seed, never fear. She has more years to come" (37). Nature's economy depends on the snowballing of production and circulation: to spend is to indebt, to entail and interlink, to proliferate in connections.[16] So seeds are also for squirrels, redpolls, mice, jays, foxes—each the center of a universe, pursuing business, unwittingly complicit in the overall design: "the consumer is compelled to be at the same time the disperser and planter, and this is the tax which he pays to Nature" (114). Even "the most ragged and idle loafer or beggar may be of some use in the economy of Nature, if he will only keep moving" (97). Just keep circulating. The three basic agents of transport multiply exponentially: wind from all directions and at all speeds, water from dew to torrent, steam to snow; every living being crossing paths across every inch of ground at every height, creating a vast hum of activity just below the level of our hearing. The interpenetration of plant and animal worlds continues into Thoreau's metaphors: cherries grow thrush wings, birch seeds are "tiny brown butterflies" (42), maple seeds are veined "like green moths" (50), pitch pine seeds blow like schools of brown, "deep-bellied" fish (25).

Out of this "blooming, buzzing confusion" emerge patterns that reveal a deeper structure. Birches often grow in parallel lines, because a spring freshet carried the seeds and deposited them along its edge, or the seeds lodged "in the parallel waving hollows of the snow" (44). Elms border streams since that is where green rafts of the seeds have lodged (54), and willows cluster on muddy banks because the seed down forms a white scum on the water in which the seeds germinate, and which deposits the seedlings on bare mud as the water level drops (62). Clumps of maples, elms, or ash grow around rocks in river meadows because the rock "first detained the floating seed, protected the young trees, and now preserves the very soil in which they grow" (55). Maple seeds will not spring up in the grass of a bare pasture, but plow the ground and the seeds which land there can catch and germinate; and so a plowed field can suddenly turn into a maple wood (52). Similarly, "lay bare any spot in our woods, however sandy, by a railroad cutting for instance," or where frost keeps other trees out, and a willow or poplar will "plant itself there" (59). Their downy seeds, like those of milkweed, settle into hollows "where there is a lull of the wind" (59); causeways, fence lines, and embankments catch the seed and so grow natural willow hedges along their length (60). The river shores of the barren northern prairies are populated by willows and aspens—the very trees that are the first to spring up on burnt lands:

> It is remarkable that just those trees whose seeds are the finest and lightest should be the most widely dispersed—the pioneers among trees, as it were, especially in more northern and barren regions. . . . while the heavy-seeded trees for which they may prepare the way are comparatively slow to spread themselves. (58)

So, for instance, though birch and alder are closely allied and their seeds are similar, the tiny brown winged birch seeds float high and spread everywhere, for "how many hundreds of miles" (43)—even to the far north, where a burned evergreen wood will spring up in a birch forest "as if by magic" (46). Alder seeds, however, are not winged but flattened, larger and heavier: "There is, of course, less need that they be winged, since they grow along streams or in wet places, whither their seeds may be floated in freshets; but the birches, though they have a wide range, grow chiefly in dry soil, often on the tops of dry hills" (48–49). Seed form correlates directly with plant habitat. Thoreau evidences the only alder that grows high on mountains: it alone "has winged seeds, apparently in order that the seeds may be spread from one ravine to another and also attain to higher levels" (49).

Everywhere Thoreau looks, he learns to read such relationships between seed form, plant habitat, growth patterns, and geographical distribution: each creates the other. In April 1856, George Hubbard remarked of some pines that "if they were cut down oaks would spring up, and sure enough," Thoreau confirms, across the road where Loring had just cut down his white pines, the ground was covered with oaks (VIII:315; 4/28/56). He has his theory in place within days. A few weeks later, when he was doing some surveying for John Hosmer, the old man "who had been buying and selling woodlots all his life" asked Thoreau if he could tell "how it happened that when a pine wood was cut down an oak one commonly sprang up, and vice versa" (VIII:363; 6/3/56; DI 104). It so happened by then that he *could*, and his answer became the heart of "The Succession of Forest Trees" and from there, *Dispersion of Seeds*: because of the interaction of winged seeds and heavy seeds and nuts.

The wind conveys winged pine seeds into hardwoods and onto open lands, where, as he slowly proves to his satisfaction, the seedlings survive far better than under the dense shade of their own kind. Meanwhile "the squirrels and other animals are conveying the seeds of oaks and walnuts into the pine woods, and thus a rotation of crops is kept up" (DI 106). The ramifications fill many pages of his nascent book, as he develops this fundamental insight into the key that unlocks the history of the mixed and patchwork forests and fields around Concord.[17] "Thus you can unroll the rotten papyrus on which the history of the Concord forest is written" (169).

As Thoreau teases the patterns into meaning, he realizes how careless and wasteful his neighbors and their ancestors have been, through sheer ignorance. "Our woodlots, of course, have a history, and we may often recover it for a hundred years back, though we *do* not. . . . Yet if we attended more to the history of our woodlots, we should manage them more wisely" (164). In normal usage, as pitch pines seed into open pastures, they are bushwhacked and cattle are turned out to graze on the land. Both cows and bushwhacking break down the pines, until in fifteen or twenty years, having "suffered terribly," they stubbornly push themselves above the farmer's head, finally commanding his astonished attention as trees, whereupon he drives out the cattle, fences in the trees, and declares it a woodlot—having nearly ruined it in the meantime. "What shall we say to that management that halts between two courses? Does neither this nor that but botches both?" (171–72). So in New England "we have thus both poor pastures and poor forests," while in Old England, where resources are dearer, they have taken "great pains to learn how to create forests" (172).

Thoreau's admonition bears on one of the themes of the book: that American self-involvement has bankrupted the land. There is no pure nature created gloriously apart from the American to supply him with an endless fountain of resources—be they material, aesthetic, or spiritual. We blithely assume that when one forest is cut, another will "as a matter of course" spring up, and so, "never troubling ourselves about the succession," we don't anticipate the time when succession will fail and we shall have to learn the connection of seeds with trees (23). Instead of a nature pure, boundless, and Edenic, Thoreau proposes a nature thoroughly hybrid, limited, and post-lapsarian: he who has truly *seen* the new world has seen that no world is truly new or belongs to us alone. Nature that is endlessly new is also endlessly old, imbricated with past lives and ages; Europeans who came to the "New World" were folded into American nature just like the Indians they displaced, who also burned and cultivated the land, creating a hybrid landscape that mixes nature's design with layers of human purpose, the nature of the oak and pine forest with "the necessities or whims of John and Sally and Jonas, to whom it has descended" (170). In *this* new world, knowledge is neither the evil nor the power, but the lifeline to survival. Chestnut timber, for example, is disappearing so rapidly "that there is danger, if we do not take unusual care, that this tree will become extinct here" (126). If only we understood that chestnuts do not come from God or out of nowhere but from chestnut seeds, we could replant chestnut forests. Nor is the problem confined to a single species: "The noblest trees, and those which it took the longest to produce, and which are the longest lived—as chestnuts, hickories, and oaks—are the first to become extinct under our present system and are the hardest to reproduce, and their place is taken by pines and birches, of feebler growth than the primitive pines and birches, for want of a change of soil" (130–31). We do not command, but *survive*, by obeying nature.

Accordingly, Thoreau's mission to educate is urgent, and his advice is specific. Forests must be of diverse kinds, planted in alternating bands and incorporating "countless trees in every stage of growth" (131–32), or they will not thrive.[18] Nature has a design and it is not ours—but we are having to learn it the hard way. Just as the English planters, through "very extensive and thorough experiments," have learned that the *only* way to raise oaks is by planting them under pines! And they had the audacity to declare this their "discovery," and to lay down as a principle that in cultivation art and design must regulate every step, with "'nothing whatever, or, at least, as little as possible, left to unassisted nature'" (125). Take another simple, practical problem: how to store nuts over the winter. On January 10, Thoreau buys some nuts at the store, and collects some from under the leaf mold in the forest. Over half the store nuts are spoiled; the natural nuts are every one wholesome. "Nature knows how to pack them best. They were still plump and tender"—and ready to sprout in the spring. "Would it not be well to consult with Nature in the outset?—for she is the most extensive and experienced planter of us all, not excepting the dukes of Athol" (134).

It all seems like reasonable, sound, practical Yankee advice. But Thoreau is after more than a practical handbook to resource management. Why insist so repeatedly and firmly that every plant comes from a *seed?* It would occur to few of today's readers to argue with him, but it did occur to no less a one than Horace Greeley, who published Thoreau's "Succession of Forest Trees" in the *New York Weekly Tribune* on October 8, 1860, then wrote Thoreau a letter challenging him to reconcile his theory that trees are never generated spontaneously with the fact that after fire devastates pine forest, "up springs a new and thick growth of White Birch—a tree not before known there." Greeley printed his letter, and Thoreau's reply, on February 2, 1861.[19] Thoreau framed his argument in *Dispersion* to answer just such objections. Why, for instance, if fireweed is so spontaneously generated, is it "not so produced in Europe as well as in America? Of course, the Canada thistle is spontaneously generated just as much, yet why was it not generated here until the seed had come from Europe?" (89).[20]

His insistence on the material agency of the seed made Thoreau one of Darwin's earliest and strongest allies. Even before the publication of *Origin of Species*, Thoreau was working along lines that can only be described as Darwinian.[21] The first and fundamental move toward Darwin's theory of evolution was to reject as "erroneous" the view "which most naturalists entertain, and which I formerly entertained—namely, that each species has been independently created" (*Origin* 6). Separate creation had solved the obvious problems resulting from the assertion that all life had radiated from one central point—problems like those to which Thoreau pointed in his discussion of fireweed and thistle—and did so in congruence with the dominant belief in a dead nature that had to be willed into life from outside. But Thoreau's emerging vision was of a living nature that willed *itself*, and so relied on material networks of circulation and exchange to proliferate and spread—as it did, to all parts of the globe.

The principle of life did not skip and flow independent of matter, but was transmitted by matter—through seeds: "A seed, which is a plant or tree in embryo, which has the principle of growth, of life, in it, is more important in my eyes, and in the economy of Nature, than the diamond of Kohinoor" (XIV:334; 3/22/61). "Life" became, not fragile and dependent, but sturdy and independent—profligate, stubborn, tenacious, like the willow whose twigs broke off at the least touch into Thoreau's boat, and which he pitied for being "made so brittle." Then he learned the twigs will sprout, so are "shed like seeds which float away and plant themselves in the first bank on which they lodge." Where he had once pitied the tree, "now I admired its invulnerability" (DI 63). Tradition was in error. Experience with living reality showed him the willow tree was not the symbol of sorrow and despairing love, but "an emblem of triumphant love and sympathy with all Nature. It may droop, it is so lithe, but it never weeps" (64).

The seed was an efficient explanatory principle, but it also gave Thoreau the tough, wily, independent, and astonishingly beautiful nature that could be, like a human friend, separate enough to show him mystery, yet close enough to involve him as partner and companion, in "sympathy" and even love. The seed gave Thoreau the principle of connection, of present with past and future, of all the globe with the common and local inhabitants of Concord, the designs of men with the design of nature. To Thoreau, the seed as material cause, the "little strokes" that *raise* great oaks (36–37), was still more magical than the false "magic" of spontaneous generation. The sense of the marvelous permeates his writing, and flashes into moments of joy and admiration. He finds the year's newest pitch pine seedling in a pasture, and wonders at the infant tree:

> It was, as it were, a little green star with many rays, half an inch in diameter, lifted an inch and a half above the ground on a slender stem. What a feeble beginning for so long-lived a tree! By the next year it will be a star of greater magnitude, and in a few years, if not disturbed, these seedlings will alter the face of Nature here. (27)

Not only do the mechanisms delight him, but he is moved by the way such an interwebbed nature anticipates itself, opposites interpenetrating to create beauty. The "strong, prickly, and pitchy chest" of the pitch pine cone contains about a hundred seeds in pairs. A membrane or "wing" clasps each seed "like a caged bird holding the seed in its bill and waiting till it shall be released that it may fly away with and plant it. / For already some rumor of the wind has penetrated to this cell, and preparation has been made to meet and use it." Thoreau marvels further: "This wing is so independent of the seed that you can take the latter out and spring it in again, as you do a watch crystal" (25). The wonder and delight in such precision echo Paley, except in Thoreau there is no sense that the wondrous contrivances bespeak the power and ingenuity of the Contriver, but rather the astonishing achievement wrought by the cooperation of independent agents all using each other to accomplish their ends, in a harmonious universe presided over by nature's wise design. The veined mothlike wing of the maple is essential to that design, for if the seed is to reproduce under favorable conditions, nature "must also secure it those favorable conditions" (XIV:334; 3/22/61). So the wing develops even when

> the seed is abortive—Nature being, you would say, more sure to provide the means of transporting the seed than to provide the seed to be transported. In other words, a beautiful thin sack is woven around the seed, with a handle to it such as the wind can take hold of, and it is then committed to the wind, expressly that it may transport the seed and extend the range of the species. (DI 50)

Entire worlds "spring" from the tiniest seed. Thoreau calculates the seed of the earth, were it proportional to that of the willow, "would have been equal to a globe less than two and a half miles in diameter, which might lie on about one-tenth the surface of this town" (67).

In this new order, every phenomenon in nature has a cause or connective principle. Knowing the language allowed Thoreau to understand new sentences, as for instance when he notices that sugar maples retain some of their seed so late, and "suspect[s] that their distribution may be somewhat aided by the snow" (52). Given chains of connection like this, Thoreau has a world that "springs" into meaning, and a meaning that can be read in the real world. Things are joined together not by penetration, dissolution, etherialization, by the flow of spirit precipitating into shadow forms, but by contact, a chain of contacts, multiple chains, creating an order in which man is neither ruler nor intruder but incorporated into the "economy" as casually as the beggar-tick burr borrows a pant-leg to hitchhike into town, or as the oak seedlings, coiled and ready to spring into being as a forest, borrow the material needs of the woodlot owner who logs and sells the pine trees.

Only our self-centered view has prevented us from seeing all this. Intrigued with a blue butterfly, for instance, Thoreau grumbles that the only insects of which science takes account are those that are noxious or injurious to vegetation: "Though God may have pronounced his work good, we ask, 'Is it not poisonous?'" (XII:170–71; 5/1/59). "How little observed are the fruits we do not use! How few attend to the ripening and dispersion of the white-pine seed" (DI 34). Since this is not a simple world in which we are lawgivers, but a complex world we cannot predict, we pay it the less attention:

> When lately the comet was hovering in our northwest horizon, the thistle-down received the greater share of my attention. . . . Astronomers can calculate the orbit of that thistledown called the comet, conveying its nucleus, which may not be so solid as a thistle seed, somewhither; but what astronomer can calculate the orbit of your thistledown and tell where it will deposit its precious freight at last? It may still be travelling when you are sleeping. (87)

While we have been asleep, the whole world has been traveling around us.

The single thistledown, as it rises from Thoreau's hand hundreds of feet into the air "and then passes out of sight eastward" (87), becomes, briefly, a world, one that soon passes beyond his sight. In this plural universe there are new worlds everywhere, concealed only by our ignorance—the ignorance that regards squirrels as "vermin," when it would be "more civilized as well as humane" to honor them in an annual ceremony as the planters of our forests (130). So Thoreau spends some time thinking about a squirrel world: how they "strip and spoil the tree" (but perhaps the trimming is to the tree's benefit?). How they gather in their own October harvest just like the farmer, and even more "sedulously" (39–40). There is a squirrel mind, and squirrel knowledge—"But he does not have to think what he knows" (32). These other beings are utterly at home in their worlds, in a way that Thoreau finds heroic, and endlessly fascinating. He creates moments of intersection and watches them turn away from him, pursuing their own direction, like the milkweed seed he launches which "rises slowly and uncertainly at first," until it catches the strong north wind and is born off, "till at fifty rods off and one hundred feet above the earth, steering south—I lose sight of it" (93).

Not that their worlds are always easy. Thoreau digs up oak and hickory seedlings and learns to his surprise that frail seedlings conceal thick, vigorous roots, "lusty oaken carrots" (117–18). Examination reveals the stubs of old shoots, repeated attempts to lift a tree into a hostile world; one little hickory seedling turns out to have been "at *least* eleven years old" (138). Their perseverance earns Thoreau's respect:

> There are those who write the lives of what they call *self-educated* men, and celebrate the *pursuit* of knowledge under difficulties. It will be very suggestive to such novices just to go and dig up a dozen seedling oaks and hickories, read their biographies, and see what they here contend with. (142)

Many of their difficulties are caused by the farmers who are so careless of "little" things and material causes. "The history of a woodlot is often, if not commonly, here a history of cross-purposes, of steady and consistent endeavor on the part of Nature, of interference and blundering with a glimmering of intelligence at the eleventh hour on the part of the proprietor" (170). Thoreau ends his draft with a story of a landowner who logged his pinewood in the winter of 1859. Thoreau stopped by the following October to "see how the little oaks" were doing. To his "surprise and chagrin," the field had been burned over and planted with winter rye: "What a fool! Here Nature had got everything ready for this emergency, kept them ready for many years—oaks half a dozen years old with fusiform roots full charged and tops already pointing skyward, only waiting to be touched off by the sun—and he thought he knew better. . . ." And so, instead of "an oak wood at once," he will get from his woodlot a bare field, "pine-sick" and shriven of oaks. "So he trifles with Nature," harumphs her closest student (172–73).

One of Darwin's operant principles was "Nature non facit saltum"—there are no gaps in nature. This was an old saw in natural history, one which had aided Lyell's gradualist geology, had driven Jefferson to seek for mastodons in inland North America, and Goethe to seek in the leaf the principle for all vegetable life. But Darwin's question was new: "Why, on the theory of Creation, should this be so? Why should all the parts and organs of many independent beings, each supposed to have been separately created for its proper place in nature, be so invariably linked together by graduated steps?" (*Origin* 194). His answer was that all organisms are related by descent—not metaphorically or metaphysically, but by actual, physical, lineal descent. To communicate this idea, Darwin borrowed an old figure of language and literalized it:

> As buds give rise by growth to fresh buds, and these if vigorous, branch out and overtop on all sides many a feebler branch, so by generation I believe it has been with the great Tree of Life, which fills with its dead and broken branches the crust of the earth, and covers the surface with its ever branching and beautiful ramifications. (130)

By the "Tree of Life," all living beings are related through common ancestry. But Darwin's nature doesn't really look like a noble tree. It looks more like a tangled bank:

> It is interesting to contemplate an entangled bank, clothed with many plants of many kinds, with birds singing on the bushes, with various insects flitting about, and with worms crawling through the damp earth, and to reflect that these elaborately constructed forms, so different from each other, and dependent on each other in so complex a manner, have all been produced by laws acting around us. (489)

The laws are simple and discoverable by anyone. Darwin's great argument builds through the careful wrangling of commonplace and familiar facts, such as the variability of domestic dogs and pigeons; the family resemblance of child to parent; fecundity in a world where so few can survive. Given the plain and commonplace, Darwin like a good traveler defamiliarizes it, then with surefooted steps leads the reader through this strange new landscape to the inescapable and extraordinary conclusion: life created itself. Its own power responds to the force of circumstances by proliferating, in an unbroken chain, into all the beings in and on the earth.[22] Darwin closed his book by proclaiming the beauty of his vision, even as he dreaded the way it would be received:

> There is grandeur in this view of life, with its several powers, having been originally breathed into a few forms or into one; and that, whilst this planet has gone cycling on according to the fixed law of gravity, from so simple a beginning endless forms most beautiful and most wonderful have been, and are being, evolved. (*Origin* 490)

Thoreau had already intuited something like Darwin's Tree of Life in the *Walden* sandbank passage, where the descending sand mixes, proliferates, and transforms itself into all the forms of life, in a continuous, unbroken flow. Yet the pre-Darwinian evolutionary ideas on which Thoreau's vision drew were not physical but *meta*physical: Life assumed and discarded form after form, in its upward yearning to reach man and return to God. What shocked nineteenth-century readers was not Darwin's theory of evolution (already an old idea), but his assertion that the old *metaphor* of evolution was the literal truth. A continuous chain of physical contact linked every single organic form. Thus our classifications express true genealogies: "community of descent is the hidden bond which naturalists have been unconsciously seeking, and not some unknown plan of creation, or the enunciation of general propositions, or the mere putting together and separating objects more or less alike" (*Origin* 420). In Thoreau's terms: there was *always* a seed. Make *that* a fact of your understanding, and the whole world would "spring" into being. For Thoreau, the metaphor of the seed had to be a "radical" metaphor, had to be fully spiritual and fully material, simultaneously. In literalizing the metaphor of the seed, Thoreau paralleled Darwin's move in literalizing the metaphoric Tree of Life: both created communities of descent. Thoreau saw science as metaphor; Darwin saw metaphor as science. Having exposed the metaphorical nature of science, they both go on to insist that the metaphor is real after all. Darwin alludes to Goethe: "Naturalists frequently speak of the skull as formed of metamorphosed vertebræ; . . . [and] the stamens and pistils of flowers as metamorphosed leaves"; however, they use such language "only in a metaphorical sense. . . . On my view these terms may be used literally" (438–39). Gillian Beer points out that "in the process of Darwin's thought, one movement is constantly repeated: the impulse to substantiate metaphor and particularly to find a real place in the natural order for older mythological expressions."[23] As Thoreau would put it, to so state facts that they may be mythological. If only we truly understood it, the material world would fabulate, and the truly fabulous would exist in fact. Thus would the whole world signify.

Thoreau read *On the Origin of Species* within weeks of its publication in London on November 24, 1859.[24] By February 1860 he was copying extracts from it into his notebooks, and he spent the next several months expanding his research in directions inspired by Darwin. "Never had Thoreau been so captivated by a project," notes William Howarth.[25] That September he distilled his researches into "The Succession of Forest Trees" for public presentation, and for the next two months he worked daily on gathering still more new material and expanding the "Succession" lecture into *The Dispersion of Seeds*. The work was slowed by the cold he contracted on December 3, 1860, after a day spent counting tree rings in the rain. From that point on his health steadily deteriorated until his death on May 6, 1862; yet even during the journey he took to Minnesota from May to July in 1861, in hopes of regaining his health, he was taking notes that expanded the field of his observations to the northern plains.[26] All this time he was juggling work on *Dispersion* with "Wild Fruits," the enormous project which had been ongoing through the late 1850s and toward which he was amassing, from March 1860 to January 1862, over 750 pages of lists and charts correlating his years of minute seasonal observations into a "Kalendar" of Concord. Exactly how these various projects were related or what might have come of them is unclear, and given the state of the manuscripts, may always remain so. It is more clear that his reading of Darwin changed the tenor of his work, giving it a context, a direction, and a dynamic focus that were taking Thoreau beyond old-fashioned seasonal chronology and rank-and-file natural history. The revolution that Humboldt had started, Darwin was completing.

The *Dispersion* manuscript several times alludes to Darwin on technical matters, but only once discusses more theoretical questions, in a passage musing on ponds; while Darwin needed to account for the origin of species on ocean islands, Thoreau's concern lay in the immigration of species to inland waters. The immediate problem was to account for the lilies and the fish (pouts and pickerel) that had appeared seemingly out of nowhere in Sleepy Hollow Cemetery's newly dug pond. Thoreau, directly applying Darwin, asserts the fish "had undoubtedly come up from the river, slight and shallow as the connexion is," while the lily seeds had been conveyed "by fishes, reptiles, or birds which feed on them" (DI 100–101). Thus he can assert: "If you dig a pond anywhere in our fields you will soon have not only waterfowl, reptiles, and fishes in it, but also the usual water plants, as lilies and so on. You will no sooner have got your pond dug than nature will begin to stock it" (100). Yet he continues to pursue the question: then how did *Pontederia* and spatterdock get to "the little pool at the south end of Beck Stow's," which *lacks* a stream? Perhaps carried by reptiles and birds—but the real question is broader: "*Indeed*, we might as well ask how they got anywhere, for all the pools and fields have been stocked thus, and we are not to suppose as many new creations as pools." How, then, did *any* plant get anywhere? By the same process:

> I think that we are warranted only in supposing that the former was stocked in the same way as the latter and that there was not a sudden new creation, at least since the first. Yet I have no doubt that peculiarities more or less considerable have thus been gradually produced in the lilies thus planted in various pools, in consequence of their various conditions, though they all came originally from one seed. (DI 101; cf. XIV:146–47; 10/18/60)

Thoreau is running Darwin's theory—descent of all life from a single "seed," speciated by variation across isolated populations—through his own observations and conclusions, and finding them perfectly congruent. Darwin provided the theoretical framework toward which Thoreau had been groping, and confirmed Thoreau's intuitive vision of the world: "We find ourselves in a world that is already planted, but is also still being planted as at first" (101). What he learned of the living lilies at Beck Stow's would be just as true of any fossil lilies geologists might unearth: the newest connects with the oldest, and the old is ever new: "The development theory implies a greater vital force in nature, because it is more flexible and accommodating, and equivalent to a sort of constant new creation" (102). Thoreau concludes the discussion by quoting one of Darwin's experiments in which Darwin allowed every seed in three tablespoons of pond mud, collected in February, to germinate and grow. At the end of six months "the plants were of many kinds and were altogether 537 in number; and yet the viscid mud was all contained in a breakfast cup!" (DI 102; *Origin* 386–87).

Darwin had designed this experiment to show how a bit of mud on the foot or beak of a waterbird could convey seeds across great distances, over land or water. As Thoreau said, given the seed, the next question was "how the seed is transported from where it grows to where it is planted" (DI 24). Darwin conducted many investigations of this kind, for he believed such questions were some of the most important, and least well understood, in contemporary natural science:

> No one ought to feel surprise at much remaining as yet unexplained in regard to the origin of species and varieties, if he makes due allowance for our profound ignorance in regard to the mutual relations of all the beings which live around us. Who can explain why one species ranges widely and is very numerous, and why another allied species has a narrow range and is rare? Yet these relations are of the highest importance, for they determine the present welfare, and, as I believe, the future success and modification of every inhabitant of this world. (*Origin* 6)

Questions regarding "the mutual relations of all the beings which live around us" had already preoccupied Thoreau for nearly a decade. He was deeply engaged in meticulous and broad-ranging field research devoted to addressing some of the most vexing, most important, and least comprehended questions of his time (and of ours): the causes of speciation, patterns of distribution and migration, habitat separation, the phenomena of coevolution. These form the center of Darwin's work and writing, a fact overshadowed by the consensus that equates Darwin's name only with the brutal struggle to survive. The consensus overshadowed as well the true method and intent of Thoreau's late work—what Emerson famously called his "broken task."

Stephen Jay Gould, considering the puzzle of Humboldt's eclipse, finds the most telling cause in Darwin's *Origin*—in a painful irony, the student superannuated the teacher, only a few months after Humboldt's death in 1859. Why? In the darker post-Darwinian view, Humboldtian higher harmony became "a scene of competition and struggle." Humboldt's confidence in universal progress became opportunistic local adaptation with no intrinsic upward direction, and Humboldt's multiple harmonious forces became random internal forces in precarious balance with the caprices of environmental change.[27] Yet Darwin's own language of "checks" and "struggle" regularly transmutes into visions of sublime nature. As his discussion verged on Thoreau's own field, Darwin emphasized not Thoreauvian creation, cooperation, or exchange, but the order generated by so "many different checks" that must come into play in the life of any species. But what an extraordinary order this becomes:

> When we look at the plants and bushes clothing an entangled bank, we are tempted to attribute their proportional numbers and kinds to what we call chance. But how false a view is this! Every one has heard that when an American forest is cut down, a very different vegetation springs up. . . .

Yet the second-growth forests growing on ancient Indian burial mounds display the same "beautiful diversity" as the virgin forests:

> What a struggle between the several kinds of trees must here have gone on during long centuries, each annually scattering its seeds by the thousand; what war between insect and insect—between insects, snails, and other animals with birds and beasts of prey—all striving to increase, and all feeding on each other or on the trees or their seeds and seedlings, or on the other plants. . . . Throw up a handful of feathers, and all must fall to the ground according to definite laws; but how simple is this problem compared to the action and reaction of the innumerable plants and animals which have determined, in the course of centuries, the proportional numbers and kinds of trees now growing on the old Indian ruins! (*Origin* 74–75)

Darwin's language of war and defeat sounds glum enough next to Thoreau's language of triumph, joy, and renewal, but even Darwin is swept away by a vision of the beauty and complexity created by the vying forces of life. Conversely, Thoreau does admit the presence of death, from the war of nature in *Walden*'s battle of the ants, to asides about struggling hickories, or the puzzling loss of so many acorns to frost (XIV:149). But in the unending cycle of destruction and renewal, Thoreau refuses death as nature's ultimate truth, instead throwing his rhetorical and imaginative weight into images of vitality. Death, in a move reminiscent of Whitman and Wallace Stevens, is only the necessary mother of beauty.

As Gould himself acknowledges, the bleakness of the Darwinian worldview was not shared by Darwin himself, who accepted that "Nature simply is what she is; nature does not exist for our delectation, our moral instruction or our pleasure."[28] Indeed, *Origin*, especially in its unbridled first edition, brims with life and wonder. Despite the regular intonements of "the war of nature . . . famine and death" (490), its metaphoric and emotional heart is Darwin's awe before the world's beauty. He reiterates, typically, "I can see no limit to this power, in slowly and beautifully adapting each form to the most complex relations of life" (469). The lesson Darwin himself carries away is a sublime confidence:

> As all the living forms of life are the lineal descendants of those which lived long before the Silurian epoch, we may feel certain that the ordinary succession by generation has never once been broken, and that no cataclysm has desolated the whole world. Hence we may look with some confidence to a secure future of equally inappreciable length. (489)

Thoreau launches his thoughts after the milkweed seed that has sailed aloft and is "steering south":

> Thus, from generation to generation it goes bounding over lakes and woods and mountains. . . . I am interested in the fate or success of every such venture which the autumn sends forth. And for this end these silken streamers have been perfecting themselves all summer, snugly packed in this light chest, a perfect adaptation to this end—a prophecy not only of the fall, but of future springs. Who could believe in prophecies of Daniel or of Miller that the world would end this summer, while one milkweed with faith matured its seeds? (DI 93)

For Thoreau, nature's manifold views held no room for apocalyptic visions.

Where Darwin saw war and the feeding of all upon all, Thoreau detailed cooperation, collusion, a community bound together by exchange and prospering through extravagance and excess. Thoreau read nature, even Darwin's nature, through Humboldtian eyes. To speculate: is it possible, just conceivable, that had Thoreau lived he could have offered to his time a thoroughly up-to-date and Darwinian view of nature that might have defied the canonical determinism that cut Darwin's joy to its own grim and dualistic measure? Could Thoreau, in some slight degree, have complicated the accelerating rupture that drove the chill and competitive world of "Darwinian" nature apart from the warm and womanly hearth of the humanists and artists?

Speculations aside, Thoreau did enter upon a campaign to inform and enlighten the public in general, and the masters of the land in particular. If the forests he loved were being bound over by ignorance to loss and depletion, he could intervene and educate, teach others how to read the history of the forests. The end would be community action for the improvement of all, just as he recommended in the essay salvaged from "Wild Fruits" as "Huckleberries." What to do with the farmer fool who ruins his land? "That he should call himself an agriculturalist! He needs to have a guardian placed over him. . . . Forest wardens should be appointed by the town—overseers of poor husbandmen" (DI 173). Thus concludes his long-unpublished first draft: this Thoreau would have fathered the United States Forest Service, in the Department of Agriculture![29]

Though the idea died with him, he did succeed in presenting publicly the "first fruit" of his researches, and if dispersion be any measure, he had embarked in a most promising way: "Succession" had the widest circulation of any essay within Thoreau's lifetime.[30] This is significant, for the situation he faced in writing and presenting "The Succession of Forest Trees" had specific social demands which were new to Thoreau, and which put him in an anomalous position. He who had exiled himself so grandly from the social network of Agassiz and American science now needed to reconnect with it, for this was the network of knowledge and power, and Thoreau wanted his new knowledge to acquire the power to change the face of nature in America—like his little pitch pine stars—beginning like them with the battered pastures of Concord. Thoreau had to present himself believably as a scientist, or else stand by and watch his forests die.

The Transcendentalist at the Cattle Show: Thoreau's Ironic Science

In Thoreau's new science, ironies abound: nothing is as it seems. Pine forests are really oak forests. Willow, from time immemorial the symbol of sorrow, is really the emblem of triumph and joy. The "downy atoms" of willow seed, "which strike your cheek without your being conscious of it, may come to be pollards five feet in diameter" (DI 57). When his neighbor expressed a wish for a quantity of birches, Thoreau dropped by and extracted "one hundred birch trees" from his pocket (47). On a rainy day in September 1860, during the annual cattle show, when Thoreau stood in the Concord Town Hall before the assembled membership of the Middlesex Agricultural Society to present a "serious scientific subject," he knew many in his audience thought of him rather as the town crank, an eccentric, a poet, even a woods burner.[31] In a moment steeped in ironies, not the least is that the text in which he most artfully negotiated the difficult passage between poetry and science has fallen between them into obscurity.

Thoreau opened his lecture by squaring off directly against the disciplinary expectations of his immediate audience: "Every man is entitled to come to Cattle-Show, even a transcendentalist. . . ." Just how is the transcendental poet and peripatetic town surveyor to address the substantial landowning citizens of the Middlesex Agricultural Society? Thoreau began by deflecting their mutual awkwardness through a strained joke at his own expense. He recalls to his audience another festival regular, that "weak-minded and whimsical fellow" who for a cane carries a crooked stick, when, we all know, "a straight stick makes the best cane. . . . Or why choose a man to do plain work who is distinguished for his oddity?" Perhaps, he jests disarmingly, his audience thinks they have made the same mistake. Yet he then gently and firmly turns the joke on them: it is *he* who knows the straight path in the woods, who has "several times shown the proprietor the shortest way out of his woodlot" (NHE 72–73). Having thus established his own authority as the local land surveyor (with "title" to speak, if not to property), Thoreau invites their attention to "a purely scientific subject": namely, why is it "that when a pine wood was cut down an oak one commonly sprang up, and *vice versa*" (73). Slowly and carefully, Thoreau unpacks and pieces together his answer: wind conveys pine seeds into the hardwoods and open land, while squirrels, jays, and other animals convey hardwood seeds into the pinewoods. The seedlings of neither will mature in the shade of the other, but cut the woods, and the seedlings, there but unseen all along, will spring up into their own.

As far as Thoreau's science-minded audience was concerned, their speaker negotiated the divide successfully. For all his witticisms, Thoreau took the occasion seriously and delivered a solid lecture. The society's president, ex-governor of Massachusetts George S. Boutwell, concluded the event by congratulating the audience on hearing an address "so plain and practical, and at the same time showing such close and careful study of natural phenomena."[32] So Thoreau's immediate, science-inclined audience agreed that Thoreau's representation of the forest would count as science: they reprinted it as such and even today it is so acknowledged. Accordingly, when it was reprinted in the *Eighth Annual Report of the Massachusetts Board of Agriculture*, the literary frame was deleted,[33] turning Thoreau's work into a serious scientific paper with throwaway literary asides. The deletions helped stabilize this text *as* science, making it legible within the context of other scientific reports.

It has proven more difficult for literary critics to integrate this essay into the context of other literary works. The science tends to be bracketed as "just" science, stabilizing the lecture as a serious literary essay with throwaway scientific asides, which however dominate to such a degree that they distract from the great domain of *imaginative* truth. Howarth remarks, "Ironically, [Thoreau's] contribution to succession theory was ignored by scientists for almost a century, and many later readers have assumed that this 'purely scientific subject'. . . has little value as literature." John Hildebidle, in the midst of careful consideration of "The Succession of Forest Trees" and of *The Dispersion of Seeds*, finds the coexistence in "Succession" of Thoreau the "moralist" and Thoreau the "observer" to be "rather stale and unprofitable"; "Whatever its virtues as science, the lecture as art is something of a disappointment." Joan Burbick, who gives this essay a serious and thoughtful reading, sees it as a failure in "the synthesis of natural causes and of imaginative associations." The seed, Thoreau's most (so to speak) pregnant metaphor of all, is in her words "demetaphorized into a natural cause." The seed is "redefined" as limited and material until the end of the essay, when Thoreau adds "a countermanding postscript" that reassigns the seed to the realm of imaginative, not material cause: "separate systems for the two realms," precisely what Thoreau is trying to avoid.[34]

The essay's dubious status, and Thoreau's own dubious tone in the opening, suggest in small scale the problems this larger project faced: how to make a single, hybrid text coherent to readers with double vision? To make his most radical metaphor viable in *both* realms? If Thoreau failed, it may be because his first priority seemed the more urgent: to convince his audience to listen to him, to accept science from a transcendentalist. Sitting before him in the lecture audience were the very people who manipulated the forests he so loved, and in whose decisions he could hope to intervene. Their expectations of "science" demanded from Thoreau the role of neutral and modest expert, delivering objective scientific truth, and though he might, instead, have read them a poem, that would have sacrificed the occasion to a principle in which he disbelieved, that discourse was merely discourse. That is, this was not a relativistic space of mere words, in which any words would have sufficed; Thoreau, caring deeply about forests, needed words that would influence how his neighbors would construct, literally not just literarily, the forested land which they owned and exploited. Thus he was concerned, as he wrote to Blake, with getting "more precision & authority" into the work that would lead to his 1860 lecture.[35] Facing this powerful constraint, Thoreau disciplined himself as a scientist, and addressed them with authoritative empirical data and precise cause-and-effect explanations.

But Thoreau's larger task was to intervene in the "deadening" discourse of science, and to demonstrate an alternative mode which nevertheless would be successful *as* science—successful, that is, in generating agreement among a particular and powerful community who would then accept and integrate his statements as "fact." This second task might explain why this essay is, after all, such an odd specimen of "scientific" writing, full of jokes and wordplay, asides and parables. One could assume that Thoreau simply didn't know how to reproduce the well-established genre of the scientific paper, or that he simply couldn't help himself and must leaven his text with humor, producing a science so sugarcoated even a transcendentalist could deliver it. I would like, instead, to ask how this essay would look if we take his words seriously, as if he meant what he said.

How is the power of scientific discourse constructed? Through conventions of discourse which Thoreau deliberately attempts to disrupt by employing two techniques, "feedbacking" and "inversion."[36] The first convention of scientific discourse requires that the author suppress her "personal" view and eliminate any suggestion of a connection between the author and the object of study. This convention had already been in place for over 150 years, and as a regular reader of science writing, Thoreau was not only familiar with it but regularly chafed at it: "Ah what a poor dry compilation is the Annual of Scientific Discovery. I trust that observations are made during the year which are not chronicled there" (3:354; 8/5/51). In "feedbacking," the author attempts to disrupt the putative "objectivity" by visibly reconnecting object and subject, emphasizing rather than suppressing her own presence both in the text and in the process of science. Second, the construction of objectivity assumes that the object gives rise to its own representation; it "speaks for itself," and the author is but the transparent channel of truth. "Inversion" disrupts this coupling by reversing it, insisting that it is the representation which precedes and gives rise to the object: we see what we look for. As we have seen, Thoreau was familiar with this principle, calling it the "intentionality of the eye" whereby our knowledge constructs its supporting facts, and basing on it his major critique of conventional scientific practice. By employing these two techniques, Thoreau ran the risk of defining his work as not scientific at all, but wholly literary, undercutting the effectiveness of his critique even before it achieved a hearing.

"Feedbacking" focuses on the role of the putative "discoverer," in this case Thoreau himself. In "objective" science, this role would be that of transparent intermediary between the scientist's object and ourselves, the readers and witnesses. The act of discovery being essentially passive, anyone, the story goes, could have stumbled across it; I just happened to be the one, and I merely convey my finding to you. The narrating "I/eye" we expect in scientific rhetoric claims merely to record what is there all along for anyone to see, staying rhetorically out of sight, suppressing any sense of its own agency—for, recall, there has been no agency. The very power of this view rests on this premise: command by obedience. But if objectivity is undercut, one can no longer claim simply to channel the docile body of the discovered to its interested onlookers, nor posit oneself as the passive vehicle of intelligence, pure, unmarked, invisible, neutral, and uncontaminating. Feedbacking, therefore, disrupts this fictive role by foregrounding agency. The discoverer/scientist/author will emphasize rather than suppress individual presence, action, and circumstance, through the use of what Steve Woolgar calls "modalizers" which "draw attention to the existence and role of an agent in the constitution of a fact or factual statement."[37] Or even more dramatically, the author may put in a sudden and unexpected appearance—not an easy thing to do, I've noticed—revealing the convention that has kept her "silent."

I suggest this is precisely what Thoreau does. Throughout "Succession" he insists on foregrounding his own role as an agent, both in the field and at the podium; furthermore, all the other operative elements are active agents as well, co-producers of both the forest and of his own process of discovery. Each one of his major narratives of deduction, quest, and discovery is framed by his actions and local circumstances. For example, an observation of a squirrel "planting" a hickory nut begins, "On the 24th of September, in 1857, as I was paddling down the Assabet, in this town, I saw a red squirrel run along the bank under some herbage, with something large in its mouth" (78–79). The passive, agentless statement is here nowhere to be found; his statements are all active: "I saw," "I approached," "Digging there, I found," "I walked," "I selected." Nor is Thoreau the only one actively engaged about his business. He is surrounded by equally active entities, who are as aware of *his* presence as he of theirs: "One of the principal agents in this planting, the red squirrels, were all the while curiously inspecting me, while I was inspecting their plantation" (80). He must even compete with these "agents" for specimens: "The jays scream and the red squirrels scold while you are clubbing and shaking the chestnut trees, for they are there on the same errand, and two of a trade never agree" (82). The squirrels cast down the green chestnut burrs, "and I used to think, sometimes, that they were cast at me" (83). He mocks the English foresters who were "discovering" the secret of cultivating oaks: "but they appear not to have discovered that it was discovered before, and that they are merely adopting the method of Nature, which she long ago made patent to all. She is all the while planting the oaks amid the pines without our knowledge," and so our woodchoppers cut down the pines and "rescue an oak forest, at which we wonder as if it had dropped from the skies" (82). But of course it didn't—the absurdity of his image adds the backstroke to the edge of Thoreau's argument. The oak forest was *produced*, by the actions of pines and cones and winged pine seeds, wind, sun, shade, oaks, acorns, squirrels and jays and mice, the woodlot proprietors, the watchful town surveyor, the woodchoppers and their axes, the Middlesex Agricultural Society, British agriculturalists, "Nature" herself.

The essay thus demonstrates a number of feedbacking techniques for disrupting the illusion of an invisible omniscient "I" whose all-seeing "eye" has laid bare the truth of the universe. The first is, instead of gliding smoothly and silently down the halls of discovery, to make lots of narrative noise. This essay, as already noted, opens with a strained and extended joke. It bumps along through extravagant assertions ("that I can tell,—it is no mystery to me"), asides ("As I have said . . ."), circumlocutions ("I think that I may venture to say . . ."), and gratuitous remarks ("The ground looks like a platform before a grocery, where the gossips of the village sit to crack nuts and less savory jokes" [86]), to a hyperbolic fable about a squash. We are never suffered to forget that this essay is being voiced by a single idiosyncratic personality.

The second is to reassert the fictive character of the narrative by inserting long rhetorical flights, which should be totally unacceptable in a "factual" context. Thus, from a straightforward description of a pine seed, Thoreau leaps by way of its ingenious construction to a conceit about the "patent office at the seat of government of the universe" which oversees such things (75). He returns to this notion at the essay's close, expanding it to suggest the millennial potential of a seed, and of a patent office that distributed such seeds (91). Such transcendental interruptions elaborate on that distinctive authorial voice, which refuses to operate invisibly but instead ostentatiously takes control to remind us who is speaking, and in what context.

The third is to reiterate the active agency, at every level, of the author who is constructing this narrative. After announcing that he, alone, has the answer to the "mystery," Thoreau takes us collectively by the elbow: "*Let me* lead *you* back into *your* wood-lots again" (74; emphasis added). Having insisted on being his listeners' guide, he reminds them regularly that *he* is the one doing the leading, walking, deducing, seeing, probing, connecting. It is he who fashions the myriad data, and links them into a story designed for the consumption of his particular audience, namely, the legal owners of the woods in question, who are also his employers, those who in some sense "own" him—or not, any more than they "own" the squirrels and jays who also inhabit and make their living from the woodlots, just as, when you come down to it, the proprietors themselves do, in cutting and marketing the timber. Thoreau foregrounds his own agency in constructing this convoluted tale, and in so doing, acts as the solvent dissolving the boundaries between all the other elements, which are also revealed as actors, agents in their own right, co-producers of the story Thoreau is seen to be fashioning.

Thus the fourth technique is to emphasize the voices of all the other participants, from proprietors who question, to jays who scold, to scientists who make competing claims. Thoreau treats all the participants in his narrative as actors, "agents." This is not a passive and docile field subject to his manipulations, or a rational cosmos revealed by his God vision, or even a world evoked into meaning by his Adamic power of naming. The various actors here not only question and scold, but hide, and hide evidence, from him; they collect, disseminate, and plant, carry, bury, swallow, design, fly, choke and overshadow, spring up, thrive, die. Thoreau's world here is a very busy place, and Thoreau lets us see just how hard he must work to compose a coherent story out of its confusion.

Both Thoreau and Woolgar discuss "intentionality," or "inversion," in the context of "discovery," wherein, according to the ideology of representation, the object was there all along and the discoverer more or less manages to stumble across it. Thoreau focuses on our propensity to *miss* what we stumble across routinely—as he writes in 1842 in "Natural History of Massachusetts," "it is much easier to discover than to see when the cover is off. . . . We must look a long time before we can see" (NHE 29). Here he illustrates the principle by revealing all the unseen oak seedlings living in a particularly dense pine grove: "Standing on the edge of this grove and looking through it . . . you would have said that there was not a hardwood tree in it, young or old. But on looking carefully along over its floor I *discovered*, though it was not till my eye had got used to the search," that the forest floor was scattered regularly with little oaks, from three to twelve inches high (80; emphasis added). That is, once he knew what to look for, Thoreau knew, first, to seek out a particular kind of forest; and once there, he taught his eyes to see what no one had ever seen, but had been there all along. Once one knows what to look for, one finds it.

"Discovery" defines something both novel and significant. Previously, the oak seedlings were neither; once constructed otherwise, they leapt into view. The discovery of America by Columbus was a social process, beginning with an orientation and setting out, ending with consolidation and assertion. The process is successful when it is socially acknowledged: we agree (or did until recently) that it was Columbus who "discovered" America, not the Vikings —and certainly not the "Indians" (an object of discovery can hardly discover itself!).[38] Thoreau's point is that the "cover" is always "off"; the act or process of discovery constructs the unseen into the seeable, and then finds and declares it for all to see, just as he was doing in his Middlesex Agricultural Society lecture. Two months later he added shrewdly in his *Journal*, "How is any scientific discovery made? Why, the discoverer takes it into his head first. He must all but see it" (XIV:267; 11/25/60).

If discovery proceeds by making visible those objects in the natural world that cooperate with our knowledge, the *narrative* of discovery is constructed by separating nature from knowledge, fact from truth, and insisting that instead of being complexly intertwined they are wholly independent. This dualism allows the inversion by which the now "freestanding" fact is imagined to have given rise to the hypothesis or document, and the preceding steps are denied, minimized, or simply forgotten.[39] Once we imagine that the fact or object gave rise to our knowledge, we can point back to the object we now imagine as independent and uncontaminated, as the unanswerable legitimation of what we have decided is the essential and incontrovertible truth. Thoreau provides an example: everyone knew that seeds lay dormant in the ground for many years. So when an old pine grove was cut and oak trees sprang up, that fact proved what everybody knew, that for generations acorns had lain dormant under the pines. The strength of the social work that constructed this circle of "knowledge" could be measured by the resistance which Thoreau met when he sought to offer an alternative explanation. It did take some degree of effort to show that acorns do not preserve their vitality for very long, but the acorns cooperated nicely with Thoreau: he could show that by November virtually every acorn left on the ground had either "sprouted or decayed" (88).[40] Obviously there could be no dormant survivors. He intervened in what "everybody knew" by showing them what *he* knew (and that, incidentally, the acorns themselves—who should know, after all—were on *his* side).

Thoreau's larger claim, that the forests and fields of Concord were the fleeting expression of an interactive system of innumerable agents acting across various paths of contact, even yet meets with a degree of resistance; for it assumes that "phenomena" are, as he put it, *not* independent of us but related to us, and that our knowledge can exist only in that shifting field *between* ourselves and our "objects." What Thoreau was arguing against was the "ideology of representation," as Woolgar calls it, which objectifies the world as thing, then asserts that what is out there can be correctly, i.e., objectively, seen. Lapses are due to contamination or failure which can, at least in theory, be corrected to yield accurate vision: "The fire of scientific scrutiny burns away from the idea—the hypothesis or the theory—the stain of its origin."[41] But Thoreau's point is that no vision can be simply accurate. As he writes, "There is no such thing as pure *objective* observation." All observation is necessarily "*subjective*" (VI:236). To the problem of representation, he proposes the solution of relational knowing:

> I think that the man of science makes this mistake, and the mass of mankind along with him: that you should coolly give your chief attention to the phenomenon which excites you as something independent on [*sic*] you, and not as it is related to you. The important fact is its effect on me. . . . With regard to such objects, I find that it is not they themselves (with which the men of science deal) that concern me; the point of interest is somewhere *between* me and them (*i.e.* the objects). . . . (X:164–65; 11/5/57)[42]

"Between" subject and object: that is, in their conjunction or relationship, which denies that "objective" knowing is even possible, let alone desirable. Science eliminates that very point of interest: "You would say," he grumbles, "that the scientific bodies were terribly put to it for objects and subjects. A dead specimen of an animal, if it is only well preserved in alcohol, is just as good for science as a living one preserved in its native element" (XI:360; 11/30/58). Thoreau criticizes science repeatedly for its lack of "relation," to us and to its objects—for its presumption, as George Levine says, "that to observe a thing carefully, one must not care about it." Self-annihilation is necessary to protect science from the consequences of knowledge and so authenticate it.[43]

No science will ever preserve a forest in alcohol. Thoreau's sarcasm and dismay turn on his crucial paradox of independence and relational knowing: everything is sovereign, therefore nothing is passive and docile, "subject" to our "objective" knowing. Or, everything is related, therefore nothing hangs on us and our designs alone. We cannot impose, only invite. So in his complex moral vision, knowledge is authenticated not by protection from, but by exposure to, its consequences: we should believe him because he *does* care, not because he doesn't. In Thoreau's alternative science, authority comes from individual involvement and experience. What is more, we should care too, because we are all similarly involved, implicated. As our designs tangle with those of willows and squirrels and oaks and beggar-ticks, we all become co-producers of Concord, and by extension, the "environment" around us, wherever we are.[44] Hence we can no longer rationalize treating oaks and pines as just a commodity, of value only as we can insert them into our social-economic system. Thoreau's essay becomes a prescient plea for what later readers would call environmentally based consciousness and action.

Within various registers—from dormant acorns to village religious dogma to the conquest of Mexico and the American institution of slavery—Thoreau worries about the way a posteriori rationalization operates to make advocacy appear "objective," a social construction appear "natural" and inevitable. As Woolgar puns, such a construction becomes a thing, a *res* which "*res*ists" our efforts to deconstruct it.[45] Against this form of social resistance Thoreau throws up a form of counterresistance: "things," objects, which refuse to be passively manipulated into our constructions, but actively "*res*ist" us; we can, if we wish, imagine that acorns lie dormant, but they resist, and sprout or rot despite us. (Or as Latour puts it: "reality as the latin word *res* indicates, is what *resists*").[46] We may dismiss squirrels, jays, thrushes, and mice as inconsequential, even as vermin, but they too resist and go about planting and shaping our forests. They perform in effect their wild woodland version of "civil disobedience"—or as Thoreau's original title had it, "*Res*istance to Civil Government."

Among these various acts of resistance and cooperation, Thoreau fabricates a narrative out of his own actions, which include not only looking and counting (239 pitch pine cone cores in one pile alone! [84]) but surveying the land for its proprietor's economic use, during the course of which he too is inspected and even scolded. By foregrounding all this Thoreau is advertising the status of his representation *as* a representation, not some clairvoyant reading of the truth of the universe. That is, the nature of representation itself is an issue here, and his critique is meant on principle. Yet Thoreau is faced with the problem of mounting "an adequate and effective resistance in a situation where adequacy and effectiveness are defined by the ideology (representation) under critique."[47] Thoreau does attempt to reconstitute the moral order of representation, by disputing the role of science as "the sealer of weights and measures," sole arbiter of what "counts." That is, he, a poet and a transcendentalist, "counts" too—as do the other participants, two-, four-, and non-legged altogether. There are many measures; they should all be welcome at the cattle show.

But representation which reflexively interrogates the status of representation is defined by the ideology of representation as poetic or literary. So when Thoreau mounts a critique of representation by offering an alternate story and an alternative means of representation, one which acknowledges its status as storytelling, his critique, meant as principled, runs the risk of being sidelined as marginal to the great domain of scientific truth. If science offers "the" representation of the woods, has not the poet merely offered a secondary one, forming at best a kind of supplement? Is this not after all a quirky and marginal text, neither mainstream science nor very good literature? Clearly this essay rejects the assumptions encoded in such a statement, asking instead that we attempt to recover Thoreau's writing in all its extravagant intertextuality. His attempt to remake science was more broadly an attempt to interlace the separating domains of science, society, and poetry at exactly the historical point when the three were precipitating out of a common culture into specialized disciplines. In other words, Thoreau was not just pioneering a "discovery" or a new scientific concept. He was arguing for a new concept of science, a nonmodern science in which the subject and object are not split into separate and independent entities but caught mutually in a web of relationship. He shows us how we might continue to make science, while recalling all the while that it is finally *ourselves making* science.

Thoreau closes his essay by releasing the metaphor that has been lurking just below the surface of his prose, concealed in plain sight like his little sheltered oak seedlings coiled under the pines, waiting to "spring up":

> Though I do not believe that a plant will spring up where no seed has been, I have great faith in a seed,—a, to me, equally mysterious origin for it. Convince me that you have a seed there, and I am prepared to expect wonders

—or the millennium and the reign of justice, even, "when the Patent Office, or Government, begins to distribute, and the people to plant, the seeds of these things" (91). Then he tells us an odd story about turning loose in a corner of his garden a "brace of terriers," six squash seeds (dispersed to him straight from the Patent Office). A little hoeing and manuring, and "*abracadabra presto-chango* . . . lo! true to the label, they found for me 310 pounds of *poitrine jaune grosse* there, where it never was known to be, nor was before." Why, look at the squash my seeds have discovered! Thoreau chortles, mocking the tag-phrase magic of spontaneous generation. Not only that: "But I have more hounds of the same breed. . . . Other seeds I have which will find other things in that corner of my garden, in like fashion, almost any fruit you wish . . ." (91–92). Other seeds in an imaginary garden, indeed: the seed-hounds of a thought will point us to a discovery—the discovery of oak trees where you would have said there were none, of a community where you would have said were only trees, of literary coproduction where you would have said was only a lone naturalist in the woods. The two orders, material and imaginary, do not fall away from but interpenetrate and define each other.

Truly, interdisciplinarity is (as Stanley Fish's title has it) "so very hard to do"—but perhaps not quite impossible. It requires multiple and simultaneous acknowledgment of manifold factors and perspectives, keeping material and imaginary realms in focus at once in a kind of material semiosis, both letting a representation stand and querying the doubleness of its stance. Thoreau is dazzled by the magic of this simultaneity: that a representation could so unfailingly find its object, in a universe so infinitely undiscovered. Nature always answers to our conceptions. Given such seeds as this, "the corner of my garden is an inexhaustible treasure-chest. Here you can dig, not gold, but the value which gold merely *represents* . . ." (92; emphasis added). Yet, he concludes, even in the presence of such a garden, the fertile source of infinite representations, men still prefer the sterile trickery of isolated and perpetual self-production: "Yet farmers' sons will stare by the hour to see a juggler draw ribbons from his throat, though he tells them it is all deception." Disciplines too seem to prefer this magic trick of producing themselves endlessly out of themselves.

This is why I find so sad Thoreau's choice of a closing statement: "Surely, men love darkness rather than light" (92). Given the seed, the earth, and the mind, he asks us to hope for more than sterile self-reproduction, yet stands already in the anticipation of being misunderstood. In the work most deeply influenced by Darwin, Thoreau like Darwin had to defend his vision, had to assert the wonder and beauty of the merely material, the "little *things*" that were seeds: "I have great faith in a seed,—a, to me, equally mysterious origin for it. Convince me that you have a seed there, and I am prepared to expect wonders. . . ." The "mysterious origin" of a seed is, of course, another seed: the most prosaic of magic. Just as Darwin's "Tree of Life" fills the earth and covers its surface "with its ever branching and beautiful ramifications," so does Thoreau's seed, if only his hearers will attend:

> Other seeds I have which will find other things in that corner of my garden, in like fashion, almost any fruit you wish, every year for ages, until the crop more than fills the whole garden. You have but little more to do than throw your cap up for entertainment these American days. Yet farmers' sons will stare by the hour to see a juggler draw ribbons from his throat, though he tells them it is all deception.

Against the sterile deception of endless self-reproduction, the carnival trickery of "spontaneous generation," Thoreau interposes the chain of real connection, of real causes and their astounding effects. Yet he closes without optimism: "Surely, men love darkness rather than light" (92). Time and again in his late essays Thoreau effectively says "Yes!" in thunder until compelled to say "No."

Thoreau's closing reference to his "garden" links this essay to his other cross-grained and hybrid seeds—"Autumnal Tints," "Wild Apples," "Huckleberries"—in which the garden, *his* garden, is the wild. Not some faraway Eden, but here under our noses, in our living vision: "These are *my* China-asters, *my* late garden-flowers. . . . Only look at what is to be seen, and you will have garden enough, without deepening the soil in your yard. We have only to elevate our view a little, to see the whole forest as a garden" (NHE 172). This "garden" will be not a warehouse of resources for translation into the networks and centers of commerce, but an ultimate value in itself—that which "gold merely represents." Because it is no one's and everyone's it is "ours" for the *seeing*, offering us a kind of abundance not consumed with use. He who "shoots at beauty," who has "dreamed of it, so that he can *anticipate* it," then, indeed, "flushes it at every step, shoots double and on the wing, with both barrels, even in corn-fields." Nor will his desire ever fail him: "If he lives, and his game spirit increases, heaven and earth shall fail him sooner than game" (NHE 175–76).

To our quiet desperation, nature offers spiritual bounty and abundance: a profusion of berries, abandoned apples, forests full of acorns and chestnuts, "Slight and innocent savors which relate us to Nature, make us her guests, and entitle us to her regard and protection" (NHE 241). Here is the weight behind his oft-quoted line from "Walking": "in Wildness is the preservation of the World" (NHE 112). Against the coming of the evil days, in which men who love darkness better than light run after carnival tricks and the husks of experience, Thoreau invokes the real redeemer: Nature wild and unappropriated, the muck that will fertilize our withered fields, Chaos that does not just burrow through but abolishes altogether the "dead dry life of society," engendering each new Cosmos. At the close of *Walden*, cosmos emerges not from the sterile frost of winter but from that thawing, dissolving, flowing, self-organizing chaos of spring, even to sacrifice and destruction and the rain of flesh and blood. We should not forget that Thoreau built his Eden not only on cooperation and community but on waste and excess, vultures, carrion, and decay. His Eden will incorporate the Fall: it is a paradise erected on transgression.

6

Walking the Holy Land

—What a novel life to be introduced to a dead sucker
floating on the water in the spring!— Where was it spawned pray?
The sucker is so recent—so unexpected—so unrememberable
so unanticipatable a creation— While so many institutions
are gone by the board and we are despairing of men & of ourselves
there seems to be life even in a dead sucker—whose fellows at least
are alive. The world never looks more recent or promising—
religion philosophy poetry—than when viewed from this point.
To see a sucker tossing on the spring flood—its swelling
imbricated breast heaving up a bait to not-despairing gulls—
It is a strong & a strengthening sight.
Is the world coming to an end?—
Ask the chubs.

—Henry David Thoreau, *Journal*

The past three chapters have argued that Thoreau built his worlds—of *Walden*, the *Journal*, the late essays—from a universe that was not fixed and determinate but fluid and open-ended. These worlds were organized by his perceiving mind, "central" to be sure, but only to himself: the observer, he wryly noted, always seemed central to himself, not suspecting that a thousand other observers are equally central (3:298; 7/10/51). Hence the value of that "meteorological journal of the mind— You shall observe what occurs in your latitude, I in mine" (3:377; 8/19/51). And like good meteorological scientists, you and I shall combine our observations toward a community of knowledge: "A wise man will avail himself of the observation of all" (3:91–92; after 7/1/50). But at the heart of such an insight is contingency: "Who shall say what *is*? He can only say *how* he *sees*." No man inhabits the same world, then, as another (2:355; 12/2/46). But if what you saw was contingent on who you were and what you were looking for—not to mention where and when you looked—truth could be spun only tentatively out of chance, which another chance might undo. Thoreau thus planted in his worlds the seeds of their own unmaking. Yet instead of concluding that his worlds were thereby doomed, he imagined a new kind of worlding which wove creation and decreation together: a new world in which cosmos and chaos were not antagonists, but in which cosmos, "Beauty," was chaos by another name.

Contingent Wholes: A Few Herbs and Apples

In his poem "Days" (1857), Emerson imagines the "hypocritic Days" filing past with their gifts, "diadems and fagots," to each "after his will"; he, watching the pomp,

> Forgot my morning wishes, hastily
> Took a few herbs and apples, and the Day
> Turned and departed silent. I, too late,
> Under her solemn fillet saw the scorn. (*Poems* 228)

What was metaphor for Emerson was for Thoreau a literal act, as he spent his days after *Walden* multiplying observations on "herbs and apples," gathered under the provisional title "Wild Fruits." Though the shape Thoreau finally intended for this work is unclear, when his fatal illness confined him to the house he did turn the two lectures "Autumnal Tints" from 1859 and "Wild Apples" from 1860 into essays; both were published in *Atlantic* shortly after his death. A third essay, "Huckleberries," has been reconstructed by Leo Stoller. Apparently the longer work was to have been a collective whole, composed of an assemblage of essays drawing together groups of facts, unified by the mind of the author in interaction with the natural elements of his environment—not the cosmos entire, but partial and contingent *Ansichten*, passages of order collected and connected "after his will" from his own Days. In each of the three published segments, themes of decline and fragmentation are countered with pleas for collective strength and achievement: these late "nature" essays are finally some of Thoreau's most social writing.

"Autumnal Tints" opens by insisting that the fall of the year represents not a decline but a jubilant flowering and ripening, all the finer for addressing, not our gross senses, but our taste for beauty. To establish this, Thoreau had thought to offer us, he says, a book of leaves; but offers instead a chronological series of descriptions of "tints": purple grasses, red maples, yellow elms, scarlet oaks. His primary theme is asserted early: a high-colored stripe of purple under a wood turns out, upon examination, to be composed of "a kind of grass in bloom," with "a fine spreading panicle of purple flowers, a shallow, purplish mist trembling around me." Single plants seem dull, thin, colorless; but together they are "a fine lively purple, flower-like, enriching the earth. Such puny causes combine to produce these decided effects" (NHE 140). As the weeks advance, each species, each tree, even each single leaf, declares itself. A small red maple, having "faithfully discharged" its duties—husbanding its sap, sheltering birds, and ripening and discharging its seeds—"runs up its scarlet flag on that hillside," flashing out its virtue and beauty: "We may now read its title, or *rubric*, clear" (148–49). Each maple chooses its own set of colors and its own timetable:

> It adds greatly to the beauty of such a swamp at this season, that, even though there may be no other trees interspersed, it is not seen as a simple mass of color, but, different trees being of different colors and hues, the outline of each crescent treetop is distinct, and where one laps on to another. (150)

Equally beautiful are the fallen leaves: "For beautiful variety no crop can be compared with this. . . . The ground is all parti-colored with them" (157). But for all their variety, they combine to the same purpose: "How they are mixed up, of all species, oak and maple and chestnut and birch! But Nature is not cluttered with them; she is a perfect husbandman; she stores them all." For this is "the great harvest of the year," when the trees repay "the earth with interest," adding "a leaf's thickness to the depth of the soil." Together they prepare "the virgin mould for future corn-fields and forests, on which the earth fattens." Building the soil, they "stoop to rise" by "subtle chemistry, climbing by the sap in the trees," so that the sapling's first fruit comes at last to adorn the crown of the monarch of the forest (156–57).

Years earlier, Thoreau had noted how a flag set on a hillside tames and "subdues the landscape to itself" (3:72; 5/12–5/31/50)—like Stevens' "jar in Tennessee," "The wilderness rose up to it, / And sprawled around, no longer wild."[1] But in "Autumnal Tints," the imperial vision collapses, and it is the flags of the leaves that control the human landscape: "There is an auction on the Common to-day, but its red flag is hard to be discerned amid this blaze of color" (NHE 160). Nature herself is holding her annual fair, and "every tree is a living liberty-pole on which a thousand bright flags are waving." We have only "to set the trees, or let them stand," and nature will fly "flags of all her nations." Does not this suggest, Thoreau asks mildly, that man's spirits should rise as high, "—should hang out their flag . . . ?" (162–63). Thoreau depicts here a segment of nature, in which we can witness how phenomena of nature interrelate—including, significantly, the village itself. To Humboldt, "Nature is free because its order is determined within itself as a whole."[2] The anarchy of the autumn hillside produces a similar order, as fall becomes a celebration of collective identity, a utopian vision of the entire countryside as a flag of flags.

Access to such vision is available to anyone who chooses to see: "Only look at what is to be seen," advises Thoreau; "All this you surely *will* see, and much more, if you are prepared to see it,—if you *look* for it" (NHE 172–73). The final pages of the essay become a plea for vision, for understanding that it is not the eye but the mind that sees. Nature is concealed from us only because we look with our eyes and not our minds: if we are not prepared to see, for "threescore years and ten" we will think that "all the wood is, at this season, sere and brown." "A man sees," he continues, "only what concerns him"—and thus does the scope or limitation of your dreams determine the world you live in, and the manner of your living in the world. His prize exhibit is the scarlet oak, which "requires a peculiar alertness, if not devotion to these phenomena," to appreciate. Seen unsympathetically it is dull and dark, but to his receptive eye it shows "an intense, burning red" in the low sun, turning the whole forest into a flower garden "in which these late roses burn" (169–71). Their fire is partly borrowed from the sun, and partly from the eye, making the scarlet oaks the purest instance of the interactiveness of all phenomena: "You see a redder tree than exists."[3] His vision cultivates such a garden, not the "petty garden" of a few asters and weeds, but "the great garden" that is all the world, were we just to elevate our vision.

The duality implied by this darkly burning color is also signaled by the "rich and wild beauty" of the leaf's shape. The only drawing Thoreau ever commissioned for his work was an outline of a scarlet oak leaf for this essay (Fig. 1). With their deep scallops these leaves have "solved the leafy problem," how to combine terra firma, "earthy matter," with light and "skyey influences." They "appear melting away in the light," or "dance, arm in arm with the light—tripping it on fantastic points, fit partners in those aerial halls," intermingled so closely that "you can hardly tell at last what in the dance is leaf and what is light" (NHE 166–67). In this figure the old problem of dualism, the heavy oscillation between earth and spirit, is at last "solved": each resolves into the other in a moving cosmic dance. The leaf, as Sattelmeyer notes, is Thoreau's "hieroglyph" of creative energy (NHE xxviii); Thoreau's drawing also parallels another hieroglyph, his map of Walden (Fig. 2), in which the pond's outline maps the morality of life. In "Autumnal Tints," the leaf-map suggests first water, "a pond," then land, "some fair wild island in the ocean" whose "smooth strand" and "sharp-pointed rocky capes" mark it as a home for mankind, "a centre of civilization at last." Is the leaf

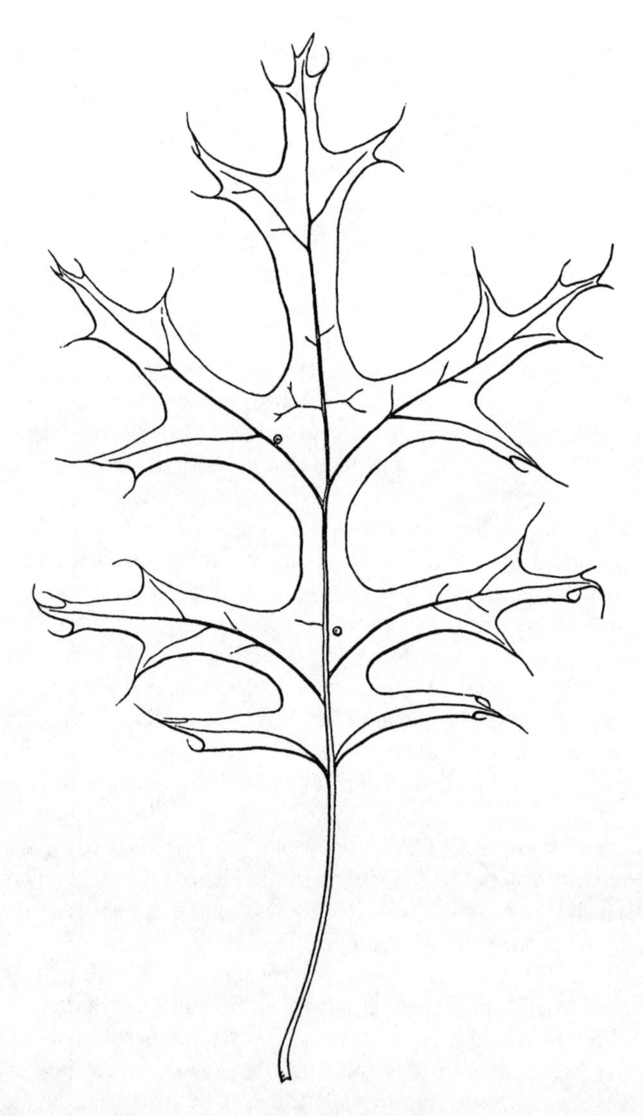

Figure 1. The engraving of a scarlet oak leaf, which Thoreau commissioned for the October 1862 publication of "Autumnal Tints" in *Atlantic Monthly.* On February 20, 1862, he wrote Ticknor & Fields: "In my lecturing I have always carried a very large & handsome one displayed on a white ground, which did me great service with the audience." He requested a simple outline engraving so "that the readers may the better appreciate my words—I will supply the leaves to be copied when the time comes." He sent the specimen leaf on March 1: "I wish simply for a faithful outline engraving of the leaf bristles and all," he reiterated (CO 637, 639).

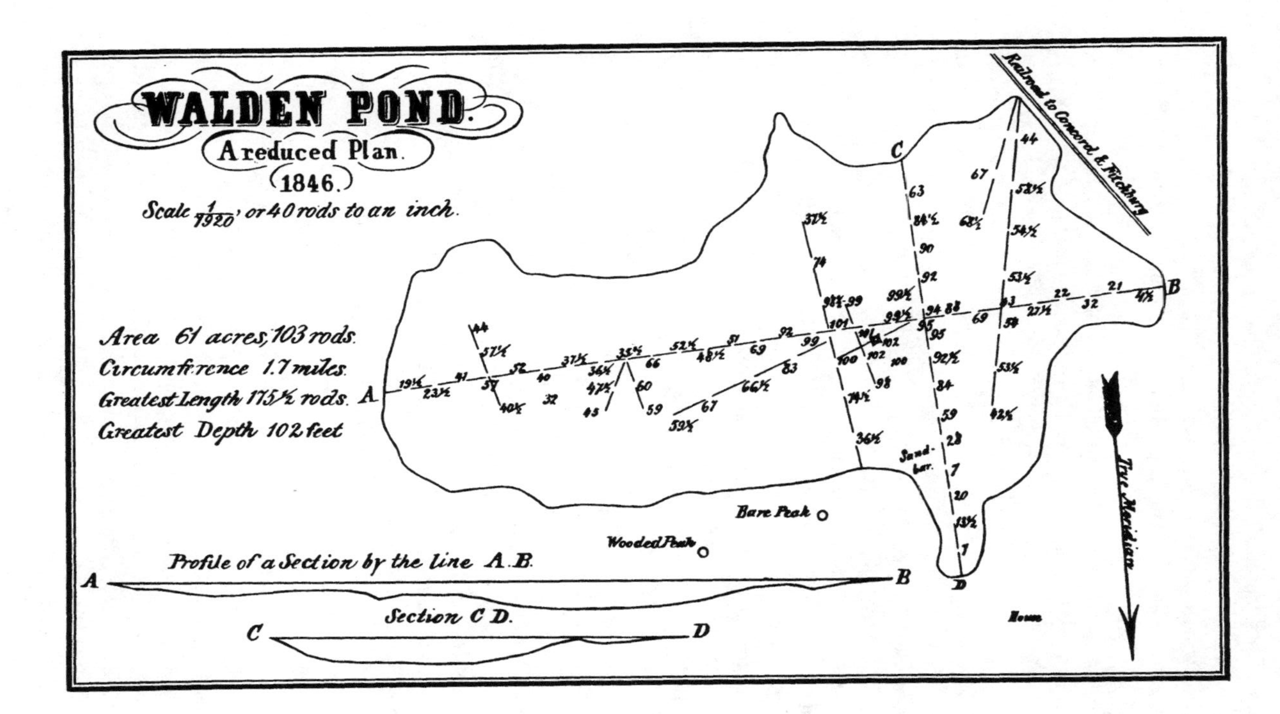

Figure 2. The map of Walden Pond which Thoreau supplied for *Walden*, "The Pond in Winter." The professional lithograph was based on two drawings of Thoreau's, a large one (16 1/4 x 20 1/2 inches) he made in 1846, and a second, "reduced" version probably drawn in early 1854. The accuracy "has been confirmed with modern instruments"; see Stowell, *A Thoreau Gazetteer* 5–9.

not, "in fact, a shore to the aerial ocean, on which the windy surf beats?" (168). In the figure of the leaf Thoreau discovers his own yang-yin sign of dualism without hierarchy. Instead of Emerson's "stupendous antagonism," Thoreau offers us his vision of the dance of a leaf in the wind.

To Humboldt, contemplation of the beauty of nature moved the susceptible observer to perceive, and act on, moral good.[4] To Thoreau, the falling of the autumn leaves "teach[es] us how to die," as they "lie down" with an "Indian-summer serenity"; when they fall, "the whole earth is a cemetery pleasant to walk in" (158). But more important, the flaring of the leaves teaches us how to *live*, to come to a perfect ripeness and maturity, to see the beauty of collective creation and the fitness of returning our labor to the "soil" of our culture. By their extravagance, they declare to us what we miss in our haste and carelessness. As he ends the essay, faithful observation will show us that "each humblest plant" has its own "peculiar autumnal tint" (177); the leaves' showiness finally leads us to observations less spectacular and more subtle. Above all, in autumn the trees are not "dying" but at the height of their vitality. The sap in the November scarlet oaks is flowing fast: "They are full of life," full of a "strong oak wine" of a "pleasantly astringent, acorn-like taste" (169–70). This is the idea developed in "Wild Apples": the joy and exhilaration of ripeness, the vitality expressed in the trees and their fruit, even through the months of winter.

"Wild Apples" is, like "Autumnal Tints," an essay compacted of radical metaphors, but even more intensely: the immigrant domestic apple tree is identified with New World Europeans, and their misuse signals the way America is bringing the evil days upon itself. Thoreau himself is a "wild apple," for though there are native wild apples out in Michigan, "*our* wild apple is wild only like myself, perchance, who belong not to the aboriginal race here, but have strayed into the woods from the cultivated stock" (NHE 189). He liked to think of himself as a slow-maturing poet, whose fruits "may be less delicious, but are a more lasting food and are so hardened by the sun of summer and the coolness of autumn that they keep sound over winter"—not "June-eatings" but "russets, which last till June again" (VI:190–91; 4/8/54). Thus what Thoreau gives us in this essay is his own spiritual autobiography: the growth of his respect for the rank and wild, and his savoring of their "spirited and racy" flavor (NHE 199), scorned by the farmer; his dying quest for the "half-fabulous" native crab apple; his elegy for their joint passing and his prophetic warning of the coming of the evil days. "How the Wild Apple Grows" narrates the joint biography of their difficulties and their resistance. The sprouts, browsed by cattle, spread "low along the ground" until, "regularly clipped by cows," they form perfect little cones, sharp "as if trimmed by the gardener's art" (192), already old but gaining in power what they lost in time. At last, after twenty years or more,

> they are so broad that they become their own fence, when some interior shoot, which their foes cannot reach, darts upward with joy: for it has not forgotten its high calling, and bears its own peculiar fruit in triumph.
>
> Such are the tactics by which it finally defeats its bovine foes. (193)

Released, the plant shoots upward, forming an hourglass on its pyramidal base; "The spreading bottom, having served its purpose, finally disappears," and the generous tree now admits the cows to its shade and its fruit. The cows thus "create their own shade and food, and the tree, its hour-glass being inverted, lives a second life, as it were" (193–94). The despised tree, once offering only shelter for birds, at last blossoms and offers its harvest, "sincere, though small," and perchance the sweeter "for the very difficulties it has had to contend with" (194–95). Lest the reader overlook his oblique communication, Thoreau reads the lesson clear:

> Every wild apple shrub excites our expectation thus, somewhat as every wild child. It is, perhaps, a prince in disguise. What a lesson to man! So are human beings . . . browsed on by fate; and only the most persistent and strongest genius defends itself and prevails, sends a tender scion upward at last, and drops its perfect fruit on the ungrateful earth. Poets and philosophers and statesmen thus spring up in the country pastures, and outlast the hosts of unoriginal men.
>
> Such is always the pursuit of knowledge. . . . (195)

There are, he adds parenthetically, less remarkable ways in which wild apples are propagated; but of course the story they tell is not *his* story: a border life, its vitality sculpted by the contending forces of wild and cultivated, persisting on the outskirts of the civil—like his beanfield and his cabin, "the connecting link between wild and cultivated . . ." (WA 158).

Part of the relish of the essay is Thoreau's elaborate connoisseurship of the flavors and textures of these scrubby, shrunken, sour, bitter, and cidery wild fruit, emphasizing that frontier contacts need not lack exquisite refinement. His discriminating sensualism guides him from taste through fragrance to spiritual effluence: "There is thus about all natural products a certain volatile and ethereal quality which represents their highest value, and which cannot be vulgarized, or bought and sold" (NHE 183). Yet apples are after all primarily a market commodity. The driver transporting apples to market "begins to lose his load" the moment he sets off; though he feels them and thinks they are all there, "I see the stream of their evanescent and celestial qualities going to heaven from his cart, while the pulp and skin and core only are going to market" (184). The farmer handling his apples "rubs off all the bloom, and those fugacious ethereal qualities leave it" (185).

Thoreau has been able to cheat the farmer, by turning to apples left neglected, and inhaling, tasting, and drinking in them "a fruit never carried to market" (208). Yet such days are nearly gone:

> The era of the Wild Apple will soon be past. It is a fruit which will probably become extinct in New England. . . . I fear that he who walks over these fields a century hence will not know the pleasure of knocking off wild apples. Ah, poor man, there are many pleasures which he will not know! (209)

In an era of grafted purchased trees, platted and fenced, "we shall be compelled to look for our apples in a barrel" (209). To avert doom, our second fall from Eden, we must not avoid but eat again of the tree of knowledge, the knowledge of the wild. For the loss of wild apples is the lost possibility of harmony between wild and cultivated, and Thoreau ends with a biblical prophecy of a land laid waste: "'Be ye ashamed, O ye husbandmen. . . . The vine is dried up, and the fig tree languisheth; . . . and the apple tree, even all the trees of the field, are withered: because joy is withered away from the sons of men'" (210).

Not just he, but we all are borderers, living on the ragged edge of time in a degraded landscape of "shreds and fringes" (XI:296). Thoreau offers a requiem for the cosmos. In seeking to know "the habits of the wild animals, my brute neighbors," he has learned what he has lost: "the cougar, panther, lynx, wolverene [sic], wolf, bear, moose, deer, the beaver, the turkey, etc., etc.,—" all exterminated here; "Is it not a maimed and imperfect nature that I am conversant with?" He listens "to [a] concert in which so many parts are wanting," reads a poem to find "that my ancestors have torn out many of the first leaves and grandest passages, and mutilated it in many places." What if some demigod had picked out the best of the stars? "I wish to know an entire heaven and an entire earth" (VIII:220–21; 3/23/56). But in Concord, Massachusetts, in 1856, he finds he is much too late—though not, happily for him, too late to enjoy what remains:

> At present, in this vicinity, the best part of the land is not private property; the landscape is not owned, and the walker enjoys comparative freedom. But possibly the day will come when it will be partitioned off into so-called pleasure-grounds, in which a few will take a narrow and exclusive pleasure only. . . . (NHE 104)

In his unfinished essay "Huckleberries," Thoreau continues the lament of decline and extinction: the private property markers which confine the saunterer to the highway are ordering pickers out of the huckleberry fields.

By a similar law, every boy used to rush to the hillside, speaking for his own spot and "indicating its bounds"; by a similar law do we take "possession of the territory of Indians and Mexicans." The spreading networks of commerce defined not only the tracks of opportunity but the fences of exclusion. Thoreau could step off the tracks and make his own pathways, bypass the market; "But ah we have fallen on evil days! . . . What becomes of the true value of country life—what, if you must go to market for it? . . . The wild fruits of the earth disappear before civilization," or leave only their husks in the markets (248–49). Thoreau adds, alluding to a more spectacular instance of commodification, that in the "constitution of our society," we "compromise" and "permit the berries . . . to be enslaved" (251). But to these laments he offers a solution, which is, of course, collective: in each town there should be "a committee appointed, to see that the beauty of the town received no detriment." In this country "precious objects of great natural beauty should belong to the public," so that, instead of making the new world into the old, we can try to "keep the new world new." Most men do not care for nature and would happily sell their share in her beauty; for that very reason "we need to combine to protect all from the vandalism of a few" (254–56). Therefore "it would be wise for the state to purchase and preserve" a few primitive forests, for example; each town should have a park, "of five hundred or a thousand acres," not for any business or military use, "but stand and decay for higher uses—a common possession forever, for instruction and recreation." Walden wood might have been preserved, he laments (259) —assuring, by unexpected irony, that the state *would* buy and restore it, even though he assumed it would all soon go to summer houses.

Such collective proposals are common in the work of Concord's notorious individualist. Years afterward, Thoreau confronted in his *Journal* the careless, every-man-for-himself behavior that resulted in the damaging spread of the fire he had accidently helped set in his beloved woods; then he sketched out an elaborate plan for a program of cooperative burning, fighting forest fires, and even an original scheme for a volunteer fire department (3:79–85; 6/4–6/9/50). In *Walden* he proposed a system of adult public education and even government funding of the arts and education. After all, "To act collectively is according to the spirit of our institutions," and surely in a republic we can combine to combat our own ignorance (NHE 108–10). Even more pointedly, his invective in "Slavery in Massachusetts" is directed against the majority who turn away from justice to "peaceably pursue [their] chosen calling on this fair earth," acting like good Americans first, and men never: "The law will never make men free; it is men who have got to make the law free" (RP 102, 98).

In these late essays, even his engagement with nature is a communal act: the profusion of wild fruits is spread for man and animals alike. "They seem offered to us not so much for food as for sociality, inviting us to a pic-nic with Nature. We pluck and eat in remembrance of her. It is a sort of sacrament—a communion—the *not* forbidden fruits, which no serpent tempts us to eat" (NHE 241). We are, finally, eating the body of Nature, our true savior, sharing the ceremonial bread and wine of the Eucharist. All of his sensual celebrations have this savor: it "excites him" to behold the purple poke, "all on fire with ripeness"; he loves "to press the berries between my fingers, and see their juice staining my hand," to "walk amid these upright, branching casks of purple wine," tasting with his eye "Nature's vintage" (142–43). The headiest pleasure of all is offered by frozen and thawed wild apples, whose "rich, sweet cider" seems "to have borrowed a flavor from heaven through the medium of the air in which they hang." He and his companion "both greedily fill our pockets with them,—bending to drink the cup and save our lappets from the overflowing juice,—and grow more social with their wine" (207–8).

It is precisely these ineffable, shared joys that are destroyed by the demands of the market. The real enemy, it turns out, is not science but business. He complains he cannot find blank books to write in, not because they are all laboratory notebooks, but because men are such "slaves of business" that they are all ruled with red or blue lines "for the dollars and cents" (4:4; 8/21/51; also 4:54; 9/7/51). "In my experience nothing is so opposed to poetry—not crime—as business. It is a negation of life" (IV:162; 6/29/52). Men tend, when "the husk gets separated from the kernel," to "run after the husk and pay their respects to that." Just as men revere only the husk of Christianity (NHE 212), only the "husks" of wild fruits make it to the market. "It has come to this, that the butcher now brings round our huckleberries in his cart. . . . Such is the inevitable tendency of our civilization, to reduce huckleberries to a level with beef-steaks"—and we "all know" what is it to go "a-beef-steaking. It is to knock your old fellow laborer Bright on the head to begin with . . ." (249). The tracks of capitalist networks are, like the Fitchburg Railroad, laid across the bodies of laborers. To commodify is to cannibalize.

Finally, it is business—not the internal logic of science—that separates the hapless scientist from his subject. "A." hires "B.'s" field, to pick the huckleberries (with a patented rake), which he sells to "C.," who cooks them into a pudding for "Professor D.," who "sits in his library writing a book—a work on the *Vaccinieae* of course." And, of course, the book will be worthless, "with none of the spirit of the huckleberry in it. . . ." Thoreau recommends "a different kind of division of labor—that Professor D. should be encouraged to divide himself freely between his library and the huckleberry field" (NHE 250). Professor D. is welcome to continue writing his book —*if* he uses his science to get at the *kernel*, not just the husk. Science which consents to be mobilized is, like the mass of men, only chasing after the husk—worse, it acts as the agent which separates the husk from the kernel to begin with, declaring the latter worthless and enhancing the marketability of the former, fitting truth to the measure of the market. But in the fable of the wild apple, the "spreading bottom" of slow and patient accumulation of knowledge establishes the foundation for the "tender scion," defending it from "bovine" enemies; once having served its purpose, it "disappears." To achieve useful ignorance, learn science then forget it. What Thoreau warns against is science without "reverence," which presumes to explain living beings with mechanical laws, as if they were "some machinery of his own making" —science which cannot forget itself. Employing the same metaphor of fruit and fragrance, Thoreau reminds himself not to confuse surface with the "fine effluence" perceived only at a reverent distance, for cause and effect are equally evanescent: "Science is often like the grub which, though it may have nestled in the germ of a fruit, has merely blighted or consumed it and never truly tasted it. Only that intellect makes any progress toward conceiving of the essence which at the same time perceives the effluence." (XII:23; 3/7/59). The "effluence" is that "ripeness," which is not instrumental to some other end but ultimate in itself; "The ripeness of a leaf, being perfected, leaves the tree at that point and never returns to it" (XII:24). One might almost say it then becomes a seed.

Chance and Necessity: The Laughter of the Loon

Thoreau's understanding of contingency emerged from his own experience. "Anticipation" was enormously powerful, but it could fail, too; expectation might strengthen possibility, but could not secure the improbable. Concord insisted on producing the same old round of phenomena. And some items in that round particularly insulted his longing for control. Upon hearing of a friend's suicide, for instance, he peevishly tried to disown reality: "I did not know when I had planted the seed of that fact that I should hear of it" (IV:280; 8/5/52). His field studies, instead of proving the predictability of nature, documented the opposite: "I can hardly believe that there is so great a diffirence [*sic*] between one year & another as my journal shows," he marvels in 1851, adding a few months later, "How admirable it is that we can never foresee the weather—that that is always novel" (4:76; 9/12/51 and 4:227; 12/29/51). Nor did the years regularize the universe. In 1860 he still marvels that the only certainty is surprise: "Though you walk every day, you do not foresee the kind of walking you will have the next day" (XIII:108; 1/26/60). Outer weather was closely attended by inner weather: "Our moods vary from week to week, with the tides and the temperature and the revolution of the seasons" (IV:44; 5/9/52). How could our own temperament succeed in sculpting the atmosphere, when our only certainty was its liability to change?

For Thoreau was still attracted to the sense of mastery offered by Emerson's view of fate, that every human being was the artificer of his own destiny: "Events, circumstances, etc., have their origin in ourselves. They spring from seeds which we have sown" (VI:226; 4/27/54). But Thoreau's reaction to the world's uncanny responsiveness—even the "most foreign scrap of news" seems but a hue of his inmost thought—is to refine his posture of invitation, guarding against "idleness" lest it invite sickness and distraction (VI:226). That is, he did not attempt to exclude chance categorically, but wove it into the fabric of his life and journal, amassing happenstance toward its accumulation into pattern. By contrast, Emerson's tone in "Fate" (1860) is strenuous, commanding, nothing if not categorical. In its conclusion he chants, "Let us build altars to the Beautiful Necessity," precisely to drive home his assertion that there *is* no chance, "that there are no contingencies; that Law rules throughout existence . . . " ("Fate" 968). Accordingly we must give our life to the law, which is also circumstance, nature, fate, necessity. Arrayed against this pantheon of unified truth is man, with his intellect, freedom, will, and power: the power of thought that "dissolves the material universe, by carrying the mind up into a sphere where all is plastic" (956). In this "stupendous antagonism," "Everything is pusher or pushed"; pushed by nature or fate, we push back with the power of thought, which is invincible. "Every solid in the universe is ready to become fluid on the approach of the mind, and the power to flux it is the measure of the mind." Your failure is merely your weakness. There is no luck, only causation (964).

For his strongest supporting evidence Emerson turns to "the new science of Statistics," which proves it is "a rule, that the most casual and extraordinary events—if the basis of population is broad enough—become matter of fixed calculation."[5] Chance is tamed by law, caprice ruled by science:

> Doubtless, in every million there will be an astronomer, a mathematician, a comic poet, a mystic. . . . In a large city, the most casual things, and things whose beauty lies in their casualty, are produced as punctually and to order as the baker's muffin for breakfast. Punch makes exactly one capital joke a week; and the journals contrive to furnish one good piece of news every day. ("Fate" 951)

Even disaster, disease, and suicide are not fortuitous but show "a kind of mechanical exactness." Our life is "walled up" by such "pebbles from the mountain" (951). Emerson eloquently advances a popular contemporary argument detailed by Ian Hacking: while at first measurements and statistics registered only numerical regularities, after 1830 such regularities were "enthroned" as statistical "law," which "passed beyond a mere fact of experience. . . . The law of large numbers became a metaphysical truth." Out of this "truth" arose "a metaphysical quandary, which was called statistical fatalism": human behavior was lumped into probabilistic laws which were equated with the laws of physics, ruling therefore in favor of determinism, and ruling out free will.[6]

As Hacking continues, the real knot was not metaphysical, but political: "The issue that was hidden was not the power of the soul to choose, but the power of the state to control what kind of person one is." As instances of state control, Hacking points to both utilitarian philanthropic attempts to modify a population, and the slightly later eugenics movement, their common theme being the management of class and race.[7] Humboldt's science, which sought pattern through measurement, collection, and connection, had produced a view of scientific truth as probabilistic and contingent and of political progress as collective individual action. But there were powerful social reasons why such probabilistic truth led not to widespread celebrations of freedom and choice, but to assertions of determinism and social control. This direction was encouraged in America by Emerson, who expressed the determinism authorized by mainstream natural science and put to social use by the organicist state, in line with Emerson's monism of mind and the law. In effect, then, rational holism swallowed up empirical holism by adopting the empiricist's descriptive methodology, the gathering and collating of interconnected facts, into its own a priori assertion of proscriptive and all-governing law.

The backlash to this imperial monism took the form of what Hacking calls "a sort of statistical nihilism," which restored the prerogatives of sheerest chance: Hacking instances Dostoyevsky's call for untrammeled freedom under the standard of "caprice."[8] Thoreau's anarchy stops well short of nihilism, for he found a way to involve chance and order together. So his response to the notion of contingent truth drawn from an unpredictable universe was not to condemn the universe or tame it, nor to recoil from it in anxiety or terror, but to play in it and with it. He exudes relish, joy, exultation, even in aspects we find disturbing or repulsive. His response is grounded in the conviction that "nature" is not inherently malignant, nor even necessarily indifferent, but fundamentally "innocent." Moments of alienation from a nature not definable in human terms could evoke strong reactions in Thoreau—dismay on Ktaadn, revulsion on Fire Island—but these moments mark a limit, allowing us to characterize his more typical response. In one of his most famous aphorisms, nature becomes "the Wild," a figure for absolute freedom and goodness unsubdued by social (but not necessarily *human*) design. As developed in his late essay "Walking," "the wild" ultimately promises to redeem human society, saving it from the sterile inbreeding of man interacting only with man. Thoreau's optimism grows from this belief. It is inaccurate to claim that Thoreau "knew not evil,"[9] for "evil" haunted him throughout the 1850s until his death in 1862. But he could still hope that "the wild"—the socially unconstructed, whether in individual or outer nature—would yet redeem us.

A key moment occurs in one of *Walden*'s most famous passages, in which Thoreau engages a loon in a watery game of hide and seek. He is paddling along the north shore of Walden Pond on a calm October afternoon, looking "in vain" for a loon, when one sails out from the shore to meet him. For an hour the loon leads the man, diving and rising, never letting Thoreau get near: "It was a pretty game, played on the smooth surface of the pond, a man against a loon" (WA 235). H. Daniel Peck finds in this scene a prime example of Thoreau's "worlding," the interaction, "—the 'dance'—of the creative self and the world," brought into being and sustained by the imagination in a tableau of difference against the backdrop of the pond.[10]

This worlding does not just continue, or even just end: it is deliberately unmade. "At length" the loon breaks off the game, by uttering

> one of those prolonged howls, as if calling on the god of loons to aid him, and immediately there came a wind from the east and rippled the surface, and filled the whole air with misty rain, and I was impressed as if it were the prayer of the loon answered, and his god was angry with me; and so I left him disappearing far away on the tumultuous surface. (236)

Shall we emphasize Thoreau's "worlding"? Or the "unworlding" that follows, the entry of turbulence and the wild? Thoreau has come down to us as the advocate of both, the cabin builder who made Walden into a world for our contemporary imagination, and the oracle who first warned us that "in Wildness is the preservation of the World" (NHE 112). He situated himself on the outskirts of town, as the active mediator of civil and wild. To civilization belonged art, poetry, science—the organizing powers of the human mind; but the wild was the source, the generative basis of the world. Accordingly, civil and wild, order and chaos, emerge in Thoreau not as antagonists but as reciprocal forces. He ranged ever more deeply across the Concord landscape, ever encountering more "facts" which he recorded daily in faith that their amassing would reveal pattern at last. This recursive project opened before him the study of an enormously complex system, the village and environs of Concord, and the flow of its infinitely various, nonrepeating but patterned phenomena gave him a crucial insight: that the chaos of wild nature was not meaningless or a void, but information which he could process into meaning.[11]

Thoreau could argue that all nature is important only as it relates to man, but he also holds the door open to unpredictability, "the wild," that which relates to man only casually and incidentally if at all. His world-making game, on that October afternoon, lasts just as long as the loon permits it to. Long after the event, Thoreau tells us how he organized the loon's behavior, as for instance how he "concluded" that the loon's "laughter" was in "derision" of Thoreau's efforts. But at this very moment of apparent reciprocity, when he as the sole agent of his world has just concluded the essential meaning of the loon, the loon howls; "immediately" the wind rises from the east, and "ripples" the "smooth" surface of the pond, forcing Thoreau in his canoe off the water altogether and leaving the loon "disappearing far away on the tumultuous surface" (WA 236). Initially, that is, he shows us the familiar story: all is harmony in nature and all is harmonious between man and nature. But harmony is static, and exists only in a reified universe. Such a universe lasts as long as Thoreau alone centers and completes the process of world making, enjoying the show, gaming with the amusing loon, whom Thoreau thinks he has well in hand. But what Carolyn Porter calls the "cognitive abyss" opens when the observer's "disembodied eye" is "faced with its own presence in the world it presumes to observe."[12] Thoreau stepped over this abyss daily. Only nature as reified "object" can be labeled and cataloged, but he will instead experience nature as *process*, possible only when the observer understands his participant status. This awareness is triggered by the observer's startled recognition that nature is *looking back*. Thoreau's *Journal* is peppered with moments of fascinated shock when he finds himself the object of an animal's (or even a plant's) gaze, reminding him that he too is visible, that he is not some undercover agent. His *own* "cover" is always "off." It is this very awareness that makes true reciprocity possible, and Thoreau signalizes it by allowing, at key moments of vision, "the wild" to enter. The "pretty game" with the loon ends at the moment the loon calls on "the god of loons to aid him," and the smooth playing board of the pond is roughened into tumult, forcing the man to call off the game—and call it for what it was: a fictive creation, dependent on reciprocity. The independent action of the loon reopens the void between them, and through the void rushes chaos—the wild.

In this instance, typical of his craft in *Walden*, Thoreau's mediating act of writing the scene into coherence, making it "poetic," invokes chaos only to contain it firmly within the narrative frame, as potential, interesting, or even fascinating, but not as threat. But there were moments when the gentle play of reciprocity broke down, and the smooth surface of his poetry broke apart. His reiterated love of "the wild" presumed a partnership in sympathy between his own innermost nature and the outer "wildness" of forest and meadow. But in the one notorious moment when he sought out "real" wilderness, nature not yet constructed or mediated by man, he encountered, and established, an absolute limit where no basis for reciprocal knowing existed. Personified "Nature," on Ktaadn, had nothing to say to him but a challenge: "why came Ye here before your time—" (2:339–40). Nor did he have any means of approach. "Unhandselled" nature offered nothing "to hand," nothing for him to grasp. In this state of detachment her gaze looks not at but straight through him, penetrating his own hollow and fragile ribcage in a rhetorical reversal of Emerson's all-dissolving mind. Absolute matter resists the mind absolutely, in a moment of frustrated revelation which Thoreau dramatized in the published version as his cry of "*Contact! Contact!*" (MW [646]). This was, as I have argued, the payoff of the Emersonian apocalypse of the mind, and Thoreau answered it not by reasserting the absolute power of the mind against matter, but by reestablishing means of "contact," and arguing the importance of mutuality such that contact might be sustained, not broken.

Another moment, far less polished, centered not on the dissolution but on the objectification of his own bodily flesh. In this moment, the "Not-Me" of nature embraced the repellent "Not-Me" of his own body, in another fallback to a dividing and indifferent Emersonianism. In July 1850, Thoreau was dispatched to the site of the Fire Island shipwreck to recover the bodies and effects of Margaret Fuller Ossoli and her family. He arrived five days after the event, to find little left. All he could bring back was a button he ripped off the Marchese Ossoli's coat.[13] His recorded reaction mixes revulsion and awe. The evident insignificance of the "actual" posed a puzzle in its starkest form: How could there be any "real" connection between the "actual," governed by chance or accident, and the ordering imagination?

> I find the actual to be far less real to me than the imagined. Why this singular prominence & importance is given to the former I do not know. . . . I have never met with anything so truly visionary and accidental as some actual events. They have affected me less than my dreams Whatever actually happens to a man is wonderfully trivial & insignificant—even to death itself I imagine. (3:94–95; after 7/29/50)

He concludes that such casual and meaningless "actual" events cannot *really* touch us: "He complains of the fates who drown him that they do not touch *him*. They do not deal directly with him." The button itself represents the implacable absurdity of this contradiction:

> I have in my pocket a button which I ripped off the coat of the marquis of Ossoli on the sea shore the other day—held up it intercepts the light & casts a shadow, an *actual* button so called— And yet all the life it is connected with is less substantial to me than my faintest dreams[.] (3:95)

After writing in this vein for several lines, Thoreau veers into revulsion:

> I do not think much of the actual
> It is something which we have long since done with. It is a sort of vomit in which the unclean love to wallow[.]

He then tries to brush off the sight (of an unidentified body) with contempt: "There was nothing at all remarkable about them they were simply some bones lying on the beach. They would not detain a walker there more than so much sea weed. I should think that the fates would not take the trouble to show me any bones again. I so slightly appreciate the favor[.]" His final move in this sequence is to upbraid himself, incite himself to worthier effort: "Do a little more of that work which you have sometime confessed to be good—which you feel that society & your justest judge rightly demands of you—" (3:95). Similar questions and resolutions continue for several more pages.

Then, three months later, he returns to the experience, this time casting it in the form of a narrative: "I once went in search of the relics of a human body—," he begins. What he finds on the "smooth & bare" beach this time has a kind of dignity: "there lay the relics in a certain state—rendered perfectly inoffensive to both bodily & spiritual eye by the surrounding scenery—." The body establishes a curious community, alone with the beach and the sea, "whose hollow roar seemed addressed to the ears of the departed—articulate speech to them. . . . It reigned over the shore—that dead body possessed the shore as no living one could— It showed a title to the sands which no living ruler could.—" (3:127–28; after 10/31/50). The living Thoreau is not entitled to share this community, yet his own "possession" of a body gives him a point of contact which enables him to feel respect rather than alienation or revulsion. Two kingdoms front each other, dead and living, natural and spiritual, linked in a moment of recognition by Thoreau's intuition of fatality. In other words, three months after the actual event Thoreau can reimagine physicality as a channel of communication between differences, instead of being repelled by it as a kind of "vomit" polluting the spiritually unclean.

Nevertheless, there is a cosmic coolness to his later reaction that shocks us, strikes us as inhuman, certainly less sympathetic than the spontaneous horror of his initial response. Two and a half years later Thoreau had occasion again to meditate on chance and the body, and he exhibits an even more unnerving detachment. One morning in January 1853 the powder mills four miles from his house blew up, and he immediately hitched a ride to the scene. In his *Journal* he describes what he saw in careful, journalistic detail: the shape and color of the cloud of smoke, the order and location of the explosions, the appearances of the various buildings, remnants of timber, clothing, and, last of all, bodies:

> Some of the clothes of the men were in the tops of the trees, where undoubtedly their bodies had been and left them. The bodies were naked and black, some limbs and bowels here and there, and a head at a distance from its trunk. The feet were bare; the hair singed to a crisp. I smelt the powder half a mile before I got there. Put the different buildings thirty rods apart, and then but one will blow up at a time. (IV:453–55; 1/7/53)

Richard Bridgman finds the opening sentence an intolerable attempt at mordant wit, though it may be an artifact of Thoreau's compressed and "dry" style in the retelling.[14] The objectification of the body is partly Thoreau's, but more is suggested by the closing punchline. The real objectification is that instituted by the factory system, which first renders workmen into bodies, which then take care of the business of "the absent proprietor"—who might, Thoreau pointedly adds, have so spaced the buildings that the explosion of one need not have set off, in sequence, each of the others.

Two days later, the sublime music of the "telegraph harp" causes him to reflect less coolly on the incident:

> Day before yesterday I looked at the mangled and blackened bodies of men which had been blown up by powder, and felt that the lives of men were not innocent, and that there was an avenging power in nature. Today I hear this immortal melody, while the west wind is blowing balmily on my cheek, and methinks a roseate sunset is preparing. Are there not two powers? (IV:459; 1/9/53)

In a similar vein, shortly after John's death he had observed, "There seem to be two sides to this world presented us at different times—as we see things in growth or dissolution—in life or death," in God's eye, or the eye of memory; "if we see nature as pausing immediately all mortifies and decays—but seen as progressing she is beautiful" (1:372; 3/14/42). Part of what he explores with his concept of "the wild" is the paradox of these two "powers" or "sides," for it becomes apparent to him that they are not separable, that a nature "progressing" and hence "beautiful" must simultaneously be mortifying and decaying.

Months after the explosion the progress of nature unexpectedly casts up a reminder: a neighbor happens to show Thoreau "quite a pile" of fragments of blackened timber which he had been collecting as they drifted by on the river. Thoreau reads them as they drift down to the sea, slowly dispersing and collecting as driftwood, and the one image leads him to the other that still haunted him: "To see a man lying all bare, lank, and tender on the rocks. . . ." The actual event sinks into history and is forgotten by men, or turned into anecdote, but nature remembers. "It is long before Nature forgets it. How slowly the ruins are being dispersed!" The human mind may pretend to the power to dissolve and disperse, but nature carries forward the material reminders of human error and miscalculation, "still capable of telling how and where they were launched, to those who can read their signs" (V:211–12; 6/1/53). The most telling sign of all is the human body itself, draped like sea-drift on the rocks, revealing without sight its vulnerability to the eyes of all: "To see a man lying all bare, lank, and tender on the rocks, like a skinned frog or lizard! We did not suspect that he was made of such cold, tender, clammy substance before." "A man," not just "the body," is made of "substance," and he cannot banish materiality without banishing himself; there is something troubled and lingering in Thoreau's repetition of the word "tender" against the old granitic truth of the rocks.

Chance does govern the body, as Emerson understood. His response was to abolish chance through the power of the mind. Thoreau's response was similarly detached or "philosophical" but less aggressive: first, to rationalize or ameliorate misfortune by positing social or spiritual causes, which suggested the possibility of control through improved engineering or intensified aspiration. Failing that, to embrace the uncontrollable as part of life's "progress" and therefore beautiful, an astringent but bracing tonic: the laughter of the loon who engages us in a game, but then calls down the wild "god of loons" to his aid.

Chance recalls for us the limitations of our control: the illusion of the contemplative eye, secure in the objectivity of its God vision, is shattered when it is abruptly confronted by its status as a participant. The smooth surface of the social arrangements by which explosives are manufactured, distributed, and deployed is torn apart by the "accident" that reveals the underlying contradictions. Nature cooperates with our attempts to master and control only up to the point at which "she" reasserts her native rights to resist and withdraw, like the loon on the pond, leaving the smooth surface of our social narrative in tumult and disorder. Thoreau records these moments—whether playful like his game with the loon, serious like Ktaadn, or dangerous like Fire Island—as "frontier" experiences, those moments when the mind "fronts" a fact, and he must save his skin if he can. We—the industrializing United States—can delude ourselves into believing that by pouring a mold of the social and civil over nature we have "tamed" it, preserving ourselves from change and from chance. But Thoreau insists that, though human society may thus fortify itself against the wild, individual human beings are still both members of society and inhabitants of nature. He models his own life as a "border" existence between the civil and natural, demonstrating to his readers that they too are borderers. Social arrangements cannot "tame" nature, and a society which withdraws from its source is false, sterile, and doomed.

"Walking, or the Wild"

To heal this incipient division between nature and man, Thoreau recommends that we all learn to live as borderers. His "frontier" encounters with inhuman nature inscribe a necessary outer limit, and accordingly he does not exclude but cultivates and ceremonializes them through language. But he lives not out on the American frontier nor on the edge of the official (and rapidly receding) "wilderness" but, like the rest of us, in a compromised and cultivated landscape. It is within this middle earth that he inscribes his border life, what he calls a "living way" (with the religious overtones intact) in his poem "The Old Marlborough Road." The old and disused road goes nowhere really—or, if you can follow it, "round the world." This poem was published in June 1862 as part of the long essay "Walking, or the Wild," which was composed of Thoreau's two favorite lectures from the 1850s. From it the Sierra Club borrowed its motto: "In Wildness is the preservation of the World." In "Walking" Thoreau poses what was for him a crucial question: if one surrenders the destination to the journey, how does one weave the civil and the wild together in a life, a "living way," sacrificing neither, allowing each to complete the other?

Thoreau's focus in "Walking" is not on a place or a "center" but on a process, "walking," which conducts him to a state of being: "the wild." His working dichotomy is established and characterized in his opening sentence: "I wish to speak a word for Nature, for absolute freedom and wildness, as contrasted with a freedom and culture merely civil,—to regard man as an inhabitant, or a part and parcel of Nature, rather than a member of society" (NHE 93). He would establish a perspective from which he can "regard" the civil and social, and the natural and wild, not as the antagonists they might appear to be, but as allies, linked by the "free" individual who is both a member of society and "part and parcel" of nature. This will mean "speaking for Nature," since our ears are less attuned to her voice than his; but by the end, he hopes to have conducted us to the ability to listen not just to him, but to her.

The first misconception Thoreau must dispel is that "walking" means either progress toward a destination, or a mode of "taking exercise" (97). To trade the destination for the journey means turning off the "track," or the mainstream networks of commodification and exchange—as he comments, most property is not private yet, but the day is coming when it will be, and the true walker will be prohibited as a trespasser. His word for this walker without bounds precipitates one of his loveliest puns: the "Saunterer," derived, he proposes, "'from idle people who roved about the country, in the Middle Ages, and asked charity, under pretense of going *à la Sainte Terre*,' to the Holy Land. . . ." Those who never go to the Holy Land are indeed idlers and vagabonds, he agrees, but those who do "are saunterers in the good sense, such as I mean" (93). Or perhaps, he speculates further, the word derives "from *sans terre*, without land or a home, which, therefore, in the good sense, will mean, having no particular home, but equally at home everywhere. For this is the secret of successful sauntering" (93). Rejection of property and a goal releases one into the state of at-homeness everywhere, where everywhere is Holy. *All* the Land is Holy, and the true saunterer will feel himself at home in every part of it, in a state of "absolute freedom" that can be (and often is) denied by the government, but can be conferred only by the self.

This freedom partakes of "wildness," a notion he nowhere defines but shapes over the course of the essay (if not, indeed, of his career) to stand in or for, not a certain bounded place, but a life principle vital to all places: "Life consists with wildness. The most alive is the wildest" (114). "Walking" becomes a metaphor for right living, and as he asserts, every walk/life is "a sort of crusade": the Holy Land may lie at our feet and all around us, but it will not be found without effort and peril. This is, quite rightly, "an extreme statement," one of Thoreau's extra-vagances: to recover the holy is to *see* that the familiar, known, and ordinary is unknown and strange, that familiarity is but the old incrustation we must break through to approach something "for the thousandth time" as "totally strange" (XII:371–72; 10/4/59). The world that we think is discovered and named is eternally undiscovered and unnamed: "I walk out into a nature such as the old prophets and poets, Menu, Moses, Homer, Chaucer, walked in. You may name it America, but it is not America; neither Americus Vespucius, nor Columbus, nor the rest were the discoverers of it" (NHE 102). It is one of his favorite wordplays again: you cannot discover what was never "covered" to begin with. Look around! The only "covers" are in the mind. Names are the most effective covers of all:

> Whatever aid is to be derived from the use of a scientific term, we can never begin to see anything as it is so long as we remember the scientific term which always our ignorance has imposed on it. Natural objects and phenomena are in this sense forever wild and unnamed by us. (XIII:141; 2/12/60)

The "wild and unnamed" recalls the wild and un*tamed.* Naming is the attempt to tame, unnaming is wilding—uncoupling the old representational dyad between subject and object to release the object out of our world, into its own.

Hence Thoreau can call up the unfamiliar with ease, even after walking his neighborhood daily for a decade: "An absolutely new prospect is a great happiness, and I can still get this any afternoon. Two or three hours' walking will carry me to as strange a country as I expect ever to see." Does he have in mind some wonder of nature, some unsuspected scenic vista? "A single farmhouse which I had not seen before is sometimes as good as the dominions of the King of Dahomey." To *this* "king of the homey," there is more in the landscape even within the twin limits of a ten mile radius and "three-score years and ten" than can ever be known: "It will never become quite familiar to you" (NHE 99–100).

To the skeptical he closes the essay with three instances, demonstrating the transfiguration of the ordinary. "I took a walk on Spaulding's Farm the other afternoon," the first casually begins; nowhere special, we all know the place. "I saw the setting sun lighting up the opposite side of a stately pine wood." We have all seen such, have we not?

> Its golden rays straggled into the aisles of the wood as into some noble hall. I was impressed as if some ancient and altogether admirable and shining family had settled there in that part of the land called Concord, unknown to me,—to whom the sun was servant,—who had not gone into society in the village,—who had not been called on. I saw their park, their pleasure-ground, beyond through the wood, in Spaulding's cranberry-meadow. The pines furnished them with gables as they grew. Their house was not obvious to vision; the trees grew through it. I do not know whether I heard the sounds of a suppressed hilarity or not. . . .

This "family"—so familiar, so strange—might be the neighbors who inhabit the great utopian hall Thoreau envisioned in *Walden* (243–44). Their "coat-of-arms is simply a lichen. I saw it painted on the pines and oaks," Thoreau attests. They have no politics, they do not labor, but he detects "the sound of their thinking," a sweet musical hum. It is difficult to remember them, he ends, but he can after serious effort; their "cohabitancy" seems to redeem all Concord (NHE 131–32).

Or perhaps we are village skeptics who require material proof? On another day, near the end of June, Thoreau climbed a tall white pine, to find "on the ends of the topmost branches only, a few minute and delicate red cone-like blossoms, the fertile flower of the white pine looking heavenward." He carried away to the village the "topmost spire" and showed it to the other villagers, "and not one had ever seen the like before, but they wondered as at a star dropped down" (133). But yet more wonderful still is the fact that such things are not rare, but routine, as he shows in his third instance: "We had a remarkable sunset one day last November. I was walking in a meadow. . . ." Perhaps you remember the day, he seems to invite us. Just before setting, after a cold, gray day, the sun

> reached a clear stratum in the horizon, and the softest, brightest morning sunlight fell on the dry grass and on the stems of the trees in the opposite horizon and on the leaves of the shrub oaks on the hillside, while our shadows stretched long over the meadow eastward, as if we were the only motes in its beams. It was such a light as we could not have imagined a moment before, and the air also was so warm and serene that nothing was wanting to make a paradise of that meadow. When we reflected that this was not a solitary phenomenon, never to happen again, but that it would happen forever and ever, an infinite number of evenings, and cheer and reassure the latest child that walked there, it was more glorious still. (134–35)

Furthermore, the sun lavishes the same splendor on "some retired meadow" as on the city, on a "solitary marsh hawk" as on ourselves. Every least corner of the land is susceptible to transfiguration: "The west side of every wood and rising ground gleamed like the boundary of Elysium, and the sun on our backs seemed like a gentle herdsman driving us home at evening." "So," he ends the essay, "we saunter toward the Holy Land," till one day the sun shines into our minds, hearts, and lives "with a great awakening light, as warm and serene and golden as on a bankside in autumn" (135–36). Then at last we will see we were there all along: this is not *Walden*'s "morning work" but *evening* work, when the clear late band of sun, "morning light," awakens at last the walker at the end of day.

The "West" into which the sun sets is the defining myth of this essay. Where the sun sets is evening, aging, ripeness, the burning glow of scarlet oaks in autumn, the mellow and golden light of sunset, Elysium, paradise just beyond the western hills:

> Every sunset which I witness inspires me with the desire to go to a West as distant and as fair as that into which the sun goes down. He appears to migrate westward daily, and tempt us to follow him. He is the Great Western Pioneer whom the nations follow. We dream all night of those mountain-ridges in the horizon, though they may be of vapor only, which were last gilded by his rays. (107)

Toward the West is where the sauntering Thoreau directs his steps, for the West is also freedom, not enclosure, an unexhausted landscape, destiny, the future, enterprise, adventure, aspiration. All life tends toward this West: "The needles of the pine, / All to the west incline" (CP 22). "'Westward the star of empire takes its way'" (NHE 111). The course of life as a productive process conducts toward the West: "From the East light; from the West fruit," runs his epigram (109). Thoreau reclaims the rhetoric of Manifest Destiny to his own use: a mythology of life's journey which conducts us to no place on the map but everywhere on the globe, everywhere toward the sunset, "west" of our past and cultural inscriptions. "We go eastward to realize history and study the works of art and literature, retracing the steps of the race; we go westward as into the future, with a spirit of enterprise and adventure" (106).

Despite his universalizing metaphors, the Myth of the West does apply with special force to his own homeland and century: he universalizes the local, but nevertheless claims unique status for his own locale, "America." This special, historical authority enables him to make his next, famous move, which reiterates the peculiar burden of America but reads from this local fact a global truth:

> The West of which I speak is but another name for the Wild; and what I have been preparing to say is, that in Wildness is the preservation of the World. Every tree sends its fibres forth in search of the Wild. The cities import it at any price. Men plow and sail for it. From the forest and wilderness come the tonics and barks which brace mankind. (112)

The paradox of the West is its simultaneous movement toward freedom, future, and unexhausted landscape—and to sunset, closure, and exhaustion. "The Wild" both intensifies and expands this paradox. "The Wild" means not only the unsettled West but healing tonics and barks, wolf's milk and hemlock tea, the violence of "devouring" raw flesh: "Give me a wildness whose glance no civilization can endure," Thoreau intones, "—as if we lived on the marrow of koodoos devoured raw." "Walking" becomes an extended meditation on death and generation. "I believe in the forest, and in the meadow, and in the night in which the corn grows," he declares (113). The sources of life lie not in the sunny open fields, but in dark forests, the muck of meadows,[15] the night of growth and restoration. Images of compacted dualities pile one on another. Nature would perish of too much "informing light," for "actinism," the chemical action of the sun, is repaired mysteriously in the night, which is thus proven necessary even to inorganic creation (126).[16] As day depends on night and growth needs the dark, the future rests on decay in the present. So not every man, nor every part of a man, should be cultivated, any more than every acre of earth: "part will be tillage, but the greater part will be meadow and forest, not only serving an immediate use, but preparing a mould against a distant future, by the annual decay of the vegetation which it supports" (126). The fertility of a farm requires muck from the meadows; "A township where one primitive forest waves above while another primitive forest rots below,—such a town is fitted to raise not only corn and potatoes, but poets and philosophers for the coming ages" (117). The raw marrow of life is found in the rotting heart of the darkest wood, the thickest swamp: "I enter a swamp as a sacred place, a *sanctum sanctorum*. There is the strength, the marrow, of Nature" (116). Even disease may prophesy health (121). Thoreau sums up: "In short, all good things are wild and free." The paradox even informs our most intimate relations: "The wildness of the savage is but a faint symbol of the awful ferity with which good men and lovers meet" (122). An extravagant assertion? To be sure, but then it is not indifference that builds us and tears us apart.

The doubleness, the "awful ferity" of generative nature, is captured in his memorable image: "Here is this vast, savage, howling mother of ours, Nature, lying all around, with such beauty, and such affection for her children, as the leopard . . ." (125). Generation and predation figured as one, a mother with claws: the suggestion recalls the close of *Walden*'s "Spring," where Thoreau again declares for the "tonic of wildness," of marshes and the mysterious and unexplorable, infinitely wild, "unsurveyed and unfathomed by us because unfathomable" (WA 317–18). Thoreau steers straight for the outrageous:

> We can never have enough of Nature. We must be refreshed by the sight of inexhaustible vigor, vast and Titanic features, the sea-coast with its wrecks, the wilderness with its living and its decaying trees. . . . We need to witness our own limits transgressed, and some life pasturing freely where we never wander. We are cheered when we observe the vulture feeding on the carrion which disgusts and disheartens us and deriving health and strength from the repast.

The stench of a dead horse in a hollow forced him for a time out of his pathway, but he was compensated by "the assurance it gave me of the strong appetite and inviolable health of Nature. . . ." He continues:

> I love to see that Nature is so rife with life that myriads can be afforded to be sacrificed and suffered to prey on one another; that tender organizations can be so serenely squashed out of existence like pulp,—tadpoles which herons gobble up, and tortoises and toads run over in the road; and that sometimes it has rained flesh and blood! (WA 318)

Transgressions indeed: our bodily bounds and the bounds of good ethical (let alone aesthetic) taste lie in shambles. What are we to make of this weird and potent concoction of vigor and putrescence, sunlight and secrecy, violence and affection, offal and the sublime?

Thoreau has discovered chaos. According to Hacking, the American philosopher C. S. Peirce was fond of saying that he opened his eyes and chance poured in.[17] So with Thoreau: he opened his eyes and saw, in the streets, fields, and forests, chaos: not the ancient void out of which man created pristine order, but a new insight into the imbrication of all order with disorder, disorder with the emergence of order, the *self*-organizing power of a chaotic nature quite apart from human desire or even presence. The great surging flood of nature bore on her breast lilies, mud, and corrosive insects, mud turtles and musquashes, burnt timbers and the bones of men. Thoreau found exhilaration in the spectacle of it all:

> —What a novel life to be introduced to a dead sucker floating on the water in the spring!— Where was it spawned pray? The sucker is so recent —so unexpected—so unrememberable so unanticipatable a creation— While so many institutions are gone by the board and we are despairing of men & of ourselves there seems to be life even in a dead sucker—whose fellows at least are alive. The world never looks more recent or promising— religion philosophy poetry—than when viewed from this point. To see a sucker tossing on the spring flood—its swelling imbricated breast heaving up a bait to not-despairing gulls— It is a strong & a strengthening sight. Is the world coming to an end?— Ask the chubs. (4:450; 4/15/52)

Balm and penetration, generation and violence: Nature was the "wildest" fact of all, deadly and "cimeter"-sweet:

> If you stand right fronting and face to face to a fact, you will see the sun glimmer on both its surfaces, as if it were a cimeter, and feel its sweet edge dividing you through the heart and marrow, and so you will happily conclude your mortal career. Be it life or death, we crave only reality. (WA 98)

The reality he craved was, he concluded, life *and* death, nature a leopard-mother, murderous and loving.

The *Journal* becomes an extended exploration of the permutations of this idea. For instance, in the midst of a violent spring storm that lasted for days, the kind of storm that "returns nature to her wild estate," "lays prostrate the forest & wrecks the mariner" (4:473; 4/19/52), Thoreau hears a robin singing cheerily in the woods. "His song a singular antagonism & offset to the storm— As if nature said 'Have faith, these *two* things I can do'" (4:478; 4/21/52). There were indeed "two powers" throughout nature, each entwined in the other and both together underwriting "the wild." So Thoreau becomes fond of defending the wild's refinement, "the civilization that consists with wildness. The light that is in night" (4:59; 9/7/51). Wherever there is life, creation pairs with decreation. At the core of the paradox was the earth itself: "Shall the earth be regarded as a graveyard, a necropolis, merely, and not also as a granary filled with the seeds of life? Is not its fertility increased by this decay?" (VI:162; 3/11/54). Thus one of his recurring metaphors is the need of civilization for the "muck from the meadows," verifying his high opinion of real and symbolic "compost": the library may seem a "wilderness" of books, but "Decayed literature makes the richest of all soils," he reminds himself (4:392; 3/16/52).

Natural and social "culture" both amount to "wildernesses" which can never be knowable in their totality, only through selection and intentionality. But their fertile ground provides a space for the beginning of knowledge, that moment of "awakening" for which knowledge has prepared us, when at last "knowledge" is recognized as "positive ignorance," and useful ignorance as our "negative" or "Beautiful" knowledge:

> By long years of patient industry and reading of the newspapers,—for what are the libraries of science but files of newspapers?—a man accumulates a myriad facts, lays them up in his memory, and then when in some spring of his life he saunters abroad into the Great Fields of thought, he, as it were, goes to grass like a horse and leaves all his harness behind in the stable. (NHE 127)

The "higher knowledge" we come to in this second or delayed spring may be nothing more than "a novel and grand surprise on a sudden revelation of the insufficiency of all that we called Knowledge before," the "lighting up of the mist by the sun" (128). As Henry Golemba shows, this moment of useful ignorance is also the beginning of Thoreau's own language, the "wild rhetoric" that creates "the effect of meaning's perpetual immanence," and serves "to develop a style that includes gaps, dissolves, and contradictions." Such rhetoric forces the reader to participate as a co-author in creating meaning—even as Thoreau himself participated as *co*-author in creating meaning from the vastness of a nature riddled with its own gaps and contradictions.[18]

Hence, for instance, three days after exulting in the "unanticipatable" sight of dead suckers tossing on the spring flood, he returns to the image to "co-author" its meaning, which he weaves together from wild nature, Greek mythology, and ancient art:

> The sight of the sucker floating on the meadow at this season affects me singularly. as if it were a fabulous or mythological fish—realizing my *idea* of a fish— It reminds me of pictures of dolphins or of proteus. I see it for what it is—not an actual terrene fish—but the fair symbol of a divine idea—the design of an artist—its color & form—its gills & fins & scales—are perfectly beautiful—because they completely express to my mind what they were intended to express . . . I am serene & satisfied when the birds fly & the fishes swim as in fable, for the moral is not far off. . . . When the events of the day have a mythological character & the most trivial is symbolical. (4:467–68; 4/18/52)

This moment, which ushered in the realization that the year is a circle and inspired his question about the "sights and sounds" that "accompany our life," presents one of the high points in Thoreau's mythology of the material. The fish itself, the *thing* itself, speaks for him; his thought is fully expressed in It. All science is swept aside as he sees the thing in its "primitive" state, the firstness of its idea. As Thoreau says of the poet: "He would be a poet who could impress the winds and streams into his service, to speak for him; who nailed words to their primitive senses . . ." (NHE 120). Poetry finally is distinct from science in offering this mythology of the material: in poetry narrative is abandoned, to reveal that "negative knowledge" which is *not* networked and accumulative and socially useful, like science, but instead frontal and direct. As Thoreau put it, "The science of Humboldt is one thing, poetry is another thing. The poet to-day, notwithstanding all the discoveries of science, and the accumulated learning of mankind, enjoys no advantage over Homer" (119–20). As with Emerson in "The Poet," his ideal overshoots the actuality: Thoreau can quote no poetry that "adequately expresses this yearning for the Wild," though "Mythology comes nearer to it than anything" (120). Mythology carries us back to the wild source: Romulus and Remus suckled by a wolf; the migrations of birds and fish and men bound for the west. Hence "the Wild" in one stroke abolishes history and science, cuts through all the "mud and slush" of opinion, prejudice, tradition, and delusion (WA 97) to return us to the original mystery, the "rock" of truth, the holy grail of an original relation with the universe.

A theme constantly reiterated in the *Journal* is that instant when Thoreau is recalled to the prehuman universe by some sound, odor, or vision. The most reliable courier of all was the wood thrush:

> It lifts and exhilarates me. It is inspiring. It is a medicative draught to my soul. It is an elixir to my eyes and a fountain of youth to all my senses. It changes all hours to an eternal morning. It banishes all trivialness. . . . I long for wildness, a nature which I cannot put my foot through, woods where the wood thrush forever sings, where the hours are early morning ones, and there is dew on the grass, and the day is forever unproved, where I might have a fertile unknown for a soil about me. (V:292–93; 6/22/53)

The wood thrush, as he adds, purified, "preserved and transmitted" the wild to us, in civilization. *It* is Thoreau's unquotable poet, and time and again he tries to transmit its voice to us. His yearning toward it, and toward the wild it unfailingly signifies, grows out of a fundamental belief that makes all the rest—from storms to shipwrecks to the rain of flesh and blood—bearable: "With the liability to accident, we must see how little account is to be made of it. The impression made on a wise man is that of universal innocence." This is Thoreau's protection, his barrier against despair: "Poison is not poisonous after all, nor are any wounds fatal" (WA 318).

Hence he can write, in *Cape Cod*: "The sea-shore is a sort of neutral ground, a most advantageous point from which to contemplate this world." On this ground, another *point d'appui*, washed by waves "too untamable to be familiar," the lesson that man and nature are one comes home: "it occurs to us that we, too, are the product of sea-slime." Physicality connects us all and undoes us—all:

> The carcasses of men and beasts together lie stately up upon its shelf, rotting and bleaching in the sun and waves, and each tide turns them in their beds, and tucks fresh sand under them. There is naked Nature,—inhumanly sincere, wasting no thought on man, nibbling at the cliffy shore where gulls wheel amid the spray. (CC 147)

Thoreau cautions that we should not be so proud as to assume a peculiar malignancy in nature, directed at our fragile selves. *All* bodies in nature are frail, and the process of life devours itself, ourselves a merely accidental part of the picture: nature is "inhumanly sincere." To presume evil is to presume we know *why* any of this happens, and this is the wrong question to ask:

> Science affirms too much. Science assumes to show *why* the lightning strikes a tree, but it does not show us the moral *why* any better than our instincts

> did. It is full of presumption. . . . All the phenomena of nature need [to] be seen from the point of view of wonder and awe, like lightning; and, on the other hand, the lightning itself needs to [be] regarded with serenity, as the most familiar and innocent phenomena are. (IV: 157–58; 6/27/52)

To the moral "why," science answers, "'*Non scio*, I am ignorant'"—and it is unclear that the righteous man, confident of his moral rectitude, can do any better. Indeed, one of the attractions of science is its willingness to forgo the control of nature through moral judgment:

> The skeleton which at first sight excites only a shudder in all mortals becomes at last not only a pure but suggestive and pleasing object to science. The more we know of it, the less we associate it with any goblin of our imaginations. . . . We discover that the only spirit which haunts it is a universal intelligence which has created it in harmony with all nature. Science never saw a ghost, nor does it look for any, but it sees everywhere the traces, and it is itself the agent, of a Universal Intelligence. (VI:4; 12/2/53)

Dead suckers, too, might excite disgust and horror, but Thoreau views them also as "innocent" and therefore beautiful. Their annual recurrence comes to signify the exuberance and plenitude of the spring that can afford to sacrifice multitudes, and so deserves celebration over merely artificial holidays: How insignificant are "chronological cycles" and movable church festivals "compared with the annual phenomena of your life, which fall within your experience! The signs of the zodiac are not nearly of that significance to me that the sight of a dead sucker in the spring is. That is the occasion for an *im*movable festival in my church" (XII:390; 10/16/59). Relational knowledge remains Thoreau's touchstone: look not to the calendar but to the skies you walk under and the ground you walk on.

Thoreau's cycles, inscribed in the phenomena of one year and then another, give him the amplitude to write of the grand and significant cycles of human life, merging them with all natural life—with ultimate life processes; but instead of the grand fatalism of Bryant's "Thanatopsis" or the eternal recurrence of Lyell's "great year," Thoreau tenders a riskier vision. In his beatification of the dreary and the terrible with the beautiful and joyous, Thoreau is closer to Wallace Stevens' sober declaration, in "Sunday Morning," that man, not God, is the creator of his world, and therefore "Death is the mother of beauty." In this, Thoreau's contemporary would be less Emerson than the Whitman who wrote "Out of the Cradle Endlessly Rocking," in which death is the power that gives Whitman poetic speech. Stevens, to sustain this position, also concluded that the universe which contains death must be larger than our superstitions. In "Auroras of Autumn," the apocalyptic northern lights figure the poet's fears of a nature capable of blindly extinguishing the individual, but the long meditation concludes:

> So, then, these lights are not a spell of light,
> A saying out of a cloud, but innocence.
> An innocence of the earth and no false sign
>
> Or symbol of malice.[19]

Clearly both Stevens and Thoreau are in the Emersonian tradition, for the same dialectic animates each of their poetic worlds. But Emerson's world was still a monism of the mind, Reason in the full Coleridgean sense, and his apocalypses were perceptual. He was not capable, any more than Coleridge, of imagining a chaotic nature that organized itself out of disorder. Such an imagination requires that the "otherness" of nature be acknowledged, which will entail moments of existential terror:

> This is nothing until in a single man contained,
> Nothing until this named thing nameless is
> And is destroyed. He opens the door of his house
>
> On flames. The scholar of one candle sees
> An Arctic effulgence flaring on the frame
> Of everything he is. And he feels afraid.[20]

The Arctic lights will not yield to the hubris of imaginative will; the power of the mind to unmake them works only within doors. Fronting the sheer fact of their existence teaches the poet a different lesson. Thoreau's equivalent frontier moments on the rocks of Ktaadn and the sands of Cape Cod shook a similar train of transcendental assurances. Both Thoreau and Stevens found lesser but adequate satisfactions in alternate means of contact, lives lived in view of the ultimate frontier, but preoccupied with the poet's more intimate frontiers of consciousness and perception—which proved sufficiently challenging, even in the ordinary worlds of Concord and New Haven.

Linking these figures of the nineteenth and twentieth centuries might be William James, who late in life quoted with approval his friend Benjamin Paul Blood: "'Not unfortunately the universe is wild—game-flavoured as a hawk's wing. Nature is miracle all. She knows no laws; the same returns not, save to bring the different.'"[21] In a universe that is *not* wild, the jaws of "fate" and "power" close on the protesting subject; Emerson's only solution is to close all gaps, absorb all things back into the two which form a "terrific unity" (CW 4:29). To defend himself against such a "monistic" vision, William James, in *A Pluralistic Universe* (1909), threw up another: in the earliest stage of culture, "Nature, more demonic than divine, is above all things *multifarious*. . . . The symbol of nature at this stage . . . is the sphinx, under whose nourishing breasts the tearing claws are visible" (640). Confronted with the multifarious leopard-sphinx-mother of nature, the intellect awoke, and by "generalizing, simplifying, and subordinating" began those "divergences of conception" which nature deepens rather than effaces, "because objective nature has contributed to both sides impartially, and has let the thinkers emphasize different parts of her, and pile up opposite imaginary supplements" (640). James sought in chance the way to pry open the collapsing jaws of this dualism, opening the uncertainty through which turbulence, wonder, chaos, and the wild can enter. Only thus could "foreignness," the alienation bred by "the notion of the 'one,'" yield to the "intimacy" bred by the notion of the "'many'" (776); and it is in search of this "intimacy" that James defends his pluralism. Monism insists that in the reality of realities, "everything is present to *everything* else in one vast instantaneous co-implicated completeness"—Emerson's "one vast picture, which God paints on the instant eternity" (CW 1:36). By contrast:

> The pluralistic world is thus more like a federal republic than like an empire or a kingdom. However much may be collected, however much may report itself as present at any effective centre of consciousness or action, something else is self-governed and absent and unreduced to unity. (776)

"'Ever not quite'" must be said of any all-inclusive system. Yet it is still a "coherent world" after all, for every part is "in some possible or mediated connexion, with every other part however remote, through the fact that each part hangs together with its very next neighbors in inextricable interfusion" (778). Where did it come from, this glorious Humboldtianism reborn and resplendent with hope and implications for Thoreau's favorite trio, science, philosophy, and literature? Suffice it here to say that in William James the line of empirical holism reemerges and perhaps even culminates:

> I give the name of 'radical empiricism' to my *Weltanshauung*. Empiricism is known as the opposite of rationalism. Rationalism tends to emphasize universals and to make wholes prior to parts . . . Empiricism, on the contrary, lays the explanatory stress upon the part, the elements, the individual, and treats the whole as a collection and the universal as an abstraction. My description of things, accordingly, starts with the parts and makes of the whole a being of the second order. It is essentially a mosaic philosophy, a philosophy of plural facts. . . .[22]

By contrast, ordinary empiricism "has always shown a tendency to do away with the connections of things, and to insist most on the disjunctions." Four years later James calls this "the bugaboo empiricism of the traditional rationalist critics" (778). In William James's last writings, the fortunes of empirical holism would be borne into the twentieth century entwined with and subsumed by the concepts of pluralism and pragmatism.

In a passage that verges on Emerson and looks ahead to James, Thoreau toys with a new pun:

> Trench says a wild man is a *willed* man. Well, then, a man of will who does what he wills or wishes, a man of hope and of the future tense, for not only the obstinate is willed. . . . The perseverance of the saints is positive willedness, not a mere passive willingness. The fates are wild, for they *will*; and the Almighty is wild above all, as fate is. (IV:482; 1/27/53)

Emerson sought to define "Fate" as merely power unrealized, "will" still unmanifested; the presumed outcome is a will which reigns supreme. Thoreau too dissolves "fate" into "will," but the result is the opposite; for "will" is "wild," and fate is "wild above all," untamed and untamable, as saints move not lawfully but in the lawlessness of a "higher" law: "We may study the laws of matter at and for our convenience, but a successful life knows no law. It is an unfortunate discovery certainly, that of a law which binds us where we did not know before that we were bound," he adds, perhaps recalling Emerson's obeisance to the new "laws" of statistics. "Live free, child of the mist. . . . The man who takes the liberty to live is superior to all the laws, by virtue of his relation to the lawmaker" (NHE 128–29). As a "wild" man himself, doing what "he wills or wishes," Thoreau operated with the knowledge that meaning was not authorized by man or by God but "willed" through the act of perception and by the act of writing. Truth, even the truth that flowed from the carefully nurtured seeds of facts, was finally contingent and provisional, framed by the intentions, literary *and* scientific, of the writer, and assisted or frustrated by the "will" or responsiveness of wild nature.

The same particularist vision that led to an epistemology of contact, that let chaos into his universe, changed the form of Thoreau's writing; while in *Walden* he consciously sought to create a cosmos out of the chaotic, both the bulk of the *Journal* and the late essays experiment with disconnection and patterning, creating minimally ordered forms which remain responsive to chaotic processes. The emphasis is no longer on the individual, or on the "whole" (which in any case is no longer knowable), but on the generative equation—the author, Thoreau, in the act of writing the *Journal*, on a nodal point of chaos, bringing it into a new state of organization.

That is, Thoreau is not merely witnessing chaos—the generation of order out of disorder, the disorderliness within order—as an observer on the sidelines. Nor is he simply its tragic or alienated victim, nor yet the God of his world, ordering it by divine agency. He is instead participating in the process of chaotic order, himself an element in his world, a player like the loon, entering into it and entering it into his *Journal.* In the terms of "Walking," Thoreau stands between city and wilderness, nature and civilization, like the wood thrush transmitting each to the other. But as is possible only for the human being, he stands as a member of both the civil and the wild communities, a "member" of society, yet also "part and parcel" of nature, demonstrating how both civil and wild might be—indeed must be, for the "preservation of the World"—integrated throughout the landscape.

Conclusion: Disciplining Thoreau

In the view offered here, Thoreau's nature, and the texts he organized out of his participatory experience, become increasingly decentered, disjunctive, open-ended, multiply created by agents ranging from grass and acorns to pines, squirrels, and loons, mountain rocks to meadow rills, Concord farmers and German scientists to Harvard libraries and the Walden woods, surveyor and engineer to author, to reader. Thoreau opens up the familiar structuring dichotomies—civil and wild, objective and subjective, writing and reading, mind and nature, order and chaos—only to implode them through such kaleidoscopic inclusions, leaving us with a rich compost of pairings and disruptions brought together by repeated pleas for cooperation, for independence and relational knowing rather than the hierarchy of command and obedience.[23]

The terms that I have been pairing throughout this work—civil, order, design, hierarchy, transcendence, distance; wild, chaos, chance, anarchy, embodiment, participation—would seem to resonate with the dichotomies commonly invoked in the name of postmodernism. "Post": what comes after; modern: "the eternal and the immutable in the midst of all the chaos."[24] Or, *Walden* and "what comes after." Yet where shall this sequence be divided? The antagonisms, so useful in the necessary effort to harness the complexities of recent history into periods and a narrative design, are so intimately involved in defining each other that one might ask whether they truly form stages in a sequence, even one only rhetorically implied. Modernism—industrialism, the quickening of the land through the imbreathment of a human life with the divine capacity to organize natural matter into the global network—was turning nature into man. No one was more modern than Emerson, who saw in the quickening the correspondent fragmentation, who sensed impending breakdown as control ceded from the organizing central mind to the proliferation of new facts with unaccountable energies of their own: "Things are in the saddle, / And ride mankind," he wrote in 1846. Modernism and postmodernism ushered each other in, each the other's shadow twin. If modernism lays the tracks, the postmodern dances on the cars, thinking it is all the world that rides on the railroad, or the interstates, or the Internet.

I wish, with Thoreau, to step across the tracks and view them from the saving interval of a few feet to one side. For that way, I wouldn't have to worry like the modern rail rider who can go only where the train will take her. I envy Thoreau the saunterer who could go anywhere from Beck Stow's to Boston when it suited him, and gather his science from Pliny to Darwin to Joe Polis, the Indian guide. Thoreau, in short, lacked discipline.

Thoreau knew he had discipline problems. His refusal as a teacher to discipline his students had led to his resignation from Concord's Center School.[25] His vocational troubles would become part of his legend: teacher, poet, pencil maker, hermit, lecturer, surveyor. When invited to become a member of the emerging discipline of science, Thoreau haughtily mocked the questionnaire in his *Journal*: "The fact is I am a mystic, a transcendentalist, and a natural philosopher to boot. Now I think of it, I should have told them at once that I was a transcendentalist. That would have been the shortest way of telling them that they would not understand my explanations" (V:4; 3/5/53). Yet in 1859, he was appointed to Harvard's Visiting Committee in Natural History, charged with the annual evaluation of the college curriculum; perhaps a pro forma appointment, but suggesting nevertheless that he had come to be considered a member of the scientific establishment.[26] One can nod toward his indeterminate status by saying that his work "anticipated" Darwin or helped to "found" environmentalism, but as we have seen, it is more difficult to take such claims seriously, to understand what they might mean for "our" Thoreau.

The difficulty posed by the hybrid is the resulting need to discipline it, lest it challenge the putative purity of the parent categories. How could a poet end his life counting tree rings? How could a scientist write the "Spring" chapter of *Walden*? Were such boundaries merely personal, redrawing them could be done ad hoc, as needed; but they have tremendous institutional weight. Moreover, they allow us to be productive by sorting the significant from the expendable; as producers of knowledge, our disciplines both delimit and enable us, teaching us the double edge of Foucault's explication: that discipline "'trains'" the useless multitudes into useful, individual elements. "Discipline 'makes' individuals; it is the specific technique of a power that regards individuals both as objects and as instruments of its exercise."[27] Or if not Foucault, then William James: "Sensible reality is too concrete to be entirely manageable. . . . What we do in fact is to *harness up* reality in our conceptual systems in order to drive it the better."[28] For generations "we"—students and practitioners of American literature—have harnessed Thoreau, producing him as a key literary figure, object and instrument of our profession, even as we are aware that other disciplines (philosophy, political economy, rhetoric, ecology and environmental science—even the extradisciplinary realm of popular culture) also claim him. In effect, Thoreau belongs to us only as we have made him ours—as we discipline Thoreau.

The handling of "The Succession of Forest Trees" by literary critics stabilized this cross-disciplinary essay *as* literature, although of a marginal kind: a curiosity, or a mixed success at best. The success of this strategy has been in subordinating texts which are relatively uninteresting to the resources of literary study, allowing us to concentrate productively on those that are more responsive to our disciplinary practice. Yet such acts of categorization come with a cost: here, they eviscerate one of Thoreau's more important essays; more broadly, through such multiple acts of inattention, the literary and the scientific are driven back into their separate domains. While we concentrate on language and metaphor, let "them" deal with the dull collocation of facts: what, after all, do the facts of forest succession have to do with the construction of language? In a stabilized world, little enough; the benefit of ignoring the linguistic (let alone material) construction of nature is that when we wish to mobilize nature for our own needs, we do not encounter difficult questions. "They" have taken care of it for us, assuring us of a ready supply of nature processed into pencils, paper, and electric current, with which we, in sublime indifference to material causes, can inscribe our critiques of naturalization.

Thus "science" is marginalized, but in the wrong direction: not to the praiseworthy margins of race and gender, but to our real and unseen margins, the dull margins, elsewhere. Curious, since science has been the prime instrument for inscribing difference. Why not examine its tools of mastery? The ultimate costs loom like the national debt: it is science that produces the reality we inhabit; it is, as we all wish to forget, not marginal but central. Indeed, we worry it is *we* who are marginal, like the transcendental poet confronting his skeptical audience, in an arrangement that was actively being settled during Thoreau's lifetime and in which he sought to intervene, not by making literary gestures but by attempting to show in all his work how deeply implicated was the maker of "scientific" knowledge—knowledge of the nonhuman—in human knowledge, in the human structures of language, narrative, perception. Facts, Thoreau learned, are not found. Natural facts do not lie in wait to be "dis-covered" and accumulated into structures of scientific knowledge, a reservoir of power for an expanding American society. Facts are fashioned, manufactured, made (*facere*, to make) by human acts, consolidated and circulated through language and artifacts. The disciplinary divide that has helped us to get on with our work has also helped us disguise from ourselves our role in making the world we critique.

Given Thoreau's attempts to intervene in this arrangement, it would become doubly ironic if the body of his texts were to be used to further enact the double flight to the poles of nature-science, and culture-literature. As Bruno Latour demonstrates in *We Have Never Been Modern*, this double movement, which seeks to purify each from the other, proliferates in the creation of hybrids which fill the center, the world itself; but it evacuates the center conceptually, emptying it of all the mixed entities that inhabit the ground somewhere between those pure poles. This emptying precedes and is necessary to what has been called the transcendental, romantic, or even the "American" propensity toward world making, whereby the lone, seminal, percipient self generates out of this void, by the act of perception, the world of (his) desires. In the American way, this new world is then open to use, exploration, appropriation.

But Thoreau offers another style of world making: he can be seen as a Latourian mediator, not emptying but visibly and noisily filling the center where he lives, generating events and documents and entities that are all partly natural, partly social, partly discursive. Thoreau wasn't modern, like Emerson; Thoreau was "nonmodern," able to stand to one side and articulate the marvelous and disturbing hybrid phenomena that were modernism's burgeoning progeny, born of the marriage of human and natural: the farm boy building a waterwheel in a meadow rill; fish and lilies stocking an artificial pond in a cemetery; forests made by squirrels, acorns, and axes; a villager borrowing an axe and building a cabin on the shores of a glacial lake on the edge of town.

"Disciplining" Thoreau means turning him, in Latour's terms, from a bustling "mediator" into an invisible "intermediary" who translates pure, unsullied "nature" to an unnatural society. Thoreau at Walden Pond becomes the intermediary between civilization and wilderness, those two poles tragically rushing away from each other, and which Thoreau must somehow cobble together. But it was precisely that "middle" ground which Thoreau wanted to populate. Not only did he place himself, as Stanley Cavell reminds us, "just far enough to be seen clearly,"[29] but Thoreau works very hard to make us see what a heavily populated place this is and how much hard and unremitting work it takes to keep the boundaries clear, to keep "nature" and "culture" separate. Thus Thoreau postures like Chanticleer: in the middle of everything, borrowing, building, howing, weeding, working and boasting to the neighbors (present and former, human and brute), stowing his furniture out on the grass and warming his hands at his domesticated fire. Finally the point is, as Richard Grusin emphasizes, "extravagance"—the multiplication of intimate contacts, the proliferation of hybrids. Here alone, in connections and complications, lies hope of redemption.

The post-*Walden* Thoreau spent a great deal of time considering practical problems in boundary maintenance. As the town surveyor responsible for determining and mapping boundaries he knew that the boundary between nature and man, woods and fields, was not stable: the woods were constantly violating the fields. Worse, it wasn't even the same *kind* of woods: sometimes they were composed of oaks, sometimes of pines. In a disciplinary world, seeds and boundary invasions become "a purely scientific subject." In a nondisciplined world, they become a problem in dispersal, circulation, and construction: exactly who makes woods? Not God; Thoreau quickly disposes of the old but still current notion of spontaneous generation, with its easy and invisible intermediaries, in favor of real and particular seeds. How, then, are those seeds dispersed? Who grows them, circulates them, stores them, eats them? Here is a field for Thoreauvian mediation, the work of walking, counting, observing, annotating, measuring, asking, reading. These things are happening *on the ground*, and he will be on the ground himself, tracing over many years a web of continual growth and change. It is this focus on *process* rather than product, plus his insistence on material exchanges, not Idea, that will allow his studies to converge on Darwin's; hence it is that in 1860, only months after reading *Origin of Species*, Thoreau was presenting "Darwinian" ideas on which he had been working for years.

Of course his own process of work produced a discursive product, human textual inscriptions in which Thoreau attempted to articulate the nonhuman inscriptions which his walking, seeing, and writing unfolded. The end of all this work was not just to propose new scientific explanations, but to enact and then to model an alternative way of knowing, a situated, narrative science which traced all the multiple exchanges and connections which produced, simultaneously, the forest itself and knowledge about the forest. It took so much work because, as Donna Haraway stresses, "Not just anything can emerge as a fact; not just anything can be seen or done, and so told. Scientific practice may be considered a kind of story-telling practice—a rule-governed, constrained, historically changing craft of narrating the history of nature."[30] Facts and actors in this process proliferated under Thoreau's hands: he would innovate in a new genre, borrow Humboldtian tools to build a counternetwork to the sleek and powerful directives of modernism in America.

So I wish to claim a nonmodern's "view" (or "Ansicht") of Thoreau—a situation whose precarious irony he would have been the first to appreciate. His nineteenth-century project to interlace rather than dominate the cosmos acquainted him with the paradoxical demands of both Emersonian transcendence and Humboldtian immanence, inviting him to puzzle out a narrative pathway along the edge of chaos. By now, nearly a century and a half later, what he experienced perhaps in isolation, we certainly experience as a culture. Like Thoreau and the loon, we too have engaged nature in a perilous dance, and we are even less certain than he that it isn't the "god of loons" who will finally take over and leave the smooth surface of our world in turbulence, driving us—metanarratives, postmodernism, and all—back to shore.

Hence, I believe, it is all the more important to reread Thoreau as something other than the enshrined "hermit" who sought to escape from society into a pure and uncontaminated nature, who planted himself on the ground of our country's "birth" to declare for mornings and purity and new beginnings on an untouched and unexhausted land. The construction of this iconic figure has its own history, but I hope to have made clear my own grounds for skepticism that it continues to be a useful icon, or an accurate reflection of the contested, plural, and thoroughly contaminated world of antebellum American. It was that world, after all, that gave rise to Thoreau's own sense of belatedness and his long attention to the wisdoms not just of spring but of autumn and winter. His writings were posted from the brink of "the evil days" to those who are no longer just on the brink—who indeed look for our apples "in a barrel" (or—unimaginable!—prepackaged in Styrofoam) and who can saunter in search of the Holy Land only by trespassing. To those readers he does not offer the *frisson* of apocalypse or the dystopia of a bounded, framed, faraway "wilderness" to tantalize the sad dwellers in urban and suburban waste lands.

For to draw defended boundaries is to divide, not differentiate, and division so construed can result only in hostile and defended camps: an all-embracing urban network, electric with movement and urgent with intimations of cyberspace; a distant remnant postcard wilderness frozen into nostalgia for a lost Eden. For to assert that "man" can "violate" a "virginal" nature fixes all the stale and destructive dualisms back in their hostile places, and reinstalls our lives along the old metanarrative of the Fall. The Concord woods in 1845 were already the product of centuries, at least, of human "violation," yet to Thoreau they were no less "natural," no less rich with the seeds of redemption, no less every inch the "Holy Land." When Thoreau walked the land he was struck with the way all bounds, even the ones he had himself established, dissolved:

> These farms which I have myself surveyed, these bounds which I have set up, appear dimly still as through a mist; but they have no chemistry to fix them; they fade from the surface of the glass, and the picture which the

> painter painted stands out dimly from beneath. The world with which we are commonly acquainted leaves no trace, and it will have no anniversary. (NHE 130)

What was left after his survey lines faded was an interactive field of possibilities, regeneration even of land laid waste, succession rather than apocalypse, in a landscape constituted by history and preparing in the present the mold—not the determinate stamp, but the yeasty wild compost—for its own future. Indeed, if "the preservation of the World" is in the Wild, it is in the wild seen not as a demarcated zone but as a sustainable process.

At last this narrative, at least, reaches its own contingent end: on December 3, 1860, Henry was out in the rain measuring hickories and white oaks, busy reading "the rotten papyrus" of forest history, as he had put it a few weeks before (XIV:152). He had visited with Alcott, who had a bad cold or influenza, four days earlier, and after his day in the rain contracted a severe cold himself. It was the most trivial of accidents, but he never recovered. His health steadily worsened, and as he was increasingly bound within doors, his *Journal* diminished and then lapsed into silence. For his *Journal*'s final entry, crafted with as much care as his first twenty-four years before, he chose to record the patterns of sand ridges formed by the wind in a storm:

> All this is perfectly distinct to an observant eye, and yet could easily pass unnoticed by most. Thus each wind is self-registering. (XIV:346; 11/3/61)

With these quiet words he concludes the two great themes of his writing: the forms of attention brought to nature by the observant eye; the pattern of order created and registered by the turbulence of chaos, and by Thoreau himself as he reinscribes the self-inscribed form of the wind. Is nature really coming to an end? we might want to ask him—but he's already given us his answer: "Ask the chubs."

Notes

Bibliography

Index

NOTES

Introduction

1. In references to Thoreau's *Journal*, Arabic volume numbers indicate the Princeton Edition, Roman the 1906 edition.

2. "Symbolic" here, and throughout, alludes to Coleridge's formulation of the symbol as "the translucence of the eternal through and in the temporal" (*The Statesman's Manual*, quoted in Bate 164). This relation of temporal and eternal is encapsulated in Emerson's formula in *Nature*, which concludes that "Nature is the symbol of spirit," and affirms that nature can have, therefore, no significance in itself (CW 1:17, 21).

3. The image traces to Nash 94.

4. See Beer 6–7; Knight 3–9.

5. Worster, *Nature's Economy* 60; Sattelmeyer, *Thoreau's Reading* 10.

6. Whewell believed that the integrity of the scientific enterprise was threatened, and as part of the cure he proposed, at the 1833 meeting of the British Association for the Advancement of Science, the term "scientist" to "'designate the students of the knowledge of the material world collectively.'" "Philosopher" he considered and rejected on Coleridge's grounds, as "'too wide and too lofty.'" Whewell's new term both distinguished scientists from artists, and highlighted the "common enterprise" in which various specialists were engaged. Whewell raised the issue and publicized his neologism in the popular press. Men of science who resisted it included Michael Faraday and T. H. Huxley. See Yeo, *Defining Science* 5, 110–11; Knight 381.

7. According to Richard Yeo, "The word 'science' had not entirely lost its earlier meaning of systematic knowledge, or *scientia* . . . and the term was still used synonymously with 'philosophy,' even though the British Association attempted to make 'science' mean natural knowledge" (*Defining Science* 33). The general terms "natural philosophy" and "natural history" were "made redundant by the new vocabulary of specialized subjects," e.g., astronomy, chemistry, botany, geology. For a detailed discussion of the intermixing of these several terms, see Yeo, *Defining Science* 32–38.

8. Wordsworth 606.

9. See Young 126–63 ("Natural Theology, Victorian Periodicals, and the Fragmentation of a Common Context").

10. The allusion is to Foucault, *Archaeology of Knowledge* 37; although I would hardly pretend to offer a full Foucaultian archaeology.

11. Hayles, *Chaos Bound* 4, 208, 176–77.

12. George Levine, "One Culture" 3, 7, 17, 22. For the view of science as fundamentally a form of language, see David Locke.

13. Woolgar 13, 65. Others associated with the strong programme include David Bloor, Michael Lynch, and Bruno Latour.

14. Haraway, "Situated Knowledges" 197, 188–90.

15. See Fischer. The implicit suggestion that Thoreau's project may have been "feminist" is, to be sure, anachronistic, but quite deliberate in ways I suggest elsewhere (see my article "*Walden* as Feminist Manifesto"). Meanwhile, let it be said here that I assume "gender" to be an equally partial and enabling construction site.

16. At issue is the very possibility that "poetry" is what is lawful and universal, while "science" is a proliferation of particulars. This simultaneous inversion of "poetry" and "science" makes their opposition particularly unstable, and helps to confuse any correlation with the Coleridgean symbolic order, which embraces both poetry and science. In Thoreau's later work, either or both can also be forms of discourse.

17. Klein 56–57, 66.

18. Rudwick, *Great Devonian Controversy* 6, 14.

19. Young 134, 186.

20. See Sattelmeyer, *Thoreau's Reading.*

21. See Seybold; Christie; Hodder; Sayre.

22. Charles Darwin, *More Letters* II:323; quoted in Levine, *Darwin and the Novelists* 101. Thoreau wrote his comment in his *Journal* five days after his first introduction to Darwin's *Origin of Species.*

Chapter 1. Facts and Truth

1. Robert Sattelmeyer notes that the Harvard texts in religious studies and philosophy are both the most difficult, and the most important, to assess. See *Thoreau's Reading* 15.

2. See Thompson.

3. "Locke was Thoreau's philosophical father, as he has been of many another poet," concludes Joel Porte, who discusses Thoreau's troubled allegiance to Lockean "sensationalism," and the difficulties it posed for his relationship with Emerson; see *Emerson and Thoreau* 138–40. See also Burbick 8.

4. See the discussion in Thompson 84; and Sattelmeyer, *Thoreau's Reading* 18–19.

5. Charles Darwin, *Autobiography* (1887), 69–70. Both Hildebidle (48) and Richardson (*Henry David Thoreau* 376) take this statement as revelatory of Thoreau's working methods.

6. Note that Locke, too, uses this same topic of slavery falsely justified to prove the dangerous potential of general maxims, which we may use to confirm ourselves in our mistakes: a child could use the maxim "it is impossible for the same thing to

be and not to be" to show that "a negro" is, because not white, "*not* a man" (IV.vii.16).

7. The expression is quoted from Thoreau's class essay of May 5, 1837; see EE 104.

8. See Joyaux. Joyaux describes Cousin's work as a "collateral contribution" and a "catalyst" for the American assimilation of German thought (136). Emerson read Cousin's *History of Philosophy*, a series of lectures given in 1828, in French upon its publication, and again in 1832, in its English translation. Thoreau borrowed it—probably at Orestes Brownson's recommendation—from the library of the Institute of 1770 in June 1837, and renewed it in July; in following years they both read widely in the writings of Cousin's disciples, Jouffrey and Constant. See Sattelmeyer, *Thoreau's Reading* 20–22.

9. Furthermore, his American promoters defended him on those same grounds. See Orestes Brownson, "Victor Cousin" (1836): Brownson stressed that Cousin's system was founded on the facts of "psychology," hence was solidly based on the modern, experimental, sensual philosophy of "Descartes, Bacon, and Locke," which it extended and corrected (108–9). It is essential to Brownson that a philosophical system be thus "scientific," and the "facts" on which it must be based are, for him and for Cousin, *thoughts*: "nature is thought, and God is its personality" (112). For a rebuttal that lumps Cousin and Emerson together as enemies of Christianity, see J. W. Alexander, Albert Dod, and Charles Hodge, "Transcendentalism of the Germans and of Cousin and Its Influence on Opinion in This Country" (1839).

10. See also Burbick 7.

11. See Bozeman 8–10.

12. Cayton 70–71.

13. Bozeman 89.

14. Quoted in Wylie 5.

15. Cayton 17, 33–36. Cayton documents the perceived decline of community in Emerson's Boston, in which conservatives enlisted the ideology of Federalism to help "salvage the idealized hierarchical and deferential world of colonial Massachusetts" against the "chaos and potential disaster" of the liberals' proposed democratic social order. See 8–9.

16. Bloor 608.

17. See Zochert 469–70. According to Neal Gillespie, in "Divine Design and the Industrial Revolution: William Paley's Abortive Reform of Natural Theology," Paley constructed *Natural Theology* as an irresistible synthesis of Newtonian mechanics and Ray's natural history, to avert mechanism and materialism; but ironically, the next generation rejected it as overly mechanistic. Bozeman, in *Protestants in an Age of Science*, establishes how powerful, conservative Presbyterians effectively domesticated science, transforming Bacon and Newton "into exemplars of Christian piety" and scientists into admiring priests "of the beautiful Creation." Finally, George H. Daniels, in "The Process of Professionalization in American Science," explains the vogue of natural theology among antebellum scientists: both older, established scientists and an equally precarious emergent science enlisted "common cultural values" such as Protestant theology on their behalf, a justification with which they dispensed once American science had secured a position for itself.

18. See, for example, Dahlstrand. According to Russel Blain Nye's standard history, *Society and Culture in America, 1830-1860*, "Baconian" science "assumed that science was empirical, avoiding speculation and hypothecation. In general, science meant taxonomy, the creation of systems; it aimed primarily at discovering, identifying, and classifying information" (239). Both Nye and Dahlstrand quote George Daniels' paradigmatic formulation, in *American Science in the Age of Jackson*: by the 1850s, "science suffered from a 'deluge of facts,' so much new information that it could not be fitted into the Baconian taxonomical framework" (239). In Daniels' own words, "Early fact gathering, being largely random and undirected, had resulted in an overwhelming mass of undigested data that ultimately became a source of embarrassment and confusion" (102). Having collected so much information that it became "unmanageable," since it refused to "arrange itself automatically," Baconianism reached its limits and the bankrupt system collapsed in the 1850s (106)—the same decade in which Thoreau too was allegedly losing the battle with Baconian science. See also note 23, below.

19. Bacon 518.

20. Harriet Beecher Stowe, in *Oldtown Folks*, captures the spirit of the Protestant natural theologian in her portrait of the New England schoolmaster Mr. Jonathan Rossiter:

> He had a ponderous herbarium, of some forty or fifty folios, of his own collection and arrangement, over which he gloated with affectionate pride. He had a fine mineralogical cabinet; and there was scarcely a ledge of rocks within a circuit of twelve miles that had not resounded to the tap of his stone hammer and furnished specimens for his collection; and he had an entomologic collection, where luckless bugs impaled on steel pins stuck in thin sheets of cork struggled away a melancholy existence, martyrs to the taste for science. (1295)

Rossiter started his pupils on similar researches, and soon his disciples were gathering their own collections: "It was fashionable in his school to have private herbariums and cabinets, and before a month was passed our garret-room began to look quite like a grotto" (1296).

21. Thomas Jefferson, president of the American Philosophical Society from 1797 to 1815, pronounced himself a disciple of Bacon, Newton, and Locke, and perhaps his prominence accounts for some of the Baconian stereotype. Charles A. Miller, in *Jefferson and Nature*, discusses Jefferson's Lockean theory of knowledge, which, distrusting whatever could not be sensed or made useful, "set boundaries to his curiosity and therefore to his knowledge" (27). Fearing to fall into error, Jefferson held that it was "better to have no ideas, than false ones," and in his naiveté he missed the value of a *false* hypothesis (though Stewart himself stresses this point). Thus he could reject evidence of meteors, while believing mastodons still lived, because the latter claim was congruent with his doctrine of the perfection of nature (29–31). And as for his procedure as a naturalist: "he often does not define the questions he wants to answer; rather, his initial step was to accumulate data, nearly any data," at which

he was "compulsive" (41). John C. Greene, in *American Science in the Age of Jefferson*, helps fill in the particulars:

> He had . . . a passion for collecting facts, and a mania for measurement. He measured all kinds of things: the trunks of trees, the height of mules, the weight of peas and strawberries grown on his plantation, the time it took a workman to fill a wheelbarrow and roll it thirty yards, the time it took a stone dropped in the fountain at Nîmes to reach the bottom of the pool, the dimensions of the arches in the amphitheater at Arles, the size of the Roman bricks in the Bordeaux circus, how long it took to pass through the locks at Bézières, even the time it took to dig the grave of his deceased friend Dabney Carter. (30)

While it sounds like Baconianism gone mad, Miller makes the point that Jefferson's practice was finally not at all Baconian in the sense of reasoning inductively, for he forced nature "to submit to his own system, and where it would not, he preferred to leave the field rather than revise his premises. While he felt obligated to search out the pieces of nature's puzzle and fit them together properly, he did not intend to question the frame" (53–54).

22. Cannon, *Science in Culture* 73–74. A lengthy and fascinating examination of the nineteenth-century debate over Baconianism is presented by Jonathan Smith in a book published just as this volume goes to press; see J. Smith, *Fact and Feeling*. Smith is substantially in agreement with the argument presented here.

23. See also Hopkins, "Emerson and Bacon" 419.

24. In this vein, William H. Goetzmann, in "Paradigm Lost," offers an alternative interpretation of the explorations that characterized this "Second Great Age of Discovery" and the resulting "information overload" that so disturbs Daniels. The expeditions, launched by "entire nations and whole societies, merchants, dilettantes, and even state legislatures," returned an "avalanche of new data" which existing institutions and philosophical structures were inadequate to process; "No system provided in advance for the unexpected on a vast scale. So the world itself became a large 'anomaly,' with the result that knowledge gathering did not undergo one dramatic scientific revolution but a sense of continual evolving crises" (24–25). One response was the romantic "implosion" that characterized such investigators as Oken, Schelling, Emerson, and Thoreau, as they "gave up coping with the data problem on a systems level and settled for synecdoche," in a reaction central to "the Romantic movement and eventually to the self-oriented perceptions of modernism." Yet most thinkers in Western Europe and America faced the possibilities with a "zest" that resulted in a proliferation of new ways of looking at nature, and a new paradigm of exploration and discovery in which Europe and America, then civilization itself, "become ever more efficient information-processing systems" (24–25, 34). While I would question his characterization of Thoreau (who steeped himself in the writings and reports of the very expeditions Goetzmann is studying), Goetzmann's alternative paradigm coincides with the ideas I am advancing here.

25. Desmond 353; Desmond offers an exhaustive documentation of this claim. The more problematic figure here is Asa Gray, the "Baconian" natural theologian who was, famously, Agassiz's enemy and Darwin's sounding board and ally. See Dupree.

26. Goddard 161. "'Tis quite certain that . . . the dull men will be Lockists," wrote Emerson in *English Traits* (CW, V:135; quoted in Porte, *Emerson and Thoreau* 138).

27. For a discussion of the American institutionalization of nature as a prophylactic against a diseased society, see Donna Haraway, "Teddy Bear Patriarchy," in *Primate Visions* 26–58.

28. For Linnaean science in this context, and the figure of the naturalist as quizzical unconquering conquering hero, see Pratt, esp. 23–37, 56–57.

29. See Levine, *Darwin and the Novelists* 47–55, for a discussion of this view and the contrast, only implied here, with the argument from chance, change, and undesign.

30. I am indebted to Mary Louise Pratt for clarifying this point to me. See 56–57, 86–102.

31. Sattelmeyer, *Thoreau's Reading* 29. For an extended consideration of Thoreau and Lyell in the 1840s, see Rossi, "Poetry and Progress: Thoreau, Lyell, and the Geological Principles of *A Week.*"

32. Gould, *Time's Arrow* 143.

33. As Gould shows, this passage is not at all intended by the "hero of rational empiricism" as "comic relief," but is part of his strenuous defense of the cylicity of time against the emerging fossil evidence for organic progress (*Time's Arrow* 103–4).

34. Harding, *Days* 291; Richardson, *Henry David Thoreau* 184.

35. As Robert Sattelmeyer notes, "Among the natural phenomena that Thoreau spent his life studying, fish possessed a striking capacity not only to interest him but also to stir him emotionally and to provoke in him a twin sense of wonder and kinship." See "'True Industry for Poets'" 196.

36. See Sattelmeyer, "Thoreau's Projected Work on the English Poets."

37. See Henry Petroski, "H.D. Thoreau, Engineer," for a wonderful account of this too-little-known side of Thoreau; reprinted in his book *The Pencil.*

38. William Rossi has usefully explicated this oscillation through the Lyellian metaphor of flood and lapse. See "Poetry and Progress: Thoreau, Lyell, and the Geological Principles of *A Week.*" Also useful is Robert D. Richardson, Jr.'s, discussion of Thoreau's comments on science in the "Friday" chapter of *A Week*, which compacts statements and ideas drawn from over a decade of Thoreau's writing; see "Thoreau and Science" 112–15.

39. Burbick (5) also remarks that Thoreau's writings are "Janus-like." See also Michaels; Michaels discusses how the hierarchies established by the dualisms like nature/culture, finite/infinite, and literal/figurative are always breaking down, as each term argues with the other.

40. Coleridge, "Dejection, An Ode" (1802), ll. 47–48.

Chapter 2. The Empire of Thought and the Republic of Particulars

1. James 644: no one philosophy can embrace all of nature; "The most a philosophy can hope for is not to lock out any interest forever." Or as Michel Serres shows, every polar totality creates a "third man" who is locked out; shutting out the noise of his presence allows communication, while admitting him admits change.

2. Bozeman 90.

3. Turner 98: "Belief during the first two-thirds of the nineteenth century did not rest on rationalism repudiated but on rationalism romanticized."

4. Cayton 73. See Alexander et al. 234: "Every English and American reader must fail to penetrate even the husk of German and mock-German philosophy, unless he has accepted the distinction between the *reason* and the *understanding*. . . ." Jonathan Arac (83–85) supplies an incisive commentary on the New Critical maneuver which excluded "Coleridge's most crucial term, the Reason" in order to elevate the imagination, "for Coleridge a 'mediatory' power" between Reason, the senses, and the understanding. By excluding the sun of Reason, the New Critics harmonized poetry and religion in Coleridge, eliminating "the question of religion as displacing Coleridge's poetic activity"; and by setting imagination in opposition to understanding, "thus repeated the characteristic New Critical dichotomy of poetry and science," which even yet defines the problem of this project.

5. What René Wellek (202) summarizes as "the Romantic revolt against materialistic science" established itself, then, as "'a return to law in literature and the general mind.' Law seems here to mean a non-mechanistic order of the universe" (quoting Emerson, "Life and Letters in New England" [1867], *Complete Works* X:338). According to Trevor Levere, science was fundamental to Coleridge because it *related* "mind to nature, the ideal to the real, and had to be incorporated into his system" (2). Hopkins, in *Spires of Form* (122–26), acknowledges a similar course for Emerson. For more on Emerson and science, see Robinson; also Richardson, *Emerson: The Mind on Fire*, which appeared as this book went to press.

6. See Rehbock 20–21: "And equally important, phenomena which appeared to violate the laws were not of great concern, since their inconsistency could be caused by inadequate interpretation, or by the incomplete state of the science."

7. Compare Ian Wylie's remarks: "Coleridge seems to argue that the statement 'Nature is harmonious' is true a posteriori, through observation of how the world is. Thus, by assuming that nature is God's language, he deduces that God is a perfect being. In fact, like most of his age, his argument is a-priori: God is by definition a perfect being; nature is God's language (by assumption); thus nature is harmonious (by deduction). He would not however have admitted this" (87).

8. Desmond 365. For the radical dualism of law as divine decree, see also James 642; for a discussion of law as theological sanction vs. law as order and pattern, see Levine, *Darwin and the Novelists*, esp. 89–94.

9. See Bozeman 15–16, 58. Compare with the position Walter [Susan Faye] Cannon establishes for John Herschel: "the universe is, at a high level of generalization, simple, harmonious, and rational; but not necessarily so." We can fit empirical data into a rational scheme, but only because God wills it so. As knowledge reveals the infinite complexity, it destroys our hope of understanding the totality of the system, while assuring that the search can go on forever. So for Herschel, rationality

is fundamentally contingent ("John Herschel and the Idea of Science" 226). By contrast, Bozeman's Protestants could not have accepted the final consequence of their argument—the contingency of rationality—as it was developed by Herschel. Law was still Logos, and their commitment was still to Enlightenment principles: facts by themselves were "inarticulate, 'isolated,' 'incoherent,' even 'hostile' until reduced to a system of order." In this "disdain for the singular" they were conversant with "facts" only as they indicated universal "laws," as they served as fodder for truths and systems (159).

10. See Wylie 76–77.

11. See Serres 29–38. Serres finds in the duel of polar forces the nineteenth century's signature metaphor: in the Carnot cycle of thermodynamics, energy flow from hot to cold, circulation of fluids—electrical, chemical, atmospheric, oceanic, bodily—and spiritual?—through the systems defined by the poles of positive and negative, high and low pressure, equatorial heat and polar ice. "The world is a static machine, a compression engine, an electrical engine, a chemical machine, a steam engine; the world is an organism—all without contradiction" (35).

12. See Wellek 201–4.

13. Jehlen 84.

14. Ibid. 102–3.

15. For Thoreau's impatience with the imperfection of words, see Burbick, 55–57.

16. See Jehlen on the deep contradiction inherent in Emerson's conception of action: why, she asks, does he insist on oppositions, only to roll them, repeatedly, into unity? "The reason lies . . . in the capacity of a nonantagonistic dualism to generate activity despite the absence of a dialectical impulse. This activity is reproductive rather than productive, an enactment instead of a true action. Nonetheless it constitutes movement, whereas transcendent unity implies stasis" (98). According to Jehlen, then, Emerson stresses his complementary dualisms precisely to permit the "action" (or enactment) that a realized transcendental unity would prohibit.

17. The quotation is from the Library of America edition, p. 28; the *Collected Works* text reads: "More and more, with every thought, does his kingdom stretch over things, until the world becomes, at least, only a realized will,—the double of the man" (CW 1:25).

18. Hopkins, *Spires of Form* 15.

19. The allusion is to Wallace Stevens' poem "Sunday Morning."

20. This project is unable, it should be noted, to do full justice to the contradictions and complexities of Emerson's thought, which recently have engaged a number of commentators. This discussion represents a single dimension of Emerson—but, I maintain, a crucial one: that which insists that nature is, ultimately, no more than "'*scoriæ* of the substantial thoughts of the Creator'" (CW 1:23). "*Scoriæ*," from the same root as "scatological," refers to excrement or offal.

21. Cayton 77.

22. De Man 202, 210.

23. Ibid. 203; Knights 4, 35–36.

24. Knights 20, 55. Knights argues that Coleridge began by declaring subject/object dualism as the philosophy of death, and his the philosophy of freedom; but his idea of freedom is defensive, against invasion by forces outside the mind's control. He splits the noumenal self from the phenomenal, predicating the one as "free," the

other as in "bondage," because Coleridge "identified the empirical self with the fearful, cowed, fantasizing self." Reason became self-discipline and mastery over the understanding, identified with animality and vice; the system was repressive, and never achieved, but the imperative colored his work (46–51). Knights concludes that Coleridge's idealism represented a need for an immutable world whose existence would guarantee intellectual freedom from natural pressures, "from the crushing burdens of inhabiting a body." Monism appeared to offer "a bulwark against the perils of historicity" (53).

25. Jehlen 108–9; Knights 51. Ian Wylie, in *Young Coleridge*, observes that although Coleridge is occasionally credited with a special sensitivity to detail, he actually retreated from obdurate detail to a higher sphere from which he could legislate the world by thought, escaping "the menacing power of the particular by subsuming it within a larger conception" (42–43).

26. See Knights 34–35; also Hopkins, *Spires of Form* 55. Or as Wallace Stevens mourns in "Credences of Summer": "It was difficult to sing in face / Of the object . . ." (376).

27. Knights 32–33: one of the consequences of the imperative to moral purity as the triumph over the promptings of the animal self is "the rooted suspicion of material nature, including all the biological and economic processes by virtue of which life exists." The natural world was a less than suitable subject for the highest study. High speculation does not sully itself with the empirical study of nature or practicality; the rationalists, with their patronizing attitude, "did much to legitimate a situation in which natural science was (and is still, by a series of long-institutionalized suspicions and associations) debarred from its rightful place in general culture."

28. Is it necessary to remark that Mary Shelley's *Frankenstein* (1817) is still *the* paradigmatic myth of romantic science, right down to "Dr. Strangelove," Michael Crichton's *Jurassic Park*, and the dystopias of cyberpunk?

29. According to contemporary theorists like Steven Woolgar, Bruno Latour and Donna Haraway, science *is* society. Latour demonstrates how, under the pressure of controversy, literature becomes technical—not because intellectuals are disappearing down obscure tangents, but through the logic of argumentation. The literature of controversy is "made to isolate the reader by bringing in many more resources. . . . The power of rhetoric lies in making the dissenter feel lonely" (*Science in Action* 44). The more elements the author can muster on his side, the more effectively he accomplishes this isolation of the dissenter: "the more technical and specialised a literature is, the more 'social' it becomes, since the *number of associations* necessary to drive readers out and force them into accepting a claim as a fact increase [*sic*]." Science proceeds by "linkages, resources, and allies"—as Latour adds, "if being isolated, besieged, and left without allies and supporters is not a social act, then nothing is" (62).

30. Haraway, *Crystals, Fabrics, and Fields*: strictly speaking, "organicism" should exclude nonphysical entities, and seek to explain "wholeness, directedness, and regulation" without such notions (34).

31. Quoted in Knights 41, from *Collected Letters* IV:762; see also Coleridge, *Philosophical Lectures* 196.

32. Quoted in Knights 65.

33. See Cayton 42–51.

34. Cayton 63, quoting JMN 2:65.

35. Cayton 63–64, 74–75. For another version of Emerson's corporate selfhood, see Newfield 657–84.

36. The Platonic dialogues of interest here are the *Phaedrus*, in which Plato states that "any discourse ought to be constructed like a living creature," that is, with head and feet, "a middle and extremities so composed as to suit each other and the whole work" (510); the *Gorgias*, in which he makes the same point through a discussion of painters, builders, and shipwrights (286); and the *Theaetetus*, which contains an unresolved argument about whether "the whole" and "the sum" are the same (911–16). The pieces are locked in place by Aristotle, who in the *Poetics* established that, in poetry, to transpose or withdraw any one of the parts "will disjoin or dislocate the whole" (1463).

37. See "Mechanical" and "Organic" in Williams 201–2, 227–29.

38. See Meyer H. Abrams, *The Mirror and the Lamp* 170–74.

39. Jehlen 86–87.

40. Compare Kauffman 4: "For Kant, organisms were fundamentally self-reproducing, and therefore *self-organizing wholes*. . . . While this point of view might seem mere common sense, we shall see that it has dwindled from an operant role in contemporary biology." Kauffman is pioneering self-organization in evolutionary biology.

41. Foucault, *The Order of Things* 127–28.

42. Lenoir 24–26.

43. Both Anne Macpherson, in "The Human Geography of Alexander von Humboldt," and Margarita Bowen, in *Empiricism and Geographical Thought*, have developed Humboldt's connection with Kant at length. "Accepting neither the vitalist nor the mechanist terms in which such problems had been conceived throughout the last century, he pursued a different approach and resumed his preparation for a general science" (Bowen 221). This preparation began what Bowen calls his "epic" search for "an adequate theory" that would act as "a necessary counterpart to the science of his time" (258). See also Nicolson, "Alexander von Humboldt, Humboldtian Science, and the Origins of the Study of Vegetation" 170–71.

44. From *Florae Fribergensis* (Berlin, 1793), on plant physiology, especially photosynthesis in subterranean plants; quoted in Nicolson, "Alexander von Humboldt, Humboldtian Science, and the Origins of the Study of Vegetation" 174–75. Nicolson remarks that this was Humboldt's first public statement of "his ideas for a new science of plant geography"; Helmut de Terra, that it first caught Goethe's attention (*Humboldt* 57).

45. See Whitford 293. For Humboldt as the progenitor of ecology, see Nicolson, "Alexander von Humboldt, Humboldtian Science, and the Origins of the Study of Vegetation" 183–87; Donald Worster, *Nature's Economy* 131–37.

46. Quoted in Hentschel 105; see also de Terra, *Humboldt* 74; Löwenberg et al. I:188. Löwenberg claims that Schiller came to regret his censure of Humboldt, and that their friendship remained strong through Schiller's early death, in 1805 (I:220).

47. Erwin Ackerknecht goes so far as to say that Humboldt "was never a romantic," and that Humboldt's violent denunciations of *Naturphilosophie* actually helped to overthrow it (92). On the other hand, Nicolson insists that *Cosmos* is "redolent of the Romantic tradition," and contains "many passages which give high praise to the *Naturphilosophen*" ("Alexander von Humboldt and the Geography of Vegetation" 178). Humboldt is either bivalent—or nonpolar.

48. See Ackerknecht 88, 92; also Walls, "'Napoleon of Science.'"

49. Lenoir 14, 278, 55, 112.

50. Quoted in Gould, "Church, Humboldt, and Darwin" 104. Much later, in his *Autobiography* (1887), Darwin would recall reading, as a student at Cambridge, Humboldt's *Personal Narrative* "with care and profound interest. . . . This work and Sir J. Herschel's *Introduction to the Study of Natural Philosophy* stirred up in me a burning zeal to add even the most humble contribution to the noble structure of Natural Science. No one or a dozen other books influenced me nearly so much as these two" (67–68).

51. I am most indebted to Susan Faye Cannon, *Science in Culture*, who asserts that Humboldt offered an alternative view which must be understood on its own terms. In this line, see also Anne Macpherson, "The Human Geography of Alexander von Humboldt," and Margarita Bowen, *Empiricism and Geographical Thought.*

Studies structured around the traditional opposition of "philosophical" and "Baconian" science have trouble placing Humboldt. William H. Goetzmann, in *New Lands, New Men*, groups Humboldt with the idealist *Naturphilosophen*, as do the treatments by Adolph Meyer-Abich and Malcom Nicolson. A "Baconian" Humboldt appears in Janet Browne's *The Secular Ark*, which traces the study of "geographical distribution" to Humboldt, as "an inquiry of the highest philosophical order" which searched for "the meaning of nature" along a route alternative to the anatomical studies of the *Naturphilosophen* (42); yet finally, Humboldt figures more memorably as the perpetrator of perverse and pointless statistical methodology (80). Philip Rehbock, in *The Philosophical Naturalists*, also designates "distributionist studies" as the opposite of "an idealist approach," but in grouping both as two "independent" aspects of a philosophical or transcendental biology opposed to "Baconian" induction, he neutralizes their key differences, and ultimately absorbs what is distinctive about "distributionist" science back into idealism. He, too, ends with an impoverished, "Baconian" Humboldt who analyzed the earth's surface through statistics (193).

52. Browne 80.

53. Ibid. 75; see also Cannon, *Science in Culture* 96.

54. This is demonstrated by those fields in natural and social science related to population thinking. Ernst Mayr (863 n. 6) states that no one "has yet traced the connection" between empiricism and the various forms of population thinking, both of which stress the individual over the ideal type, but it seems possible that Humboldtian thinking connects the two.

55. For example, see George H. Daniels' contemptuous dismissal, in *American Science in the Age of Jackson*, of Alphonse de Candolle's 1855 work *Géographie botanique raisonnée*. This Humboldtian work (which Thoreau read) attempted an innovative use of statistics to group data into meaningful patterns. On Darwin's debt to de Candolle, see Browne 82–85; Thoreau's debt to de Candolle is most evident in

The Dispersion of Seeds. Richard Yeo's analysis of the rhetoric of Baconianism in nineteenth-century Britain concludes that "a concentration on Bacon as the theorist of induction correlated with a suppression or avoidance of eighteenth century readings of Bacon as the author of a radical philosophy capable of transforming not only natural knowledge, but established social institutions and values" ("Idol" 288). In this view, Humboldt was indeed Baconian, but in the radical eighteenth-century sense.

56. See Bowen 230, 241–42. For instance, in a population analysis of Cuba, Humboldt asks his readers to consider the fact that "the copper-coloured men" form 83 percent of the population of the West Indies; if their conditions were not improved soon, they would take political power into their own hands—and quite rightfully so: "the political preponderance will pass into the hands of those who have strength to labour, will to be free, and courage to endure long privations." Yet the planters refuse to believe that blacks can act together; "Such are the illusions which prevail amidst the great mass of the planters of the West Indies . . . " (*Personal Narrative* [1852] III:233).

57. Cannon, *Science in Culture* 76–77, 105.

58. See Bowen 215, 217.

59. Ibid. 274, 257.

60. Bowen identifies such theories as "social empiricism": "The new trend is towards a kind of empiricism that gives assent to the complex process by which scientific research is linked with a paradigm of shared concepts and theories in a social and historical context" (266).

61. Humboldt, *Essai sur la géographie des plantes* (Paris, 1807), translated by Malcolm Nicolson, and quoted in Nicolson, "Humboldtian Science and the Origins of the Study of Vegetation" 176–77. Humboldt dedicated this work to Goethe.

62. James 645. James continues: "the pluralistic view which I prefer to adopt is willing to believe that there may ultimately never be an all-form at all, that the substance of reality may never get totally collected . . . and that a distributive form of reality, the *each*-form, is logically as acceptable and empirically as probable as the all-form commonly acquiesced in as so obviously the self-evident thing." Clearly James has picked up the trail of empirical holism, although I do not yet know of any direct connection between James and Humboldt; the lines of communication remain to be worked out. See also below, chap. 6.

63. See Löwenberg et al. I:166–78.

64. The quotation is from the Sabine translation of *Cosmos* I:3. An interesting example of the "negative" pole, in the identity/severance relationship of science and morality, is offered by Robert Chambers in *Vestiges of the Natural History of Creation* (1844), a natural theology of evolution. Chambers, unable to reconcile the contradiction of a universe governed by physical laws which produce moral wrong, finally declares that God established two sets of laws, physical and moral, and "the two sets of laws are independent of each other. Obedience to each gives only its own proper advantage, not the advantage proper to the other" (376).

65. Quoted in Meyer-Abich, "Humboldt as a Biologist" 181.

Chapter 3. Seeing New Worlds: Thoreau and Humboldtian Science

1. McIntosh 23, 235, 298–99.

2. Recent biographies of Humboldt include Botting; de Terra, *Humboldt, the Life and Times of Alexander von Humboldt*; Kellner.

3. Browne 38, 41. Humboldt's book was *Florae Fribergensis specimen plantas cryptogamicas præsertim subterraneas exhibens* (Berlin, 1793), written in Latin.

4. Quoted in de Terra, *Humboldt* 87.

5. Humboldt's homosexuality stands as an intriguing subtext to any narrative of his life and any discussion of his work. As Mary Louise Pratt remarks, it continues to be treated by his commentators "in a gentlemanly fashion, which is to say, as a dirty secret." Humboldt "was sustained by a series of enduring intimate relationships," including the French physicist Louis Gay-Lussac and, later and most famously, the astronomer François Arago. De Terra portrays his attachments as passionate, turbulent, even tormented; however, any study must contend with the systematic censorship of his personal papers, by Humboldt and his family, and with their accidental destruction concluding, most dramatically, with the sacking of the family estate at Tegel during World War II. Pratt suggests a connection between Humboldt's wanderlust and his need to escape "the heterosexist and matrimonialist structures of bourgeois society. The history of travel and science is significantly shaped by the fact that they were legitimate contexts for same-sex intimacy and exclusively male society" (240 n. 10).

6. This point impressed Agassiz—see his *Address* 19.

7. *Ansichten der Natur* was published in Germany in three editions, in 1808, 1826, and 1849. There were two competing English translations of the third edition; when I wish to avoid designating either one, I will refer to this work as *Ansichten.* Mrs. Elizabeth Juliana Sabine's translation was published as *Aspects of Nature* (London: Longman, Brown, Green, and Longman's; and John Murray, 1849). It was republished in Philadelphia (Lea and Blanchard, 1850). E. C. Otté and Henry G. Bohn published their translation as *Views of Nature* (London: Henry G. Bohn, [January] 1850).

8. *Kosmos* was published in Germany in five volumes (I: 1845, II: 1847, III: 1850, IV: 1858, V: 1862). Both major English translations of *Kosmos* bore the same main title, *Cosmos.* Neither was complete; the fifth German volume has never been translated. Mrs. Elizabeth Juliana Sabine's translation of volumes I and II was published in London beginning in 1846. E. C. Otté's translation, of volumes I–IV, was published in London by Henry G. Bohn in five volumes from 1849 to 1858; it was reprinted in New York by Harper and Brothers, 1850–70.

9. See Nelkin, *Alexander von Humboldt, His Portraits and Their Artists.*

10. Meyer-Abich, *Humboldt* 46.

11. Cannon, *Science in Culture* 81, 75.

12. Hamilton 33, 29.

13. Cannon, *Science in Culture* 95, 97, 75–76; the quotation is from Humboldt's *Personal Narrative* (1818) I:39.

14. Cannon, *Science in Culture* 96.

15. The connection between Church and Humboldt is well established. A good starting point into this literature is Stephen Jay Gould's essay "Church, Humboldt, and

Darwin," along with the other essays and references in *Frederic Edwin Church* (1989). The most famous of Church's paintings was the "blockbuster," *Heart of the Andes* (1859), which caused a sensation at the 1864 New York Sanitary Fair exhibition where it was hung in competition with Bierstadt's *Rocky Mountains—Lander's Peak* (1863). Viewers rented opera glasses to compare the two earthscapes, which together "were, in their attention to both materialistic detail and romantic concerns, peculiarly American macrocosms. . . ." Bierstadt's "majestically nationalistic" image of North America, however, won hands down (Goetzmann and Goetzmann 154–57).

16. Mary Louise Pratt, in *Imperial Eyes*, explores the causes and consequences of Humboldt's reputation in South America, while I treat the course of Humboldt's reputation in the antebellum United States in my article "'The Napoleon of Science': Alexander von Humboldt in Antebellum America." The collapse of his reputation surely accounts for the otherwise inexplicable ignorance of Humboldt among contemporary literary historians (though Wilhelm von Humboldt's work has a solid place in literary theory). Such parenthetical and slighting references as do occur hardly entice the student to investigate further. One authoritative volume on Thoreau and natural history, for instance, briefly mentions (and effectively dismisses) "the great German traveler Humboldt, a civil servant and mining engineer" (Hildebidle 25).

17. Goetzmann, *New Lands, New Men* 176.

18. For the full text of the poem, see Agassiz, *Address* 86–88.

19. Ulrich-Dieter Oppitz lists and annotates them all in "Der Name der Brüder Humboldt in Aller Welt." Care to travel to Las Vegas, Humboldt? At the last minute, the new state was named Nevada instead.

20. Grefe 31; Stoddard 446.

21. Quoted from "Columbus," in the London *Quarterly Review* 64 (1839): 47. In 1860, Daniel Coit Gilman wrote that "Humboldt called himself half an American; and others designated him as the scientific Columbus, who revealed to the old world the natural wonders of the new" (280). In its own obituary of Humboldt, the *American Journal of Arts and Sciences* (2d ser., 28 [1859]: 161–65) is more careful to qualify the phrase, crediting Humboldt's journey as "substantially, the scientific discovery of Spanish America" (164); most other writers dropped the geographic modifier, allowing them to claim Humboldt as particularly "American." See, for instance, the memorial lecture of Edward Everett, whose 1823 article in the *North American Review* had introduced Humboldt to American readers: "all honor should be paid to his memory on this side of the Atlantic; for the greatest scientific achievement of his life—his American voyage—was performed on the soil of this continent" (170).

22. See Foner.

23. Bozeman observes that the conservative Presbyterians welcomed Humboldt's *Cosmos* as "a massive attempt to exhibit the interconnection of all phenomena within a single and providential 'cosmic' scheme. Old School reviewers greeted it as an ultimate demonstration of 'phenomenal harmony'" (85).

24. Ingersoll 117. The Sabine translation of *Kosmos* interpolates, at a key point, the assurance that the first impulse of creation was given by "the Creator" (I:33), a phrase that does not appear in the competing, and more easily available, Otté translation. Otherwise, though his tone is often spiritual, Humboldt's writing lacks all religious reference, except as anthropology or literature.

25. De Terra, *Humboldt* 178–201.

26. Glaab; Smith 42–46.

27. Emmons; Smith 43. An extensive application of Humboldtian science to America's continental destiny appears in Arnold Guyot, *Earth and Man* (1849). A postbellum extension is George Perkins Marsh, *Man and Nature* (1864).

28. See Bowen 260, for a discussion of Dove's influence on Humboldt's posthumous reputation. See also Macpherson 12–14; she concludes that Dove admired speculative philosophy, failed to comprehend Humboldt, and "rejects any influence of Kant on Humboldt," systematically distorting and belittling him.

29. Löwenberg et al. II:415, 359.

30. Bowen 260, 248.

31. Ibid. 261.

32. The narrative was probably written not by the captain himself but by Jessie Frémont, his wife and anonymous amanuensis: see Viola 67, 80. Once again, a woman appears in this literature as the transparent translator and intermediary of knowledge, joining Mrs. Elizabeth Cary Agassiz, Mrs. Elizabeth Juliana Sabine, Helen Maria Williams, Thomasina Ross.

33. Bruno Latour (*We Have Never Been Modern* 113) warns against conflating science with nature, "ignoring the work of instrumentation"—*all* measures "construct a commensurability that did not exist before their own calibration. Nothing is, by itself, either reducible or irreducible to anything else. Never by itself, but always through the mediation of another." The next step, then, is "to point out what instruments and what chains serve to create asymmetries and equalities, hierarchies and differences."

34. Herber 23–24.

35. Cannon, *Science in Culture* 75.

36. See Richardson, *Henry David Thoreau* 196–97. Thoreau's total debt for printing *A Week* was $290. On October 27, 1853, the unsold copies were delivered to him, resulting in the famous witticism "I now have a library of nearly nine hundred volumes, over seven hundred of which I wrote myself" (V:459). He finally settled the account on November 28, 1853 (V:521).

37. The editors of Thoreau's *Journal*, vol. 3, identify the unnamed friend as "almost certainly Emerson" (405); for an account of their estrangement, see the "Historical Introduction" (485–88). Amos Bronson Alcott also contributed some criticism about this time; see Borst 156.

38. Harding, *Days* 235, 461.

39. Ibid. 267.

40. In stating this I am disagreeing with Robert Sattelmeyer and Richard A. Hocks, in "Thoreau and Coleridge's *Theory of Life*." They are entirely correct in observing that Thoreau's attention to Coleridge took place at a "critical juncture of his career," but I believe that Coleridge offered Thoreau the solace of the familiar. He cannot have offered Thoreau the needed "bridge between the transcendental idealism of his youth and the detailed and scientific study of nature that occupied his maturity" (279), for he neither can nor wishes to provide "a theory and a methodology" (270) for linking imagination with the sort of field-based and highly particularized natural

history study Thoreau was about to undertake. Sattelmeyer and Hocks do not mention the Humboldt abstracts.

41. See the facsimile reprint of Thoreau's *Literary Notebook* (362). The extract reads in full (my transcription):

> Fragment from a lost work of Aristotle preserved by Cicero in De Natura [?] ii.37; quoted in Humboldts Cosmos [II:29].
>
> "If there were beings living in the depths of the earth, in habitations adorned with statues and paintings, and everything which is possessed in abundance by those whom we call fortunate, and if these beings should receive tidings of the dominion and power of the gods, and should then be brought from their hidden dwelling places to the surface which we inhabit, and should suddenly behold the earth, and the sea, and the vault of heaven; should perceive the broad expanse of the clouds and the strength of the winds; should admire the sun in his majesty, beauty, and effulgence; and, lastly, when night veiled the earth in darkness, should gaze on the starry firmament, the waxing and waning moon, and the stars rising and setting in their unchanging course, ordained from eternity, they would, of a truth, exclaim, 'There are gods, and such great things are their work.'"
>
> The following is quoted by Alexander von Humboldt in his Kosmos as a "doctrine of Krishna" & he refers in a note to "Wilhelm von Humboldt on an episode of the Maha Bharata, in his collected work"
>
> "Truth was originally deposited with men, but gradually slumbered and was forgotten; the knowledge of it returns like a recollection."

42. Sattelmeyer and Hocks 277.

43. The allusion is to Humboldt's *Personal Narrative*, which Emerson owned in W. MacGillivrey's severely abridged and altered edition, *The Travels and Researches of Alexander von Humboldt* (1833).

44. Humboldt provides Thoreau with a technical reason why the sound of a waterfall is louder at night, enabling Thoreau to draw the attractive conclusion that "nature sympathised with his experiments" (3:51–52; after 4/1/50).

45. Harding, *Days* 275–76.

46. Ibid. 268–69.

47. See especially Harding, *Thoreau as Seen by His Contemporaries*.

48. Richardson, *Henry David Thoreau* 339; Borst 356–57. John Hildebidle, in *A Naturalist's Liberty*, observes that this passage from Darwin, and his method in general, does "serve to tame the sublime" (44), a useful corrective; any relational approach will have a humanizing, or hybridizing, effect. Even the alienated and cruel mother on Ktaadn speaks and is heard.

49. Dupree 232; Beer 6.

Chapter 4. Cosmos: Knowing as Worlding

1. Matthiessen 86–87, 92–93, 90. See also Christie 203, 232, on the fusion of material and immaterial as Thoreau's peculiar strength.

2. Here I can review only a handful of landmarks. A "triumph" position can be seen in J. Lyndon Shanley, who argued that Thoreau's expressions of defeat were only an occasional mood, commonplace among reflective people, and data collecting was "a deeper and satisfying exploration of the world that Thoreau wanted to know"; Shanley expressed puzzlement at those who "apparently think that exact knowledge must inevitably rob one of imagination and insight" (*Making of Walden* 8–9; see also Canby and Krutch).

A large contingent weighs in on the side of failure and defeat: Charles Feidelson, Jr., attributes the "disintegration of [Thoreau's] visionary power" to his growing awareness that he can present only a relative fact, not absolute poetic experience, and so slides into the diminished role of an amateur naturalist (137–38). To Perry Miller, Thoreau's dogged respect for the thing resulted in a dangerously split or double vision. The sheer difficulty of sustaining its precarious balance "killed him. But not until, at least in *Walden*, he had for a breathless moment, held the two in solution. . . ." Thoreau's tragic dénouement came after *Walden*, when the war of Transcendentalist vs. Natural Historian became "desperate," and Thoreau sank into the abyss of scientism. (See "Thoreau in the Context of International Romanticism" 150–51; and *Consciousness in Concord* 166, 183.) Sherman Paul wrote movingly about Thoreau's loss of "communion" in the years after 1850, his genuine "feeling of emptiness" in a nature gone "barren." In the bitter end his *Journal* became "a repository of scientific facts," and he himself "a scientist" who used measuring instruments, examined the droppings of crows and foxes, "and even killed a mouse." "Every critic," Paul asserts, "has recognized this change in Thoreau and has seen in it the sign of his lapse and failure" (See 256, 264, 274, 395–97). Nina Baym, in the influential article "Thoreau's View of Science," established that Thoreau's failure was double: he approached nature through a scientific life that was conceived as a holy quest, but failed; he failed again when he learned that science was increasingly objective and he was forced to reject it as false.

Critics like Frederick Garber, William Howarth, Robert D. Richardson, Jr., and H. Daniel Peck have recently offered more positive interpretations of Thoreau's late career, loosening the assumptions that underwrite the "balance" metaphor. Most recently, Robert Sattelmeyer and William Rossi have been reconceiving the question entirely, as, rather obviously, do I.

3. Sattelmeyer, "Remaking of *Walden*" 440. See, for instance, Krutch 182: "The completely unsystematic, almost desperately pointless character of his own quasi-scientific recordings is evidence enough that he did not really grasp what slight philosophical implications the vast enterprise of collecting and cataloguing did have; and it is quite obvious, from his various derogatory references to science, that in his mind it stood, not (as it might have) for an attempt to penetrate the secrets of life, but for the mere assembling of meaningless details."

4. Richardson, *Henry David Thoreau* 248; *Journal* 3:496.

5. Beer 42; James 645; see also 826, where James justifies his term: "The directly apprehended universe needs, in short, no extraneous trans-empirical connective support, but possesses in its own right a concatenated or continuous structure."

6. Channing 35–36. Channing often went along with Thoreau on his walks and excursions.

7. Christie 42–43.

8. Richardson, *Henry David Thoreau* 246; Christie 238. For more on the Indian notebooks, see Sayre, esp. chap. 4.

9. Harding, *Days* 290, 269.

10. Channing 275–76; Cameron 63–64; Harding, *Days* 290.

11. Richardson, *Henry David Thoreau* 227.

12. The handbill is reprinted in Walter Harding, *Days* 461.

13. For another approach to this aspect of Thoreau's work habits, see Hildebidle 48–50.

14. Sanborn 65–66.

15. For Thoreau's maps, see Stowell.

16. Nikita E. Pokrovsky also pointed this out in an unpublished paper delivered at the 50th Anniversary Meeting of the Thoreau Society, July 10, 1991.

17. The title of a poem by Wallace Stevens; see *Collected Poems* 125.

18. See Lurie 184, 323–31, 341–44; Reingold 181, 200–03.

19. Kohlstedt 181, 182, 186.

20. Harding, *Days* 291.

21. The standard way to earn a living as a student of nature had been to become a parson-naturalist like Gilbert White, which is why Darwin studied to become a "country clergyman"—an intention which, as he says, "died a natural death when on leaving Cambridge I joined the *Beagle* as Naturalist" (*Autobiography* 57). While the British tradition of natural theology was an important tributary to Thoreau's studies, and virtually defined the climate within which he worked, I believe John Hildebidle, in claiming White as a model for Thoreau, has hold of only part of the truth. As Hildebidle notes, Thoreau did not read White until 1853.

22. Peck, *Thoreau's Morning Work* 123.

23. Both Henry and Edward were nephews of the Unitarian clergyman Joseph Tuckerman.

24. The extract was copied from Edward Tuckerman's *Enumeration of North American Lichens* (1845), 20–21: "A View of the Natural Systems of Oken, Fries, and Endlicher."

25. That Thoreau's suppressed sexuality finds an outlet in "nature" is a common theme in biographical studies of him. Thoreau wondered if there wasn't "a law that you cannot have a deep sympathy with both man & nature. Those qualities which bring you near to the one estrange you from the other" (4:435; 4/11/52). Walter Harding, in "Thoreau's Sexuality," suggests one cause for Thoreau's displacement was his necessarily suppressed homoeroticism, which then in some fashion enabled his intense focus on nature. It is interesting to note that the breakdown of subject/object distinctions expressed in Thoreau coincides with a relaxation of gender barriers as well: in passages such as this one, Thoreau's own sexuality is characteristically

ambivalent. Given that Alexander von Humboldt was actively homosexual, it is tempting to speculate that the boundary breakdowns that typify "empirical holism" were more readily available to writers whose own gender construction was comparatively fluid. Conversely, the need to rigidify gender differences may have helped drive the consolidation of rational holism's mainstream dualisms: male/science/intellect; female/nature/emotion.

26. Feidelson 81. "Corn grows in the night" is both Thoreau's first radical metaphor and his favorite (excepting Walden Pond). It appears in the *Journal* (I:113; 1:229; I:313), and also in *Walden* (111) and "Walking" (NHE 113).

27. However, in other moods Thoreau valued the telescope as an aid to precise vision. He himself paid "eight dollars" for one in 1854 (VI:166; 3/13/54). Views through the telescope were, he wrote, "purely visionary"—meaning they involved no other sense but the eye, and hence it was "a disruptive mode of viewing" (VII:61; 9/29/54).

28. On Channing and Thoreau, including this exchange, see Porte, 103–8.

29. I'm pleased to report that this episode contains Thoreau's very first *Journal* reference to Humboldt, who offers a reason why natural sounds seem louder at night. Humboldt too is writing about running water: the sound of the falls of the Orinoco River (*Views* 168).

30. Sattelmeyer, "Remaking of Walden" 439; Shanley, *Making of Walden* 31–33, 57–67. Shanley (104–208) reprints the text of the first version of *Walden*, which Thoreau wrote in 1846–47.

31. See Peck, "Crosscurrents" 78–79.

32. See also Fischer, esp. 98–99. On the arbitrariness of any act of foundation, see Michaels, esp. 420.

33. Cavell 11.

34. Richardson, *Henry David Thoreau* 138–39; Richardson notes that Emerson thought of moving.

35. Michel Serres's suggestion in "Turner Meets Carnot" lurks in the background here: *Walden* as a steam engine, driven by the sun and ice of the seasons.

36. Richardson, *Henry David Thoreau* 384.

37. Sharon Cameron, in *Writing Nature*, suggests that in 1852 Thoreau realized he needed two incompatible texts, one to present nature to an audience, and one to represent nature in its own right. So *Walden* and the *Journal* develop in radically different directions, one rescuing experience from details, while in the other, experience "conversely depends on the way in which phenomena amass around each other" (160, 52).

38. Michael Gilmore explores the ironies of this situation in "*Walden* and the 'Curse of Trade,'" in *American Romanticism and the Marketplace* 35–51. It's hard to disagree with Gilmore's point that Thoreau is complicit "in the ideological universe he abhors" (36)—it must be the rare author who can congratulate himself on slipping entirely out of the structures built by local time and place. I do think Thoreau's work after *Walden* complicates Gilmore's analysis.

39. Cameron 129.

40. Ibid.

41. Van Dusen 36. In *Thoreau's Alternative History*, Joan Burbick writes of Thoreau that "change so fundamentally pervades all observation that no 'eye' can literally add up its observations and induce the whole. The effect of the whole is gained only through an act of imagination" (46).

Chapter 5. A Plurality of Worlds

1. The book was published anonymously in 1853, but its authorship was an open secret; a new edition, with a sixty-page supplement in which Whewell answered his critics, was published in 1856. For the American edition, see *The Plurality of Worlds*, introduction by Edward Hitchcock (Boston: Gould and Lincoln, 1856).

2. See Brooke 272. Chambers' evolutionary natural theology was echoed by Robert Hunt, in his *Poetry of Science* (312–13), which Thoreau also read in the early 1850s.

3. Hildebidle 99: "There is always in Thoreau a strong element of the notion (heretical, for a native of Puritan New England) that insight, and along with it redemption, are earned."

4. Lewontin, Rose, and Kamin, *Not in Our Genes* 280.

5. Behind this account lies the landmark study by Steven Shapin and Simon Schaffer, *The Leviathan and the Air Pump* (1985).

6. Lewis 25–26.

7. See Latour, *Science in Action* chap.6; "Drawing Things Together."

8. Ibid. 225.

9. Robert Sattelmeyer, "Introduction," NHE xxi.

10. Added in pencil; see *Journal* 3:611.

11. Robert Sattelmeyer notes that this single passage "was the product of several years of observation and evolution in his thought"; see "Remaking of *Walden*" 442. As William Rossi notes, one of the key revisions was altering "studio of the artist" to "*laboratory* of the Artist," reflecting Thoreau's stubborn insistence that imaginative and scientific truths were fused. See Rossi, "'Laboratory of the Artist'" 200; Sattelmeyer, "Remaking of *Walden*" 443.

12. Compare the words of an anonymous reviewer who seized on Whewell's *Plurality of Worlds* to reproach "Nature worshippers": "'you are . . . worshipping ye know not what. The stars are *not* worlds, they are mere chaotic masses. Nature is not such a finished rounded thing as you dream, much less is it God; it is only a crude process, not a perfected result, far less a living cause. This Universe, glorious as it looks to *man's* imagination, is not infinite, is not beautiful even: it is but clay in the hands of an Almighty Potter.'" *Eclectic Review* 7 (1854): 527–28. Quoted in Brooke 282 n. 381. Clearly the metaphor can be molded to various ends.

13. See Robert E. Abrams 250: Thoreau's sense of immediacy "deepens, complicates, and turns problematic during his travels into it; it does not recede back into the lost, immutable authority of some unwarped primordial world."

14. Quoted in Richardson, *Henry David Thoreau* 344; Howarth, *Book of Concord* 123. The letter was to Calvin H. Greene; CO 425–26.

15. On the tangle of late manuscripts and the character of the seasonal charts, see Richardson, *Henry David Thoreau* 381–82, and "Thoreau's Broken Task" 3; also Howarth, *Book of Concord.*

16. See Grusin.

17. The process is of course somewhat more complicated in its details: squirrels transport pine seeds in the cones, for instance. Thoreau was working out an exhaustive range of qualifications and permutations. For an account of Thoreau's forest studies, see Stoller 71–107. As Stoller points out, the only new element in Thoreau's studies, as Thoreau himself understood, was seed dispersion; "the other essential components" had been described in 1846 by George B. Emerson (a distant cousin of Waldo's), in *A report on the trees and shrubs growing naturally in the forest of Massachusetts*, a book which Thoreau relied on throughout the 1850s. See also Kehr; Whitford; and Whitney and Davis. According to Whitney and Davis, in 1850 the percentage of wooded land in Concord was at its lowest point in history: 10.5 percent. Currently it is about 50 percent.

18. Nearly 150 years later, monoculture and even-age stands are resulting in declining production, and the "new forestry" is discovering the guidelines Thoreau outlined in his unpublished manuscripts.

19. Brad Dean, ed., in Thoreau, *Faith in a Seed* 228.

20. On March 22, 1961, Thoreau in his *Journal* takes up the quarrel with "a writer in the *Tribune*," who insists that cherry trees must be spontaneously generated; evidently the debate was still lively. See XIV:331–34.

21. See also Richardson, "Thoreau's Broken Task"; Richardson notes Thoreau's "curt dismissal" of Agassiz's special creationism in June 1858, well before Thoreau read Darwin. See *Journal* X:467–68. As Kichung Kim noted, "in his time insistence on naturalistic explanations was little short of revolutionary" (129).

22. On Darwin's "plain man" argument, see Cannon, "Darwin's Vision" 162. A modern biologist would hardly phrase Darwin's theory this way, so as a corrective I offer a more standard definition of Darwinian evolution: "adaptive change as the result primarily of natural selection operating over long periods on the small variations present in plant and animal populations" (Bynum et al. 132). Darwin seldom used the word "evolution," preferring the more exact phrase "descent with modification."

23. Beer 80. As Gillian Beer says, this should be understood as "part of a profound imaginative longing shared by a great number of [Darwin's] contemporaries. . . . The palpable, the particular, became not only evidence, but ideal." Conversely, ideas would "find their truest form in substance" (42, 49). For Darwin's Tree of Life metaphor from this perspective, see Beer 93. See also George Levine's discussion of Darwinian metaphor in *Darwin and the Novelists* 109–10.

24. On January 1, 1860, Asa Gray's brother-in-law, Charles Brace, arrived in Concord with a copy of Darwin's *Origin of Species*, and he joined Thoreau and Bronson Alcott for dinner with Frank Sanborn, at which they discussed the book; Thoreau borrowed a copy. See Harding, *Days* 429; Borst 550. In his journal for 1860, Emerson records an interesting exchange with Thoreau: when he told Thoreau of Agassiz's scorn of Darwin, Thoreau replied: "If [Agassiz] sees two thrushes so alike that they bother the ornithologist to discriminate them, he insists they are . . . two

species; but if he see Humboldt & Fred. Cogswell, he insists that they come from one ancestor" (JMN 14:350). Cogswell was "'a kindly, underwitted inmate of Concord Almshouse'" (*Journal* IX:270).

25. Howarth, *Book of Concord* 198.

26. Horace Mann, Jr., who accompanied Thoreau on this journey, was already on the rise in what would have been a major career in botany. Tragically, he died seven years later, at twenty-four, of tuberculosis. See Richardson, *Henry David Thoreau* 387.

27. Gould, "Church, Humboldt, and Darwin" 104.

28. Ibid. 104, 106–7.

29. For Thoreau and the history of forest management practices, see Kehr 32–33.

30. Harding lists at least five separate publications or long summaries of it (*Days* 439–40).

31. During a fishing trip in April 1844, Thoreau and Edward Hoar had lit a cooking fire that burned out of control, destroying over three hundred acres; the epithet followed Thoreau for years. See Howarth, *Book of Concord* 34.

32. Harding, *Days* 438.

33. Ibid. 439. The deletions, which are acknowledged, consist of the opening three paragraphs and the closing three paragraphs, with no editing of the body of the text; the ostensible purpose may have been to save space.

34. Howarth, *Book of Concord* 195; Hildebidle 68; Burbick 129–30.

35. Richardson, *Henry David Thoreau* 343–44; CO 423–24.

36. I am borrowing the methodology here from the "strong programme" of SSK, the sociology of scientific knowledge, and the specific terminology from Steve Woolgar, *Science, the Very Idea.* For an outstanding treatment of the rhetoric of science, which deftly joins and separates the discourses of science and literature, see David Locke, *Science as Writing.*

37. Woolgar 71.

38. Ibid. 58–61. For a non-SSK discussion of the rhetoric of discovery, see Pratt 202–4. Another aspect of this argument concerns Thomas Kuhn's concept of the scientific paradigm, which creates the very rules that make scientific facts "seeable," or the mechanism by which, in Thoreau's terms, the scientist makes his discovery by "taking it into his head first." Thoreau's identification of anomalies like the oak seedlings suggests that he is working outside "normal science," and until the paradigm was adjusted to include the anomalies so important to Thoreau, his new fact would seem "not quite a scientific fact at all." Thus part of the urgency of Thoreau's rhetoric is in his desire to persuade the scientist "to see nature in a different way." See Kuhn 52–53; also 64–65.

39. Woolgar 68–69.

40. On this notion of agents, or in Latour's terms "actants," "cooperating" with the investigator, see Latour, *Science in Action*, esp. Chap. 2, "Laboratories" (63–100).

41. David Locke 17.

42. According to Richardson, the probable target here is Louis Agassiz (*Henry David Thoreau* 362).

43. Levine, *Darwin and the Novelists* 212, 214.

44. For another, extended version of the argument that organisms actively produce their environment, see Lewontin, *Biology as Ideology*, particularly the final chapter, "Science as Social Action" (105–23).

45. Woolgar 59–60.

46. Latour, *Science in Action* 93. This would be, I believe, a Thoreauvian reply to the dilemma raised by Walter Benn Michaels, in "*Walden*'s False Bottoms" (420): in a world where interpretation is constantly undercutting its own foundation, where will one find principles for action? Yes, uncertainty is built in everywhere; therefore the one thing that cannot be doubted is the paradox that material reality is utterly independent of us and utterly susceptible to our least decisions. Thoreau rewrites Emerson: We must treat nature as if it were real; perhaps it is.

47. Woolgar 105. Steve Woolgar anticipates this will be the problem for the next generation of the social studies of science, and I would add that it is also currently a problem for interdisciplinary criticism: in its vulnerability, it needs all the more to assert clairvoyance; yet in its ideology, it is all the less able to do so.

Chapter 6. Walking the Holy Land

1. Stevens 76.

2. Macpherson 38.

3. See also Robert E. Abrams for an extended exploration of this phenomenal frontier.

4. Such is the main premise of *Cosmos*, Vol. II, which traces how susceptibility to natural beauty excites first love of nature, and then the desire to understand it and to see nature in other lands. Individual action combines, then, to create the progress of whole civilizations; see also *Cosmos* I: 52–54, and Macpherson's discussion (43–51).

5. Emerson supports this statement with a footnoted quotation from Quetelet: "Everything which pertains to the human species, considered as a whole, belongs to the order of physical facts. The greater the number of individuals, the more does the influence of the individual will disappear, leaving predominance to a series of general facts dependent on causes by which society exists, and is preserved." Hacking discusses the same passage (in a different translation) to show how for Quetelet "free acts are minuscule causes that cancel out and allow of the larger regularities" (123).

6. Hacking 104, 116. Hacking notes that one hundred years later "probability" did not preclude but made room for free will: the perceived laws of physics still pulled the laws of society in their train, but in the opposite direction.

7. Ibid. 120–21.

8. Ibid. 146–47.

9. Lewis 27.

10. Peck, *Thoreau's Morning Work* 123.

11. "Chaos" refers here both to the broad front in contemporary science that studies "chaotic," complex, or nonlinear systems, as well as the word's older literary and mythic meanings. As N. Katherine Hayles reminds us, such meanings do not

disappear but linger on, creating an aura "that even the more conservative investigators into dynamical systems methods find hard to resist . . ." (*Chaos Bound* 8–9). Conversely, the new paradigm of chaos has now charged earlier uses of the term, bringing into focus "classical texts that may not have fitted very well into older traditions" (23). Thoreau's writing is an exemplary instance of such a text, showing both the emergence of self-organization from chaos, and the hidden resources of a nature that "can renew itself because it is rich in disorder and surprise" (10–11). Some characteristics of chaotic systems include nonlinearity, whereby minute fluctuations cause large-scale changes; complex form, in which measurements increase as measurement scale decreases, and different scale levels exhibit recursive symmetries; sensitivity to initial conditions; and feedback mechanisms. For a more extended discussion see the introductions to Hayles, *Chaos Bound* and *Chaos and Order*; Kellert.

12. Porter 199.

13. Harding, *Days* 277–79.

14. Bridgman xi. Bridgman is entirely at odds with my interpretation, finding in Thoreau's fascination with the cruel and macabre a sign of the psychological strain brought on by the "severe tension" between his hostile and punishing temperament and his "acquired idealism," resulting in the "sterile contradictions" that cripple his work (xii–xiv). Bridgman makes his case through his own astonishing compendium of the macabre.

15. As an on-site visit reveals, in Concord dialect a "meadow" is not an open field but a wetland covered, at least much of the year, with standing water.

16. Thoreau derives this concept from the British physical scientist Robert Hunt, who wrote in *The Poetry of Science* (1848), in a passage Thoreau copied into his *Journal*, that granite, stone, and metal are all "alike destructively acted upon during the hours of sunshine, and, but for provisions of nature no less wonderful, would soon perish under the delicate touch of the most subtile of the agencies of the universe." Actinism was thus "one of the great powers of creation" (133, 138; *Journal* 3:196; 2/18/51).

17. Hacking 201.

18. Golemba 6.

19. Stevens 418.

20. Ibid. 416–17.

21. "A Pluralistic Mystic," in James 1312; quoted in Poirier 222.

22. James 1160.

23. This phrase is taken from Murray Bookchin, who defines "hierarchy" as "a strictly social term," which refers to "institutionalized and highly ideological systems of command and obedience," exclusive to *human* societies (xxii).

24. Harvey 20.

25. Harding, *Days* 53.

26. Richardson, "Thoreau and Science" 123; Richardson doubts that Thoreau would have taken any such task casually, and notes in any case his reappointment in 1860.

27. Foucault, *Discipline and Punish* 170.
28. James 741.
29. Cavell 11.
30. Haraway, *Primate Visions* 4.

BIBLIOGRAPHY

Abrams, Meyer H. *The Mirror and the Lamp: Romantic Theory and the Critical Tradition.* Oxford: Oxford UP, 1953.

Abrams, Robert E. "Image, Object, and Perception in Thoreau's Landscapes: The Development of Anti-Geography." *Nineteenth-Century Literature* 46 (1991): 245–62.

Ackerknecht, Erwin H. "Georg Forster, Alexander von Humboldt, and Ethnology." *Isis* 46 (1955): 83–95.

Agassiz, Louis. *Address Delivered on the Centennial Anniversary of the Birth of Alexander von Humboldt.* Boston: Boston Society of Natural History, 1869.

Agassiz, Louis. *Essay on Classification.* 1857. Rev. 1859. Cambridge: Harvard UP, 1962.

Agassiz, Louis. *Methods of Study in Natural History.* Boston: Ticknor and Fields, 1863; Boston: Houghton Mifflin, 1896.

Agassiz, Louis, and Augustus A. Gould. *Principles of Zoölogy.* Boston: Gould, Kendall and Lincoln, 1848.

Alexander, J. W.; Albert Dod; Charles Hodge. "Transcendentalism of the Germans and of Cousin and Its Influence on Opinion in This Country." 1839, 1840. In *The Transcendentalists: An Anthology*, ed. Perry Miller. Cambridge: Harvard UP, 1950. 231–40.

Anderson, Charles R. *The Magic Circle of Walden.* New York: Holt, Rinehart, and Winston, 1968.

Appel, Toby A. *The Cuvier-Geoffroy Debate: French Biology in the Decades before Darwin.* New York: Oxford UP, 1987.

Arac, Jonathan. *Critical Genealogies: Historical Situations for Postmodern Literary Studies.* New York: Columbia UP, 1987.

Aristotle. *The Basic Works.* Ed. Richard McKeon. New York: Random House, 1941.

Bacon, Francis. *Selected Writings.* New York: Random House, 1955.

Bate, W. Jackson. *Coleridge.* Cambridge: Harvard UP, 1968.

Baym, Nina. "Thoreau's View of Science." *Journal of the History of Ideas* 26 (1965): 221–34.

Beer, Gillian. *Darwin's Plots: Evolutionary Narrative in Darwin, George Eliot, and Nineteenth-Century Fiction.* 1983. London and Boston: Routledge, 1985.

Billington, Ray Allen. *Westward Expansion: A History of the American Frontier.* 4th ed. New York: Macmillan, 1974.

Bloor, David. "Coleridge's Moral Copula." *Social Studies of Science* 13 (1983): 605–19.

Bookchin, Murray. *The Ecology of Freedom: The Emergence and Dissolution of Hierarchy.* Rev. ed. Montreal: Black Rose Books, 1991.

Borst, Raymond R. *The Thoreau Log: A Documentary Life of Henry David Thoreau, 1817–1862.* New York: Macmillan, 1992.

Botting, Douglas. *Humboldt and the Cosmos.* New York: Harper and Row, 1973.

Bowen, Margarita. *Empiricism and Geographical Thought: From Francis Bacon to Alexander von Humboldt.* Cambridge: Cambridge UP, 1981.

Bozeman, Theodore Dwight. *Protestants in an Age of Science: The Baconian Ideal and Antebellum American Religious Thought.* Chapel Hill: U of North Carolina P, 1977.

Bridgman, Richard. *Dark Thoreau.* Lincoln: U of Nebraska P, 1982.

Brooke, John Hedley. "Natural Theology and the Plurality of Worlds: Observations on the Brewster-Whewell Debate." *Annals of Science* 34 (1977): 221–86.

Browne, Janet. *The Secular Ark: Studies in the History of Biogeography.* New Haven: Yale UP, 1983.

Brownson, Orestes A. "Victor Cousin." 1836. In *The Transcendentalists: An Anthology,* ed. Perry Miller. Cambridge: Harvard UP, 1950. 106–14.

Buell, Lawrence. *Literary Transcendentalism: Style and Vision in the American Renaissance.* Ithaca: Cornell UP, 1973.

Buell, Lawrence. *New England Literary Culture from Revolution through Renaissance.* Cambridge: Cambridge UP, 1986.

Buell, Lawrence. "The Thoreauvian Pilgrimage: The Structure of an American Cult." *American Literature* 61 (1989): 175–99.

Burbick, Joan. *Thoreau's Alternative History: Changing Perspectives on Nature, Culture, and Language.* Philadelphia: U of Pennsylvania P, 1987.

Bynum, W. F.; E. J. Browne; and Roy Porter. *Dictionary of the History of Science.* Princeton: Princeton UP, 1981.

Cameron, Kenneth Walter. "Emerson, Thoreau, and the Society of Natural History." *American Literature* 24 (1952–53): 21–30.

Cameron, Kenneth Walter. *Thoreau's Harvard Years.* Hartford, Conn.: Transcendental Books, 1966.

Cameron, Sharon. *Writing Nature: Henry Thoreau's Journal.* New York: Oxford UP, 1985.

Canby, Henry Seidel. *Thoreau.* Boston: Houghton Mifflin, 1939.

Cannon, Susan Faye. *Science in Culture: The Early Victorian Period.* New York: Dawson and Natural History Publications, 1978.

Cannon, Susan Faye [Walter F.]. "Darwin's Vision in *On the Origin of Species.*" In *The Art of Victorian Prose,* ed. George Levine and William Madden. New York: Oxford UP, 1968. 154–76.

Cannon, Susan Faye [Walter F.]. "John Herschel and the Idea of Science." *Journal of the History of Ideas* 22 (1961): 215–39.

Carlyle, Thomas. *Sartor Resartus.* London: J. M. Dent, 1908.

Cavell, Stanley. *The Senses of Walden.* New York: Viking, 1972.

Cayton, Mary Kupiec. *Emerson's Emergence: Self and Society in the Transformation of New England, 1800–1845*. Chapel Hill: U of North Carolina P, 1989.

Chai, Leon. *The Romantic Foundations of the American Renaissance*. Ithaca: Cornell UP, 1987.

Chambers, Robert. *Vestiges of the Natural History of Creation*. London: John Churchill, 1844.

Channing, William Ellery. *Thoreau: The Poet-Naturalist*. Boston: Roberts Brothers, 1873.

Christie, John Aldrich. *Thoreau as World Traveler*. New York: Columbia UP and American Geographical Society, 1965.

Clark, Harry Hayden. "Emerson and Science." *Philological Quarterly* 10.3 (July 1931): 225–60.

Cohen, I. Bernard. *Thomas Jefferson and the Sciences*. New York: Arno Press, 1980.

Coleridge, Samuel Tyler. *Aids to Reflection*. 1829, 1840. Port Washington, N.Y.: Kennikat Press, 1971.

Coleridge, Samuel Tyler. *Letters, Conversations, and Recollections*. 2 vols. London: Edward Moxon, 1836.

Coleridge, Samuel Tyler. *On the Definitions of Life Hitherto Received. Hints Towards a More Comprehensive Theory*. [*The Theory of Life*.] In *Miscellanies, Aesthetic and Literary, to which is added The Theory of Life*, ed. T. Ashe. London: George Bell and Sons, 1885. 351–430.

Coleridge, Samuel Tyler. *Philosophical Lectures*. Ed Kathleen Colburn. New York: Philosophical Library, 1949.

Cook, Reginald L. *Passage to Walden*. 1942. 2d ed., New York: Russell and Russell, 1966.

Cousin, Victor. *Introduction to the History of Philosophy*. Trans. Henning Gotfried Linberg. Boston: Hilliard, Gray, Little, and Wilkins, 1832.

Crompton, Louis. *Byron and Greek Love: Homophobia in Nineteenth-Century England*. Berkeley: U of California P, 1985.

Culler, Jonathan. *The Pursuit of Signs: Semiotics, Literature, Deconstruction*. Ithaca, N.Y.: Cornell UP, 1981.

Dahlstrand, Frederick C. "Science, Religion, and the Transcendentalist Response to a Changing America." *Studies in the American Renaissance 1988* 12 (1988): 1–25.

Dale, Peter Allan. *In Pursuit of a Scientific Culture: Science, Art, and Society in the Victorian Age*. Madison: U of Wisconsin P, 1989.

Daniels, George H. *American Science in the Age of Jackson*. New York: Columbia, 1968.

Daniels, George H. "The Process of Professionalization in American Science: The Emergent Period, 1820–1860." *Isis* 58 (1967): 151–66.

Darwin, Charles. *The Autobiography of Charles Darwin, 1809–1882*. 1887. Ed. Nora Barlow. New York: Norton, 1969.

Darwin, Charles. *On the Origin of Species: A facsimile of the first edition*. Ed. Ernst Mayr. Cambridge: Harvard UP, 1964.

Darwin, Charles. *The Voyage of the Beagle*. (*Journal of Researches*, 1845.) Ed. Leonard Engel. New York: Doubleday, 1962.

De Man, Paul. "The Rhetoric of Temporality." In *Critical Theory since 1965*, ed. Hazard Adams and Leroy Searle. Tallahassee: Florida State UP, 1986. 199–222.

Derrida, Jacques. "Structure, Sign, and Play in the Discourse of the Human Sciences." *Writing and Difference*. Chicago: U of Chicago P, 1978. 278–93.

De Terra, Helmut. "Alexander von Humboldt's Correspondence with Jefferson, Madison, and Gallatin." *Proceedings of the American Philosophical Society* 103 (1959): 783–806.

De Terra, Helmut. *Humboldt: The Life and Times of Alexander von Humboldt, 1769–1859*. New York: Knopf, 1955.

De Terra, Helmut. "Motives and Consequences of Alexander von Humboldt's Visit to the United States." *Proceedings of the American Philosophical Society* 104 (1960): 314–16.

De Terra, Helmut. "Studies of the Documentation of Alexander von Humboldt." *Proceedings of the American Philosophical Society*. Part I: 102 (1958): 136–41. Part II: 102 (1958): 560–89.

Desmond, Adrian. *The Politics of Evolution: Morphology, Medicine, and Reform in Radical London*. Chicago: U of Chicago P, 1989.

Dupree, A. Hunter. *Asa Gray, American Botanist, Friend of Darwin*. 1959. Baltimore: Johns Hopkins UP, 1988.

Egerton, Frank N. "Ecological Studies and Observations before 1900." *History of American Ecology*. New York: Arno Press, 1977. 311–51.

Eichner, Hans. "The Rise of Modern Science and the Genesis of Romanticism." *PMLA* 97 (1982): 8–30.

Eliot, T. S. "Tradition and the Individual Talent." *Selected Prose of T. S. Eliot*. New York: Farrar, Straus and Giroux, 1975. 37–44.

Emerson, Ralph Waldo. *The Collected Works of Ralph Waldo Emerson*. Cambridge: Harvard UP, 1971–87.

Emerson, Ralph Waldo. *Complete Works*. Boston: Houghton, Mifflin, 1903.

Emerson, Ralph Waldo. "Fate." *Essays and Lectures*. New York: Library of America, 1983.

Emerson, Ralph Waldo. *The Journals and Miscellaneous Notebooks*. Cambridge: Harvard UP, 1960–82.

Emerson, Ralph Waldo. *Poems. The Complete Works of Ralph Waldo Emerson*. Vol. IX. Boston: Houghton, Mifflin, 1904.

Emmons, David M. "Theories of Increased Rainfall and the Timber Culture Act of 1873." *Forest History* 15 (1971): 6–14.

Everett, Edward. *Orations and Speeches on Various Occasions*. Vol 4. Boston: Little Brown, 1879.

Feidelson, Charles Jr. *Symbolism and American Literature*. Chicago: U of Chicago P, 1953.

Fischer, Michael R. "*Walden* and the Politics of Contemporary Literary Theory." In *New Essays on Walden*, ed. Robert F. Sayre. Cambridge: Cambridge UP, 1992. 95–113.

Fish, Stanley. "Being Interdisciplinary Is So Very Hard to Do." *Profession 89*. MLA. 15–22.

Foner, Philip S. "Alexander von Humboldt on Slavery in America." *Science and Society* 47 (1983): 330–42.

Foucault, Michel. *The Archaeology of Knowledge*. Trans. A. M. Sheridan Smith. New York: Random House, 1972.

Foucault, Michel. *Discipline and Punish: The Birth of the Prison*. 1975. Trans. Alan Sheridan. New York: Random House, 1979.

Foucault, Michel. *The Order of Things: An Archaeology of the Human Sciences*. New York: Random House, 1970.

Frémont, John Charles. *Report of the Exploring Expedition to the Rocky Mountains in the year 1842, and to Oregon and North California in the years 1843–44*. Washington [D.C.]: Gales and Seaton, Printers, 1845.

Garber, Frederick. *Thoreau's Redemptive Imagination*. New York: New York UP, 1977.

Geertz, Clifford. "From the Native's Point of View: On the Nature of Anthropological Understanding." In *Local Knowledge: Further Essays in Interpretive Anthropology*. New York: Basic Books, 1983. 55–70.

Gillespie, Neal C. "Divine Design and the Industrial Revolution: William Paley's Abortive Reform of Natural Theology." *Isis* 81 (1990): 214–29.

Gillespie, Neal C. "Preparing for Darwin: Conchology and Natural Theology in Anglo-American Natural History." *Studies in History of Biology*, vol. 7, ed. William Coleman and Camille Limoges. Baltimore: Johns Hopkins UP, 1984. 93–145.

Gilman, Daniel Coit. "Humboldt, Ritter, and the New Geography." *New Englander* 18 (1860): 277–306.

Gilmore, Michael T. *American Romanticism and the Marketplace*. Chicago: U of Chicago P, 1985.

Glaab, Charles N. "The Historian and the American Urban Tradition." *Wisconsin Magazine of History* 47 (1963): 12–25.

Goddard, H. C. "Unitarianism and Transcendentalism." In *American Transcendentalism: An Anthology of Criticism*, ed. Brian M. Barbour. Notre Dame, Ind.: U of Notre Dame P, 1973. 159–77.

Goethe, Johann Wolfgang. *Italian Journey, 1786–1788*. Trans. W. H. Auden and Elizabeth Jayer. New York: Schocken Books, 1968.

Goetzmann, William H. *New Lands, New Men: America and the Second Great Age of Discovery*. New York: Viking, 1986.

Goetzmann, William H. "Paradigm Lost." In *The Sciences in the American Context: New Perspectives*, ed. Nathan Reingold. Washington D.C.: Smithsonian, 1979. 21–34.

Goetzmann, William H., and William N. Goetzmann. *The West of the Imagination*. New York: W. W. Norton, 1986.

Golemba, Henry. *Thoreau's Wild Rhetoric*. New York: New York UP, 1990.

Gould, Stephen Jay. "Church, Humboldt, and Darwin: The Tension and Harmony of Art and Science." In *Frederic Edwin Church*, ed. Franklin Kelly. Washington, D.C.: Smithsonian Institution Press, 1989. 94–107.

Gould, Stephen Jay. *Time's Arrow, Time's Cycle: Myth and Metaphor in the Discovery of Geological Time*. Cambridge: Harvard UP, 1987.

Greene, John C. *American Science in the Age of Jefferson.* Ames, Iowa: Iowa State UP, 1984.

Grefe, Maxine. *"Apollo in the Wilderness": An Analysis of Critical Reception of Goethe in America, 1806–1840.* New York: Garland Publications, 1988.

Grusin, Richard. "Thoreau, Extravagance, and the Economy of Nature." *American Literary History* 5 (1993): 30–50.

Hacking, Ian. *The Taming of Chance.* New York: Cambridge UP, 1990.

Hamilton, Robert. "Memoir of Baron Alexander von Humboldt." In *A History of British Fishes*, Vol. II. London: Hardwicke and Bogue, [1876?]. 17–39.

Hankins, Thomas L. *Science and the Enlightenment.* Cambridge: Cambridge UP, 1985.

Haraway, Donna Jeanne. *Crystals, Fabrics, and Fields, Metaphors of Organicism in Twentieth-Century Developmental Biology.* New Haven: Yale UP, 1976.

Haraway, Donna. *Primate Visions: Gender, Race, and Nature in the World of Modern Science.* New York: Routledge, 1989.

Haraway, Donna. "Situated Knowledges." *Simians, Cyborgs, and Women: The Reinvention of Nature.* New York: Routledge, 1991. 183–201.

Harding, Walter. *The Days of Henry Thoreau.* 1965. New York: Dover, 1982.

Harding, Walter, ed. *Thoreau as Seen by His Contemporaries.* 1960. New York: Dover, 1989.

Harding, Walter. "Thoreau's Sexuality." *Journal of Homosexuality* 21.3 (1991): 23–45.

Harding, Walter, and Michael Meyer. *The New Thoreau Handbook.* New York: New York UP, 1980.

Harvey, David. *The Condition of Postmodernity: An Enquiry into the Origins of Cultural Change.* Oxford: Basil Blackwell, 1989.

Hayles, N. Katherine, ed. *Chaos and Order: Complex Dynamics in Literature and Science.* Chicago: U of Chicago P, 1991.

Hayles, N. Katherine. *Chaos Bound: Orderly Disorder in Contemporary Literature and Science.* Ithaca: Cornell UP, 1990.

Hentschel, Cedric. "Alexander von Humboldt's Synthesis of Literature and Science." In *Alexander von Humboldt 1769/1969.* Bonn: InterNationes, 1969. 97–132.

Herber, Elmer Charles, ed. *Correspondence between Spencer Fullerton Baird and Louis Agassiz—Two Pioneer American Naturalists.* Washington, D.C.: Smithsonian, 1963.

Herschel, John F. W. *A Preliminary Discourse on the Study of Natural Philosophy.* 1830. Chicago: U of Chicago P, 1987.

Hildebidle, John. *Thoreau: A Naturalist's Liberty.* Cambridge: Harvard UP, 1983.

Hodder, Alan D. "'Ex Oriente Lux': Thoreau's Ecstasies and the Hindu Texts." *Harvard Theological Review* 86 (1993): 403–38.

Hopkins, Vivian Constance. "Emerson and Bacon." *American Literature* 29 (1958): 408–30.

Hopkins, Vivian Constance. *Spires of Form: A Study of Emerson's Aesthetic Theory.* Cambridge: Harvard UP, 1951.

Howarth, William L. *The Book of Concord: Thoreau's Life as a Writer.* New York: Viking Press, 1982.

Howarth, William L. *The Literary Manuscripts of H. D. Thoreau.* Ohio State UP, 1974.

Humboldt, Alexander von. *Aspects of Nature in Different Lands and Different Climates, with Scientific Elucidations.* Trans. Mrs. Sabine. Philadelphia: Lea and Blanchard, 1850. New York: AMS Press Reprint, 1970.

Humboldt, Alexander von. *Cosmos: A Sketch of a Physical Description of the Universe.* Trans. E. C. Otté. Vols. I and II: New York: Harper and Brothers, 1850, 1852. Vol. III–V, 1869–70.

Humboldt, Alexander von. *Personal Narrative of Travels to the Equinoctial Regions of the New Continent, During the Years 1799–1804.* Vol. I and II. Trans. Helen Maria Williams. London: Longman [Brown, Green and Longmans], 1818.

Humboldt, Alexander von. *Personal Narrative of Travels to the Equinoctial Regions of America, During the Years 1799–1804. By Alexander von Humboldt and Aimé Bonpland.* Written in French by Alexander von Humboldt. Trans. and ed. Thomasina Ross. 3 vols. London: Henry G. Bohn, 1852.

Humboldt, Alexander von. *The Travels and Researches of Alexander von Humboldt.* Ed. W. MacGillivray. 1833. London and New York: Nelson, 1855.

Humboldt, Alexander von. *Views of Nature, or Contemplations on the Sublime Phenomena of Creation, with Scientific Illustrations.* Trans. E. C. Otté and Henry Bohn. London: Bohn, 1850.

Hunt, Robert. *The Poetry of Science, or Studies of the Physical Phenomena of Nature.* London, 1848. Boston: Gould, Kendall, and Lincoln, 1850.

Ingersoll, Robert Green. "Humboldt, The Universe is Governed by Law." *The Gods and Other Lectures.* Peoria, Ill.: C. P. Farrell, 1878. 93–117.

James, William. *Writings, 1902–1910.* New York: Library of America, 1987.

Jehlen, Myra. *American Incarnation: The Individual, the Nation, and the Continent.* Cambridge: Harvard UP, 1986.

Joyaux, George J. "Victor Cousin and American Transcendentalism." In *American Transcendentalism: An Anthology of Criticism,* ed. Brian M. Barbour. Notre Dame, Ind.: U of Notre Dame P, 1973. 125–38.

Kant, Immanuel. *Critique of Judgement.* Trans. J. H. Bernard. New York: Hafner, 1951.

Kant, Immanuel. Introduction to *Physische Geographie.* Trans. J. A. May. *Kant's Concept of Geography and Its Relation to Recent Geographical Thought.* Toronto: Toronto UP, 1970. 255–64.

Kauffman, Stuart A. *The Origins of Order: Self-Organization and Selection in Evolution.* New York: Oxford UP, 1993.

Kehr, Kurt. "Walden Three: Ecological Changes in the Landscape of Henry David Thoreau." *Journal of Forest History* 27 (1983): 28–33.

Keller, Evelyn Fox. *Reflections on Gender and Science.* New York: Yale UP, 1985.

Kellert, Stephen H. *In The Wake of Chaos.* Chicago: U of Chicago P, 1993.

Kellner, L. *Alexander von Humboldt.* London: Oxford UP, 1963.

Kim, Kichung. "Thoreau's Science and Teleology." *ESQ* 18 (1972): 125–33.

Klein, Julie Thompson. *Interdisciplinarity: History, Theory, and Practice.* Detroit: Wayne State UP, 1990.

Klencke, Hermann. *Alexander von Humboldt: A Biographical Monument.* In *Lives of the Brothers Humboldt*, trans. Juliette Bauer. New York: Harper and Brothers, 1853. 13–231.

Knight, David. *The Age of Science.* New York: Basil Blackwell, 1986.

Knights, Ben. *The Idea of the Clerisy in the Nineteenth Century.* Cambridge: Cambridge UP, 1978.

Kohlstedt, Sally Gregory. "The Nineteenth-Century Amateur Tradition: The Case of the Boston Society of Natural History." In *Science and Its Public: The Changing Relationship*, ed. Gerald Holton and William A. Blanpied. Dordrecht-Holland: D. Reidel, 1976. 173–90.

Krutch, Joseph Wood. *Henry David Thoreau.* 1948. New York: William Morrow, 1974.

Kuhn, Thomas S. *The Structure of Scientific Revolutions.* 1962. 2d ed. Chicago: U of Chicago P, 1970.

Latour, Bruno. "Drawing Things Together." In *Representation in Scientific Practice*, ed. Michael Lynch and Steve Woolgar. Cambridge: MIT Press, 1988, 1990. 19–68.

Latour, Bruno. *Science in Action: How to Follow Scientists and Engineers through Society.* Cambridge: Cambridge UP, 1987.

Latour, Bruno. *We Have Never Been Modern.* Trans. Catherine Porter. Cambridge: Harvard UP, 1993.

Lenoir, Timothy. *The Strategy of Life: Teleology and Mechanics in Nineteenth-Century Biology.* U of Chicago P, 1982.

Levere, Trevor H. *Poetry Realized in Nature: Samuel Taylor Coleridge and Early Nineteenth-Century Science.* Cambridge: Cambridge UP, 1981.

Levine, George. "One Culture: Science and Literature." In *One Culture: Essays in Science and Literature*, ed. George Levine. Madison: U of Wisconsin P, 1987. 3–32.

Levine, George. *Darwin and the Novelists: Patterns of Science in Victorian Fiction.* Cambridge: Harvard UP, 1988.

Lewis, R. W. B. *The American Adam: Innocence, Tragedy, and Tradition in the Nineteenth Century.* Chicago: U of Chicago P, 1955.

Lewontin, Richard C. *Biology as Ideology.* New York: Harper Collins, 1991.

Lewontin, Richard C.; Steven Rose; and Leon J. Kamin. *Not in Our Genes: Biology, Ideology, and Human Nature.* New York: Random House, 1984.

Locke, David. *Science as Writing.* New Haven: Yale UP, 1992.

Locke, John. *An Essay Concerning Human Understanding.* Ed. Alexander Campbell Fraser. 1894. New York: Dover, 1959.

Löwenberg, J.; Robert Avé-Lallemant; and Alfred Dove. *Life of Alexander von Humboldt.* Ed. Karl Bruhns. Trans. Jane and Caroline Lassell. 2 vols. London and Boston: Longmans, Green, and Lee and Shepard, 1873.

Lurie, Edward. *Louis Agassiz: A Life in Science.* 1960. Baltimore: Johns Hopkins UP, 1988.

Lyell, Charles. *Principles of Geology.* Vol. I. 1830. Chicago: U of Chicago P, 1990.

McIntosh, James. *Thoreau as Romantic Naturalist: His Shifting Stance toward Nature.* Ithaca: Cornell UP, 1974.

McFarland, Thomas. *Coleridge and the Pantheist Tradition.* Oxford: Oxford UP, 1969.

Macpherson, Anne. "The Human Geography of Alexander von Humboldt." Ph.D. diss.: U of California, Berkeley, 1971.

Marsh, James. "Preliminary Essay." *Aids to Reflection*, by S. T. Coleridge. 1829. Port Washington, N.Y.: Kennikat Press, 1971. 9–58.

Mattheissen, F. O. *American Renaissance: Art and Expression in the Age of Emerson and Whitman.* Oxford UP, 1941.

Mayr, Ernst. *The Growth of Biological Thought: Diversity, Evolution, and Inheritance.* Cambridge: Harvard UP, 1982.

Meyer-Abich, Adolph. "Alexander von Humboldt." In *Alexander von Humboldt 1769/1969.* Bonn: InterNationes, 1969. 7–94.

Meyer-Abich, Adolph. "Alexander von Humboldt as a Biologist." In *Alexander von Humboldt, Werk und Weltgeltung.* Munich: R. Piper & Co. Verlag, 1969. 179–96.

Michaels, Walter Benn. "*Walden*'s False Bottoms." 1977. In *Walden*, ed. William Rossi. New York: Norton, 1992. 405–21.

Miller, Charles A. *Jefferson and Nature: An Interpretation.* Baltimore: Johns Hopkins UP, 1988.

Miller, Perry, ed. *Consciousness in Concord: The Text of Thoreau's Hitherto "Lost Journal" (1840–41), Together with Notes and a Commentary.* Boston: Houghton Mifflin, 1958.

Miller, Perry. "Thoreau in the Context of International Romanticism." *NEQ* 34 (1961): 147–59.

Nash, Roderick. *Wilderness and the American Mind.* Rev. ed. New Haven: Yale UP, 1973.

Nelkin, Halina. *Alexander von Humboldt, His Portraits and Their Artists: A Documentary Iconography.* Berlin: Dietrich Reimer Verlag, 1980.

Nelkin, Halina. *Humboldtiana at Harvard.* Cambridge: Widener Library, 1976.

Neubauer, John. "Nature as Construct." In *Literature and Science as Modes of Expression*, ed. Frederick Amrine. Dordrecht: Kluwer, 1989. 129–40.

Newfield, Christopher. "Emerson's Corporate Individualism." *American Literary History* 3 (Winter 1991): 657–84.

Nicolson, Malcolm. "Alexander von Humboldt and the Geography of Vegetation." In *Romanticism and the Sciences*, ed. Andrew Cunningham and Nicholas Jardine. Cambridge: Cambridge UP, 1990. 169–85.

Nicolson, Malcolm. "Alexander von Humboldt, Humboldtian Science, and the Origins of the Study of Vegetation." *History of Science* 25 (1987): 167–94.

Novak, Barbara. *Nature and Culture: American Landscape Painting, 1825–1875.* New York: Oxford UP, 1980.

Nye, Russel Blaine. *Society and Culture in America, 1830–1860.* New York: Harper and Row, 1974.

Oken, Lorenz. *Elements of Physiophilosophy.* 1809. Trans. Alfred Tulk. London: Ray Society, 1847.

Oppitz, Ulrich-Dieter. "Der Name der Brüder Humboldt in Aller Welt." In *Alexander von Humboldt, Werk und Weltgeltung.* Munich: R. Piper & Co. Verlag, 1969. 277–429.

Orsini, G. N. Giodano. "The Ancient Roots of a Modern Idea." In *Organic Form: The Life of an Idea*. London: Routledge & Kegan Paul, 1972. 7–23.

Paley, William. *Natural Theology. Works*, Vol. V. London: C. and J. Rivington et al., 1825.

Paul, Sherman. *The Shores of America: Thoreau's Inward Exploration*. Urbana: U of Illinois P, 1958.

Peck, H. Daniel. "The Crosscurrents of *Walden*'s Pastoral." In *New Essays on Walden*, ed. Robert F. Sayre. Cambridge: Cambridge UP, 1992. 73–94.

Peck, H. Daniel. *Thoreau's Morning Work: Memory and Perception in A Week on the Concord and Merrimack Rivers, the Journal, and Walden*. New Haven: Yale UP, 1990.

Petroski, Henry. "H. D. Thoreau, Engineer." *American Heritage of Invention and Technology* 5.2 (1989): 8–16.

Plato. *The Collected Dialogues*. Ed. Edith Hamilton and Huntington Cairns. Princeton: Princeton UP, 1961.

Poirier, Richard. *The Renewal of Literature: Emersonian Reflections*. New Haven: Yale UP, 1987.

Pokrovsky, Nikita E. Paper delivered at 50th Anniversary Meeting of the Thoreau Society, July 10, 1991.

Porte, Joel. *Emerson and Thoreau: Transcendentalists in Conflict*. Middletown, Conn.: Wesleyan UP, 1966.

Porte, Joel. *Emerson in His Journals*. Cambridge: Harvard UP, 1982.

Porter, Carolyn. "Reification and American Literature." In *Ideology and Classic American Literature*, ed. Sacvan Bercovitch and Myra Jehlen. Cambridge: Cambridge UP, 1986. 188–217.

Porter, Carolyn. *Seeing and Being: The Plight of the Participant Observer in Emerson, James, Adams, and Faulkner*. Middleton, Conn.: Wesleyan UP, 1981.

Pratt, Mary Louise. *Imperial Eyes: Travel Writing and Transculturation*. New York: Routledge, 1992.

Rehbock, Philip F. *The Philosophical Naturalists: Themes in Early Nineteenth-Century British Biology*. Madison: U of Wisconsin P, 1983.

Reingold, Nathan. *Science in Nineteenth-Century America: A Documentary History*. New York: Farrar, Straus, and Giroux, 1979.

Richards, Janet Radcliffe. "Hypothetico-Deductive Method." *Dictionary of the History of Science*, ed. Bynum et al. Princeton: Princeton UP, 1981.

Richardson, Robert D., Jr. *Emerson: The Mind on Fire*. Berkeley: U of California P, 1995.

Richardson, Robert D., Jr. *Henry David Thoreau: A Life of the Mind*. Berkeley: U of California P, 1986.

Richardson, Robert D., Jr. "Thoreau and Science." In *American Literature and Science*, ed. Robert J. Scholnick. Lexington: UP of Kentucky, 1992. 110–27.

Richardson, Robert D., Jr. "Thoreau's Broken Task." Introduction to *Faith in a Seed*, ed. Bradley P. Dean. Washington, D.C.: Island Press, 1993. 3–17.

Robinson, David M. "Fields of Investigation: Emerson and Natural History." In *American Literature and Science*, ed. Robert J. Scholnick. Lexington: UP of Kentucky, 1992. 94–109.

Rossi, William. "'Laboratory of the Artist': Henry Thoreau's Literary and Scientific Use of the Journal, 1848–54." Ph.D. diss., U of Minnesota, 1986.

Rossi, William. "Poetry and Progress: Thoreau, Lyell, and the Geological Principles of *A Week.*" *American Literature* 66 (June 1994): 275–300.

Rousseau, G. S. *Organic Form: The Life of an Idea.* London: Routledge and Kegan Paul, 1972.

Rudwick, Martin J. S. *The Great Devonian Controversy: The Shaping of Scientific Knowledge among Gentlemanly Specialists.* Chicago: U of Chicago P, 1985.

Rudwick, Martin J. S. *The Meaning of Fossils: Episodes in the History of Palaeontology.* 1972. 2d ed. Chicago: U of Chicago P, 1985.

Sanborn, F. B. *Thoreau the Poet-Naturalist, by William Ellery Channing.* New edition, enlarged. 1902. New York: Biblo and Tannen, 1966.

Sattelmeyer, Robert. "The Remaking of *Walden.*" In *Walden*, ed. William Rossi. New York: Norton, 1992. 428–44.

Sattelmeyer, Robert. "Thoreau's Projected Work on the English Poets." *Studies in the American Renaissance, 1980.* Boston: Twayne Publishers, 1980. 239–57.

Sattelmeyer, Robert. *Thoreau's Reading: A Study in Intellectual History with Bibliographical Catalogue.* Princeton: Princeton UP, 1988.

Sattelmeyer, Robert. "'The True Industry for Poets': Fishing with Thoreau." *ESQ* 33 (4th quarter, 1987): 189–201.

Sattelmeyer, Robert, and Richard A. Hocks. "Thoreau and Coleridge's *Theory of Life.*" In *Studies in the American Renaissance, 1985*, ed. Joel Myerson. Charlottesville: U of Virginia P, 1985. 269–84.

Sayre, Robert F. *Thoreau and the American Indians.* Princeton: Princeton UP, 1977.

Sauer, Carl O. "On the Background of Geography in the United States." In *Selected Essays 1963–1975.* Berkeley: Turtle Island Foundation, 1981. 241–59.

Serres, Michel. *Hermes: Literature, Science, Philosophy.* Ed. Josué V. Harari and David F. Bell. Baltimore: Johns Hopkins UP, 1982.

Seybold, Ethel. *Thoreau: The Quest and the Classics.* New Haven: Yale UP, 1951.

Shanley, James Lyndon. "Historical Introduction." In *Walden.* Princeton: Princeton UP, 1973. xvii–xxxv.

Shanley, James Lyndon. *The Making of Walden.* Chicago: U of Chicago P, 1957.

Shanley, James Lyndon. "Thoreau: Years of Decay and Disappointment?" *The Thoreau Centennial*, ed. Walter Harding. New York: State U of New York P, 1964. 53–64.

Slovic, Scott. "'The Eye Commanded a Vast Space of Country': Alexander von Humboldt's Comparative Method of Landscape Description." *Publication of the Society for Literature and Science* 5.3 (May 1990): 4–10.

Smellie, William. *The Philosophy of Natural History.* Boston: W. J. Reynolds, n.d.

Smith, Henry Nash. *Virgin Land: The American West as Symbol and Myth.* New York: Vintage Books, 1950.

Smith, Jonathan. *Fact and Feeling: Baconian Science and the Nineteenth-Century Literary Imagination.* U of Wisconsin P, 1994.

Stallo, John Bernard. *General Principles of the Philosophy of Nature. With an outline of some of its recent developments among the Germans, embracing the philosophical systems of Schelling and Hegel, and Oken's system of nature.* Boston: W. Crosby & H. P. Nichols, 1848.

Stevens, Wallace. *Collected Poems.* New York: Random House, 1982.

Stewart, Dugald. *Elements of the Philosophy of the Human Mind. The Works of Dugald Stewart.* Vol. I. 1792. Cambridge, Eng.: Hilliard and Brown, 1829. Vol II. Edinburgh: George Ramsey, 1814.

Stoddard, Richard Henry. *The Life, Travels, and Books of Alexander von Humboldt.* New York: Rudd and Carleton, 1859.

Stoller, Leo. *After Walden: Thoreau's Changing Views on Economic Man.* Stanford: Stanford UP, 1957.

Stowe, Harriet Beecher. *Oldtown Folks.* 1869. In *Three Novels.* New York: Library of America, 1982. 877–1468.

Stowell, Robert F. *A Thoreau Gazetteer.* Princeton: Princeton UP, 1970.

Thompson, Cameron. "John Locke and New England Transcendentalism." In *American Transcendentalism: An Anthology of Criticism*, ed. Brian M. Barbour. Notre Dame, Ind.: U of Notre Dame P, 1973. 83–102.

Thoreau, Henry David. *Cape Cod.* Ed. Joseph J. Moldenhauer. Princeton: Princeton UP, 1988.

Thoreau, Henry David. *The Collected Poems of Henry Thoreau.* Ed. Carl Bode. Enlarged edition. Baltimore: Johns Hopkins UP, 1965.

Thoreau, Henry David. *The Correspondence of Henry David Thoreau.* Ed. Walter Harding and Carl Bode. New York: New York UP, 1958.

Thoreau, Henry David. *Early Essays and Miscellanies.* Ed. Joseph J. Moldenhauer and Edwin Moser, with Alexander Kern. Princeton: Princeton UP, 1975.

Thoreau, Henry David. *Faith in a Seed: The Dispersion of Seeds and Other Late Natural History Writings.* Ed. Bradley P. Dean. Washington, D.C.: Island Press, 1993.

Thoreau, Henry David. *Journal*, Vol. 1: *1837–1844.* Ed. Elizabeth Hall Witherell, William L. Howarth, Robert Sattelmeyer, and Thomas Blanding. Princeton: Princeton UP, 1981.

Thoreau, Henry David. *Journal*, Vol. 2: *1842–1838.* Ed. Robert Sattelmeyer. Princeton: Princeton UP, 1984.

Thoreau, Henry David. *Journal*, Vol. 3: *1848–1851.* Ed. Robert Sattelmeyer, Mark R. Patterson, and William Rossi. Princeton: Princeton UP, 1990.

Thoreau, Henry David. *Journal*, Vol. 4: *1851–1852.* Ed. Leonard N. Neufeldt and Nancy Craig Simmons. Princeton: Princeton UP, 1992.

Thoreau, Henry David. *The Journal of Henry David Thoreau.* 14 vols. Ed. Bradford Torrey and Francis Allen. Boston: Houghton Mifflin, 1906; New York: Dover, 1962.

Thoreau, Henry David. *The Maine Woods.* Ed. Joseph J. Moldenhauer. Princeton: Princeton UP, 1972.

Thoreau, Henry David. *The Natural History Essays.* Ed. Robert Sattelmeyer. Salt Lake City: Peregrine Smith, 1980.

Thoreau, Henry David. *Reform Papers.* Ed. Wendell Glick. Princeton: Princeton UP, 1973. 91–109.

Thoreau, Henry David. *Thoreau's Fact Book*. 3 vols. Ed. Kenneth Walter Cameron. Hartford, Conn.: Transcendental Books, 1966, 1987.

Thoreau, Henry David. *Thoreau's Literary Notebook in the Library of Congress. Facsimile text.* Ed. Kenneth Walter Cameron. Hartford, Conn.: Transcendental Books, 1964.

Thoreau, Henry David. *Walden*. Ed. J. Lyndon Shanley. Princeton: Princeton UP, 1973.

Thoreau, Henry David. *A Week on the Concord and Merrimack Rivers*. Ed. Carl F. Hovde, William L. Howarth, and Elizabeth Witherell. Princeton: Princeton UP, 1983.

Tuckerman, Henry T. "Alexander von Humboldt." *Godey's Lady's Book* 41 (1850): 133–38.

Tuckerman, Henry T. "The Naturalist." *Characteristics of Literature Illustrated by the Genius of Distinguished Writers. Second Series.* Philadelphia: Lindsay and Blakiston, 1851. 56–77.

Turner, James. *Without God, without Creed: The Origins of Unbelief in America.* Baltimore: Johns Hopkins UP, 1985.

Tyler, Samuel. Rev. of *Cosmos* by Alexander von Humboldt. *Biblical Repertory and Princeton Review* 24 (1852): 382–97.

Van Dusen, Robert. *The Literary Ambitions and Achievements of Alexander von Humboldt.* Bern: Herbert Lang, 1971.

Viola, Herman J. *Exploring the West.* Washington D.C.: Smithsonian Books, 1987.

Walls, Laura Dassow. "'The Napoleon of Science': Alexander von Humboldt in Antebellum America." *Nineteenth-Century Contexts* 14.1 (1990): 71–98.

Walls, Laura Dassow. "*Walden* as Feminist Manifesto." *ISLE: Interdisciplinary Studies in Literature and Environment* 1.1 (Spring 1993): 137–44.

Wellek, René. *Confrontations: Studies in the Intellectual and Literary Relations between Germany, England, and the United States during the Nineteenth Century.* Princeton: Princeton UP, 1965.

Whewell, William. *Novum Organon Renovatum.* Sel. in *William Whewell: Theory of Scientific Method*, ed. Robert E. Butts. Indianapolis: Hackett, 1989. 103–249.

Whitford, Kathryn. "Thoreau and the Woodlots of Concord." *NEQ* 23 (Sept 1950): 291–306.

Whitney, Gordon G., and William C. Davis. "From Primitive Woods to Cultivated Woodlots: Thoreau and the Forest History of Concord, Massachusetts." *Journal of Forest History* 30 (1986): 70–81.

Williams, Raymond. *Keywords: A Vocabulary of Culture and Society.* 1976. Rev. ed. New York: Oxford UP, 1983.

Woolgar, Steve. *Science: The Very Idea.* New York: Tavistock and Ellis, in association with Methuen, 1988.

Wordsworth, William. "Preface to the *Lyrical Ballads*." 1802. *William Wordsworth*, ed. Stephen Gill. New York: Oxford UP, 1984. 595–615.

Worster, Donald. "Ecology of Order and Chaos." *Environmental History Review* 14.1–2 (1990): 1–18.

Worster, Donald. *Nature's Economy: A History of Ecological Ideas.* 1977. Cambridge: Cambridge UP, 1985.

Wylie, Ian. *Young Coleridge and the Philosophers of Nature.* Oxford: Oxford UP, 1989.

Yeo, Richard. "An Idol of the Market-Place: Baconianism in Nineteenth Century Britain." *History of Science* 23 (1985): 251–98.

Yeo, Richard. *Defining Science: William Whewell, Natural Knowledge, and Public Debate in Early Victorian Britain.* Cambridge: Cambridge UP, 1993.

Yeo, Richard. "Reading Encyclopedias: Science and the Organization of Knowledge in British Dictionaries of Arts and Science, 1730–1850." *Isis* 82.311 (1991): 24–49.

Young, Robert M. *Darwin's Metaphor: Nature's Place in Victorian Culture.* Cambridge: Cambridge UP, 1985.

Zochert, Donald. "Science and the Common Man in Ante-Bellum America." *Isis* 65 (1974): 448–73.

INDEX

SCIENCE AND LITERATURE
A series edited by George Levine

One Culture: Essays in Science and Literature
Edited by George Levine

In Pursuit of a Scientific Culture: Science, Art, and Society in the Victorian Age
Peter Allan Dale

Sexual Visions: Images of Gender in Science and Medicine between the Eighteenth and Twentieth Centuries
Ludmilla Jordanova

Writing Biology: Texts in the Social Construction of Scientific Knowledge
Greg Myers

Gaston Bachelard, Subversive Humanist: Texts and Readings
Mary McAllester Jones

Realism and Representation: Essays on the Problem of Realism in Relation to Science, Literature, and Culture
Edited by George Levine

Science in the New Age: The Paranormal, Its Defenders and Debunkers, and American Culture
David J. Hess

Fact and Feeling: Baconian Science and the Nineteenth-Century Literary Imagination
Jonathan Smith

The Word of God and the Languages of Man: Interpreting Nature in Early Modern Science and Medicine. Volume 1: Ficino to Descartes
James J. Bono

Seeing New Worlds: Henry David Thoreau and Nineteenth-Century Natural Science
Laura Dassow Walls